Advances in cultured meat technology

It is widely recognised that agriculture is a significant contributor to global warming and climate change. Agriculture needs to reduce its environmental impact and adapt to current climate change whilst still feeding a growing population, i.e. become more 'climate-smart'. Burleigh Dodds Science Publishing is playing its part in achieving this by bringing together key research on making the production of the world's most important crops and livestock products more sustainable.

Based on extensive research, our publications specifically target the challenge of climate-smart agriculture. In this way we are using 'smart publishing' to help achieve climate-smart agriculture.

Burleigh Dodds Science Publishing is an independent and innovative publisher delivering high quality customer-focused agricultural science content in both print and online formats for the academic and research communities. Our aim is to build a foundation of knowledge on which researchers can build to meet the challenge of climate-smart agriculture.

For more information about Burleigh Dodds Science Publishing simply call us on +44 (0) 1223 839365, email info@bdspublishing.com or alternatively please visit our website at www.bdspublishing.com.

Related titles:

Assessing the environmental impact of agriculture
Print (ISBN 978-1-78676-228-3); Online (ISBN 978-1-78676-228-3, 978-1-78676-231-3)

Climate change and agriculture
Print (ISBN 978-1-78676-320-4); Online (ISBN 978-1-78676-322-8, 978-1-78676-323-5)

Genome editing for precision crop breeding
Print (ISBN 978-1-78676-447-8); Online (ISBN 978-1-78676-449-2, 978-1-78676-450-8)

Reducing greenhouse gas emissions from livestock production
Print (ISBN 978-1-78676-439-3); Online (ISBN 978-1-78676-441-6, 978-1-78676-442-3)

Chapters are available individually from our online bookshop:
https://shop.bdspublishing.com

climate•SMART•publishing
IN AGRICULTURAL SCIENCE

BURLEIGH DODDS SERIES IN AGRICULTURAL SCIENCE
NUMBER 145

Advances in cultured meat technology

Edited by Professor Mark Post, Maastricht University, The Netherlands, Professor Che Connon, Newcastle University, UK and Dr Chris Bryant, University of Bath and Bryant Research, UK

burleigh dodds
SCIENCE PUBLISHING

Published by Burleigh Dodds Science Publishing Limited
82 High Street, Sawston, Cambridge CB22 3HJ, UK
www.bdspublishing.com

Burleigh Dodds Science Publishing, 1518 Walnut Street, Suite 900, Philadelphia, PA 19102-3406, USA

First published 2023 by Burleigh Dodds Science Publishing Limited

Library of Congress Control Number: 2023940849

British Library Cataloguing in Publication Data
A catalogue record for this book is available from the British Library

ISBN 978-1-80146-376-8 (print)
ISBN 978-1-80146-378-2 (PDF)
ISBN 978-1-80146-377-5 (ePub)
ISSN 2059-6936 (print)
ISSN 2059-6944 (online)

DOI: 10.19103/AS.2023.0130

Typeset by Deanta Global Publishing Services, Dublin, Ireland

Contents

Series list

Title	Series number
Achieving sustainable cultivation of maize - Vol 1 From improved varieties to local applications Edited by: Dr Dave Watson, CGIAR Maize Research Program Manager, CIMMYT, Mexico	001
Achieving sustainable cultivation of maize - Vol 2 Cultivation techniques, pest and disease control Edited by: Dr Dave Watson, CGIAR Maize Research Program Manager, CIMMYT, Mexico	002
Achieving sustainable cultivation of rice - Vol 1 Breeding for higher yield and quality Edited by: Prof. Takuji Sasaki, Tokyo University of Agriculture, Japan	003
Achieving sustainable cultivation of rice - Vol 2 Cultivation, pest and disease management Edited by: Prof. Takuji Sasaki, Tokyo University of Agriculture, Japan	004
Achieving sustainable cultivation of wheat - Vol 1 Breeding, quality traits, pests and diseases Edited by: Prof. Peter Langridge, The University of Adelaide, Australia	005
Achieving sustainable cultivation of wheat - Vol 2 Cultivation techniques Edited by: Prof. Peter Langridge, The University of Adelaide, Australia	006
Achieving sustainable cultivation of tomatoes Edited by: Dr Autar Mattoo, USDA-ARS, USA and Prof. Avtar Handa, Purdue University, USA	007
Achieving sustainable production of milk - Vol 1 Milk composition, genetics and breeding Edited by: Dr Nico van Belzen, International Dairy Federation (IDF), Belgium	008
Achieving sustainable production of milk - Vol 2 Safety, quality and sustainability Edited by: Dr Nico van Belzen, International Dairy Federation (IDF), Belgium	009
Achieving sustainable production of milk - Vol 3 Dairy herd management and welfare Edited by: Prof. John Webster, University of Bristol, UK	010
Ensuring safety and quality in the production of beef - Vol 1 Safety Edited by: Prof. Gary Acuff, Texas A&M University, USA and Prof. James Dickson, Iowa State University, USA	011
Ensuring safety and quality in the production of beef - Vol 2 Quality Edited by: Prof. Michael Dikeman, Kansas State University, USA	012
Achieving sustainable production of poultry meat - Vol 1 Safety, quality and sustainability Edited by: Prof. Steven C. Ricke, University of Arkansas, USA	013
Achieving sustainable production of poultry meat - Vol 2 Breeding and nutrition Edited by: Prof. Todd Applegate, University of Georgia, USA	014

Introduction

With the global population estimated to reach 9 billion by 2050, agricultural production must accommodate this growth without increasing the environmental footprint of our current food systems. More sustainable and alternative modes of production are required to ensure that we can feed billions of people healthily and sustainably.

This volume reviews the major technologies used in cultured meat product development. The chapters are split into three parts: Part 1 chapters focus on general aspects of cultured meat technology by first providing an overview of the technology currently available. Chapters also review the cultured politics of cultured meat, how a regulatory framework for cultured meat products can be created – with a specific focus on Singapore – and also how an infrastructure for cultured meat can be created. Chapters in Part 2 consider current technology developments for cultured meat, focusing specifically on cell line sourcing, cell growth media, bioreactors for cell multiplication and tissue engineering. A chapter on scaling cell production sustainably is also included. Part 3 chapters draw attention to quality and sustainability issues, such as developing cultured meat as a food product, the environmental impact of cultured meat production and the importance of understanding consumer attitudes to cultured meat products. A chapter on techno-economic analysis of cultured meat is also provided.

Part 1 General

The first part of the book provides an overview of current cultured meat technology. Chapter 1 begins by first discussing the limitations of conventional food systems, focusing on the environmental, social and economic impacts these systems can have. The chapter moves on to provide a rationale for a transition to cellular agriculture, specifically addressing the health, sustainability and economic benefits cellular agriculture could have. A section on advances in cellular agriculture is also included, which is followed by an overview of current challenges for the cellular agriculture field.

Chapter 2 provides talking points on the politics of cultured meat, to focus attention on the variety of issues requiring attention. The chapter first examines the public life of cultured meat, focusing on areas such as public and consumer acceptance of cultured meat products. It then goes on to discuss the economic life of cultured meat, drawing specific attention to areas that need to be addressed in order for cultured meat to be an economic success. A section on the political life of cultured meat is also provided which examines the potential political ramifications of cultured meat products. The chapter also examines social cultured meat and social change, highlighting some of the potential benefits of cultured meat if it were to become successful in the market.

The next chapter focuses on creating a regulatory framework for cultured meat products, focusing specifically on Singapore. Chapter 3 uses the Singapore Food Agency's (SFA) regulatory framework for cultured meat and the principal considerations behind Singapore's food safety system is presented as a case study in the chapter. A review of the regulatory framework is covered in detail, including considerations for safety assessment of cultured meat based on the concepts of Hazard Analysis Critical Control Points, Good Manufacturing Practices and Good Cell Culture Practices. An overview of collaboration opportunities with the academic and industry sectors, overseas regulatory counterparts and international organisations is also described to demonstrate SFA's efforts in fostering joint responsibility in cultured meat safety. Furthermore, since communication is key to building consumer confidence in SFA's food regulatory system, strategies to engage consumers on cultured meat products are also highlighted.

The final chapter of Part 1 draws attention to creating an infrastructure for cultured meat. Chapter 4 begins by highlighting current gaps in existing infrastructure, focusing on aspects such as the talent gap and funding gaps, where lack of experience and money within the cultured meat industry can affect cultured meat's overall impact in society. To combat these issues, the chapter moves on to review possible avenues for remedy, such as the importance of working towards a more collaborative ecosystem as well as the importance of improving community engagement and education regarding cultured meat.

Part 2 Technologies

Part 2 opens with a chapter that discusses cell line sourcing and characterisation for cultured meat product development. Chapter 5 begins by considering the potential benefits of cultured meat production, such as its environmental impact and sustainability potential, its overall impact on ecology within the livestock production chain, as well as the potential of cultured meat in combatting ethical concerns regarding animal welfare associated with meat production. The chapter also reviews the economic benefits and public perception and acceptance of cultured meat products. A section on current challenges in cultured meat production is also provided, focusing specifically on challenges such as the use of tissue engineering, cell culture media, the cost of growth factors, whilst also highlighting other technical considerations that need to be developed. Sections on cell line sourcing, adult stem cells, immortal cell lines, pluripotent stem cells and the current regulatory issues of cultured meat are also provided.

The next chapter highlights developments in cell growth media for cultured meat product development. Chapter 6 presents the historical context for and recent updates in developing serum-free media for muscle and fat cells–the predominant cell types in meat–in vitro. The chapter also examines the use of serum-free media for other potential cell types, specifically induced pluripotent stem cells. A section on basal media is also provided, followed by a discussion on the importance of reducing the cost and impact of serum-free media. The chapter reviews the media impact of cellular composition, focusing specifically on safety and nutrition. It also provides an overview of current challenges, opportunities and research directions within the sector.

Chapter 7 explores the use of bioreactors for cell multiplication in cultured meat product development. The chapter begins by first providing an overview of the principles and structure of a bioreactor, which is followed by a breakdown of bioreactor operation modes: batch cultivation, fed-batch cultivation and continuous cultivation. The chapter also highlights the important bioprocess parameters that need to be considered when creating a stem cell niche in a bioreactor, such as temperature, oxygen, pH level, lactic acid and culture mixing. The chapter considers cultivation anchorage-dependent cells in stirred tank bioreactors, the scale-up of a bioprocess in bioreactors and also highlights areas which could help reduce cost and optimise processing within bioreactors.

The subject of Chapter 8 is tissue engineering using scaffolds in cultured meat product development. The chapter first provides an overview of the key scaffold properties as 3D meat cultivation platforms, focusing on aspects such as tissue engineering-related criteria and additional scaffold constraints for cultured meat production. The chapter moves on to highlight potential materials for cultivated meat scaffolds. A section on construct types and fabrication techniques is also included, drawing specific attention to micro-carriers, porous constructs, fibrous scaffolds, hydrogels and 3D printing.

The final chapter of Part 2 focuses on scaling cell production sustainably in cultured meat development. Chapter 9 begins by first introducing the scale-up requirement of cells for the production of cultured meat. The chapter moves on to discuss current bottlenecks and challenges in scaling up the production of cultured meat, specifically the quality of starting cells, cell densities, suitable scaffolds and sustainability and techno-economic feasibility. The chapter also considers analytical requirements for scalable production, the development of digitalisation and automation as well as the skills required for the process.

Part 3 Quality and sustainability issues

The first chapter of Part 3 examines developing cultured meat as a food product. Chapter 10 first focuses on the importance of meat structure when considering the development of cultured meat products, which is then followed by an analysis of appearance and consumer attitudes. The chapter also reviews the importance of flavour in final product quality, drawing specific attention to protein degradation/maturation, lipid incorporation and content as well as flavour development from Maillard reactions and other minor components. A section on nutritional quality is also included, focusing on how protein, lipids, bioactive components, vitamins and minerals can affect this. The chapter also highlights the importance of good microbial safety and highlights key research questions that need to be addressed in future research.

Chapter 11 draws attention to assessing the environmental impact of commercial cultivated meat production. To provide some context, the chapter first focuses on the environmental impacts of livestock production, specifically the impact of land use and biodiversity, climate change as well as the effect of water use and emissions to waterways. The chapter moves on to examine the use of life cycle assessment and how this practice can be used to estimate the environmental impact of cultured meat products. This is then followed by an overview of the potential environmental impacts of cultivated meat. Similarly to the section on livestock production it discusses the impact of land use and biodiversity, climate change, water use and emissions to waterways whilst also assessing how energy is used during cultivated meat production processes.

The next chapter reviews consumer attitudes to cultured meat products, focusing on improving understanding and acceptance of these products. Chapter 12 begins by examining the challenges regarding consumer acceptance of novel food technologies and cultured meat, drawing specific attention to perceived naturalness and disgust, trust in the industry and concerns with humans tampering with nature. The chapter moves on to examine consumers' motivations to substitute conventional meat, which is then followed by a section that discusses ways to overcome the rejection of cultured meat products and the possibilities for cultured meat in the near future.

The final chapter of the book reviews techno-economic analysis (TEA) for cultured meat production. Chapter 13 first provides an overview of TEA objectives, then goes on to examine basic structures and philosophy of TEA models. This is then followed by a review of the faceplate, a control panel for the TEA model. The chapter also discusses major inputs and outputs to/of a techno-economic model, which is then followed by an overview of a step-by-step approach to techno-economic analysis development. A section highlighting three of the primary uses of TEA models is also included, followed by a review and gap analysis of published TEA's of commercial cultured meat production. A case study bringing together the concepts of building a TEA model is also provided.

Part 1

General

Chapter 1

Cultured meat technology: an overview

Apeksha Bharatgiri Goswami, James Charlesworth, Joanna M. Biazik, Mark S. Rybchyn and Johannes le Coutre, University of New South Wales, Australia*

1 Introduction

With the introduction of the cultured meat concept, the past decade has seen the remarkable rise of a new scientific field based in the food technology domain, with the potential to change our current food systems[1]. Cultured meat is made from animal cells grown outside an animal. The principal approach behind producing this food is based upon the idea to use cell and tissue culture techniques, originally developed for the medical science field, to grow all components of edible meat *in vitro*. At a minimum, this comprises muscle cells, which grow and develop into enlarged cell assemblies. Ideally, this biomass would feature as many aspects of animal-derived meat as possible.

Briefly, the technology entails obtaining a biopsy from a live animal and growing it to the desired volume in a suitable production environment with the help of specific media to feed the development. For the creation of three-dimensionality and texture, typically a bio-scaffold is incorporated at some point during the process (Fig. 1). The resulting material is meat, and it will have to be labelled as such. Genetically, cultured meat is identical at the cellular level to conventional meat, and in the public debate, it is also

* Johannes.lecoutre@unsw.edu.au

http://dx.doi.org/10.19103/AS.2023.0130.01

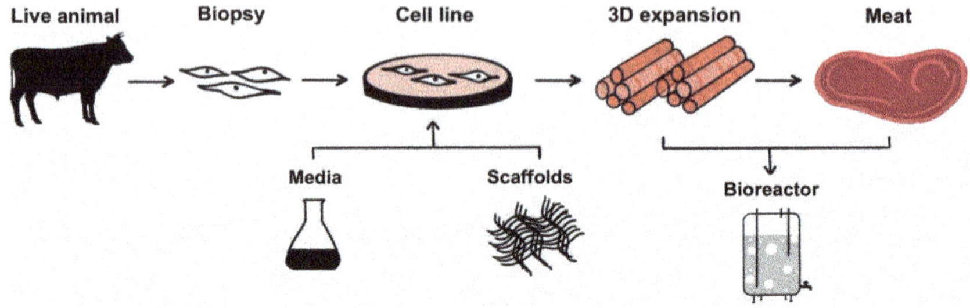

Figure 1 Overview of the cultured meat process. Initial cells are sourced via a biopsy taken from the animal of choice. Subsequently, from this source, relevant cell types are selected to develop an immortalized cell line. Finally, by using enriched media, these cells are prepared to proliferate and differentiate in a bioreactor with or without the addition of scaffolding material, which provides texture to the growing material.

referred to as lab-grown meat, clean meat, or *in vitro* meat. The scientific and commercial domain dedicated to this innovative technology refers to itself as 'Cellular Agriculture' (cellAg) to highlight the idea of introducing new domestication, i.e. the scaled production of cellular material that so far had only been obtained by higher animals or plants[2]. The term cellAg was coined in 2015 by the New Harvest non-profit organization for animal products made without animals. The technology holds the potential to improve global food security by addressing major ethical, environmental, commercial, and public concerns.

A key objective behind any industrial effort in the context of food is to provide food security. The emerging global food security gap necessitates the production of additional 60–70% calories by 2050, assuming a population of about 10 billion people globally[3]. With the intent to alleviate the burden on current food systems and to improve nutritional quality and animal welfare, the field of cellAg offers a potential solution, albeit technologically challenging to accomplish.

The technology will provide significant advantages as compared to conventional animal-based meat if the product categories are chosen well. Based upon the enclosed nature of growing biomass in bioreactors, all ingoing and outgoing material streams can be controlled.

Still, by far the largest challenge remains scaling of the technology at an affordable cost to provide nutritious, safe, and affordable material in large and impactful quantities. Current setups and approaches are nowhere near the output of the current animal-based livestock industry.

A wealth of conceptual papers, commercial reports, and anticipated consumer scenarios are being published, although we still do not know how CellAg will live up to expectations and how the related products will be perceived. The current situation is that products will be available and even

Published by Burleigh Dodds Science Publishing Limited, 2023.

are available in minute quantities to boutique restaurants and specialty niche outlets.

Cultured meat products are being envisaged at various levels of sophistication. At the simplest level, unstructured cellular material will be available that can be diluted with water to provide for a broth or soup stock. With increasing complexity, more texture can be introduced and minced meat or 'ground meat-like' structures will be available, material which at this stage already should contain additional cells or tissues such as adipocytes to provide for the organoleptic property of fatty taste or texture as well as the nutritional benefits of these cell types. Chicken nuggets or any artificially shaped meat products from other species belong in this group as well and might be available without additional cell types. The most ambitious product form will be analogous to a meat cut. To achieve meaningful meat cut analogues, it will take considerable time and significantly more technological development.

The field of food science is full of examples where one material is supposed to mimic another one, from tofu-based vegetarian meat in the tenth century Song dynasty to margarine, which is designed to mimic butter, and to specific molecules that are developed to mimic the taste of salt or sugar, such as aspartame, which imparts a sweet taste. In the cultured meat domain, it will be interesting to see if mimicking animal-based materials will be the ultimate ambition or if the derived materials will succeed to attain a food category by themselves. Why mimic animal material if you want to abandon animal-based materials?

To make cultured meat a success story, it is not just a matter of significant advances in one technology; success is dependent on making strides in an entire technology suite if the resulting products are not just meant to serve as a gimmick in a niche market.

2 Limitations of conventional food systems

Most of the food eaten by humans is derived through some agricultural and/or industrial process. Often conventional food systems are associated with deforestation, monocultures, and greenhouse gas (GHG) emissions, followed by energy consumption and emissions from transport and retail (Fig. 2). The principles of current agricultural food production are based upon the 'green revolution', which led to a significant boost in productivity post World War II[4]. A global transition took place over the past 60 years towards our current food systems, which rely on monocultures, policy, machinery, and chemical inputs, such as fertilizers, antibiotics, and pesticides. The aim behind these developments has been to satisfy the need for a more affordable, safer, and rapidly growing food supply.

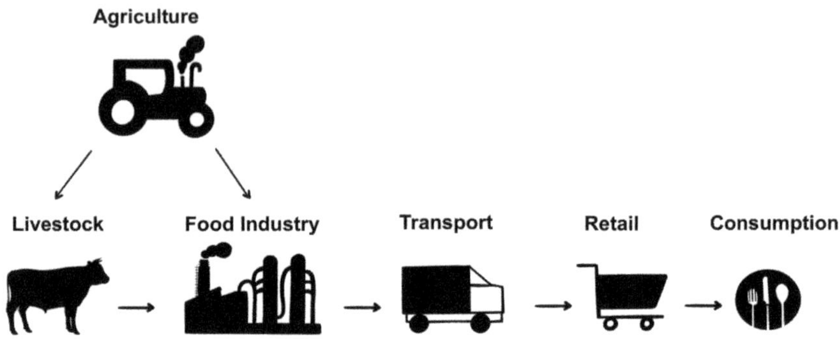

Figure 2 All stages of a food system are vulnerable to being broken, thus threatening food security. Agricultural boundaries include deforestation, GHG emissions, monocultures, resistance to pest and weed, and pollution of air and land. Vulnerability issues are exacerbated in livestock farming and the food industry. The overall sustainability of a classical food system can be improved by the responsible development of cellular agriculture.

However, as the global population increases and other challenges such as climate change occur, there is a need to address the environmental, social, and economic impacts of current agricultural technologies. We have come to a point where agricultural practice can even present a health risk - since the 1940s, nearly half of all zoonotic diseases in humans have been found to have come from livestock. Of similar concern is the usage of antibiotics in the production of meat, which has been linked to a rise in antibiotic-resistant pathogens[5]. The latter incidences are closely related to the industrialization of animal agriculture, which has increased accessibility to animal proteins and cheaper products. It also has created a perfect niche for zoonotic transmissions and has established itself as a threat to global food security.

The effects of industrial agriculture can be categorized into environmental, social, and economic impacts.

2.1 Environmental impact

Current farming practice increasingly is ecologically unsustainable. The intensive farming necessary to produce the high yields of nutrition that support global food needs has led to increased environmental impact. This includes higher use of fertilizers, irrigation, and land expansion. Our industrial agriculture successfully satisfies food demand and brings economic benefit but impacts the environment. Deforestation, pest and weed resistance, soil degradation, destruction of natural habitat, and water pollution are some of the major drivers of climate change. Emissions from the livestock industry can be categorized by source. Enteric fermentation, manure management, feed production, and energy consumption are the four main processes that contribute to GHG emissions (Fig. 3).

Published by Burleigh Dodds Science Publishing Limited, 2023.

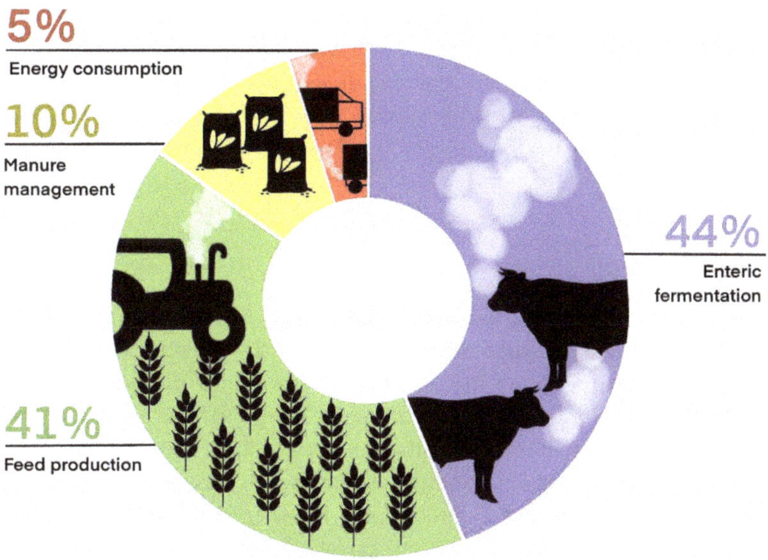

Figure 3 Global greenhouse gas emissions from livestock supply chains based upon 2015 data[30]. Enteric fermentation accounts for 3.5 gigatonnes CO_2-eq for the methane generated from ruminants and non-ruminants. Manure management includes methane and nitrous oxide released during the anaerobic decomposition of manure, which accounts for 0.8 gigatonnes CO_2-eq. Feed production accounts for 3.3 gigatonnes CO_2-eq from feed crop expansion and fertilizer production. It also includes nitrous oxide from nitrogenous fertilizers and use of manure in crop fields. Energy consumption accounts for 0.4 gigatonnes CO_2-eq attributed to the entire supply chain system, including crop management, processing, transport of feed, production site, milking, packaging, and product transport.

Millions of hectares of land have been dedicated to growing monoculture crops to produce livestock feed. This contributes to deforestation, soil degradation, and the increase in the use of herbicides, insecticides, and fertilizers. Hundreds of millions of tonnes of manure are produced every year to sustain livestock. Often, fields that rely on manure, store it in large open lagoons which upon flooding contaminate the soil and water systems with antibiotics, bacteria, pesticides, toxic chemicals, etc. A significant factor regarding climate change is the production of methane by rumen digestion of cattle, particularly cows (Fig. 3). Regarding heat accumulation, methane is a 25-times stronger effector as compared to carbon dioxide[6]. A wide range of issues are linked to the production of meat, and there is great interest in potential solutions. This includes the use of sustainable feed[7] or feed that limits greenhouse gas emissions[8] as well as clear incentives for the industry to reduce environmental impact. However, implementing these changes will take time and requires strict compliance. It is conceivable that cellAg could avoid or help limit a range of these emissions. The degree to which the change to cultured meat will have an impact does depend on how the industry improves key metrics such as power generation from renewable sources to drive bioreactors, or diverse mixtures of plant materials to serve as a supply chain. If unchanged, the environmental impact of current food systems will intensify with the growth of the meat industry as the population increases to the predicted 9.8 billion by 2050.

2.2 Social impact

To ensure global food security, a sustainable food system is required. Animal products such as meat, aquaculture, eggs, and dairy use about 83% of the world's farmland and contribute to about 57% of the different food-based emissions, while providing only 37% of our protein and 18% of our calories[2]. In addition to other environmental matters, there are increasing ethical concerns about a variety of traditional agricultural practices such as the use of 'factory farming'. Factory farming has been indicated to place undue stress on animals and may also have a role in promoting disease[9]. A recent study indicated that consumers have begun to link factory farming with cruelty and the potential for new zoonotic diseases[10]. While the social impact of a transition to cellAg products such as cultured meat is still being studied, there is a clear emerging consumer desire for non-animal-based products reflected by the rise of alternative meat and milk products[11].

2.3 Economic impact

The livestock market contributes 40% to the global agricultural output value and supports approximately 1.3 billion people in terms of food security and livelihood[2]. This reality is supported not only by consumer desire but also by governmental policies. An example of this is the lower cost of animal feed owing to government subsidies with profit margins being significantly lower if these subsidies are removed[12]. Further, tax exemptions are provided over other supplies such as equipment and employment structures. The current framework does not consider these externalized costs and the impacts on public health and on natural resources are not considered as well. These hidden costs of the current agricultural model need to be considered when modelling the transition to cellAg products. The emerging concept of 'true cost' is helpful in illustrating the overall picture[13].

It is not the first time the food industry needs widespread innovation. To meet food demands, already during the industrial revolution small farm layouts were developed into intensely packed breeding grounds for livestock. Another boost to productivity occurred during the green revolution in the second half of the twentieth century[4]. Raising livestock for food has its limitations. For example,

Published by Burleigh Dodds Science Publishing Limited, 2023.

in the US, the average number of days for chicken to reach the consumer has been reduced from 112 to 48, while the average weight has increased from 1 kg to about 3 kg[12]. To meet the growing food demands of an increasing human population, continued efficiency improvements are required[12]. The consumption of meat is engrained in all cultures and may have been responsible for shaping human development as well as providing significant nutrition and pleasing taste to consumers[14]. However, with the ever-increasing demand on land, water, GHG emissions, and ethical concerns, alternative solutions should be considered. It is a goal of the nascent cellAg field to best address as many of these points as possible.

3 Rationale for a transition to cellular agriculture

Conceptually cultured meat is an extremely appealing way to address the current shortcomings of the industrial livestock production system. So far, alternative meat products, some of which are based on recombinant and plant proteins, have already started replacing animal-based meat. The same will likely be observed for cultured meat as the field develops and the general public better understands the technology. Some current studies have indicated reluctance in the population to try cultured meat products[15,16]. Most respondents do support further research into the field[15,16] to improve a range of concerns such as worries about the taste or quality of the final cultured meat product. As the field develops, it is hoped the production of meat can be more sustainable without animal suffering and at a realistic economic return[17,18].

Driven by the limitations of animal-based meat, there have been advancements in plant-based meat as well. Other than the conventional plant-based alternatives such as tofu, novel plant-based meats are developed with enhanced meat-like sensory characteristics. Both plant-based and cultured meat consider animal welfare, human health, greenhouse gas emissions, land and water conservation, and economics to provide an effective alternative to traditional agriculture. Both types of alternatives have their benefits, and neither should be positioned to solely monopolize the market. However, cultured meat has the advantage of being sensory and nutritionally equivalent to conventional meat. It will be key to keep an eye on its attributes as pilot plants are getting bigger and production facilities are being implemented.

Importantly, advances in the cellAg field remain the subject of controversial debate, and this debate is conducted with occasionally sobering views on both sides[19]. Some clarity is provided via two studies commissioned by the Good Food Institute to be conducted by CE Delft. In a Techno-Economic Analysis (TEA) for the production of cultured meat, the overall feasibility and chances for economic success are carefully laid out[20]. In a prospective Life Cycle

Assessment (LCA) study, the same group investigates the environmental impact of a cultured meat industry operating at commercial levels[20].

As with all innovation, it will be of the essence to drive technology development in the cellAg space with rigorous and independent scientific exploration. This can be achieved by providing sufficient time and funding to ensure the sustained market introduction of safe and nutritious cultured materials. Moreover, some of the targeted health benefits might convince non-meat consumers or picky eaters to eat cultured meat if a viable nutritional option is provided without the involvement of livestock animals[21,22].

3.1 Health

Cultured meat products are being described as holding great potential to improve health and nutrition. This idea remains to be proven and will likely require further research, particularly as more products are developed. As with all dietary principles, it will be important for consumers to adhere to a balanced diet. There is a possibility to replace unfavourable saturated fatty acids with healthier alternatives, such as omega-3 fatty acids[23]. Other modifications such as inclusion of dietary fibre from cell scaffold materials or direct addition of nutrients such as vitamins could lead to cultured meat being a healthier choice than traditional meat[23]. Cultured meat is intended to be free of antibiotics, which are heavily used in intensive farming. Animal-based food production is an incubator for antimicrobial resistance, and about 70–80% of antibiotics worldwide are used on farm animals[24]. Cultured meat presents the opportunity to produce meat lacking pathogen contamination and environmental contaminants such as antibiotics, microplastics, and heavy metals like mercury in fish products. A Harvard study published 90 reasons to consider switching to cultured agriculture[24]. The study rationalizes the use of cultured meat as it would reduce antibiotic resistance, zoonosis, and GHG emissions and how this could result in nutritious clean meat. Apart from zoonosis, various abattoirs and meat packaging plants across the globe have been a risk factor throughout the COVID-19 pandemic, which occasionally has been attributed to poor hygiene conditions[25]. Without high-density animal farming, zoonotic outbreaks like swine and avian flu might be less prevalent.

3.2 Sustainability

By definition, sustainable food meets the need of the current population without compromising future generations. Cultured meat could meet the growing demands of 10 billion people by 2050, and it can have a positive impact on global hunger and climate[26]. Moreover, it has the potential to meet climate goals without affecting consumption patterns by sustainably emitting less

GHG compared to conventional meat[27]. In terms of cumulative environmental impacts, cellAg production of chicken, pork, and beef will reduce these by 17%, 92%, and 52%, respectively[28]. Ambitious modelling also suggests that 95% of the land and 78% of water currently used for agriculture can be freed and global warming could be reduced by 92%[29]. Approximately, 70 billion terrestrial farm animals are raised and killed for food every year, a figure that could be reduced upon transition to cellAg[24].

The so-called carbon miles of food seriously impact the emissions cost of food when transportation costs are added, accounting for a total of 41% of the global food emissions[30]. Another area that could be improved through cellAg is the production of food closer to major population centres, reducing the need for long-distance transportation. Overall, cellAg is resolving some of the most pressing environmental problems such as deforestation, loss of biodiversity, global warming, and pollution. If we combine the potential environmental, social, and health impacts of cellAg, the picture of a sustainable way for growing meat is beginning to emerge. Environmental impacts are extensively explained elsewhere in this book.

3.3 Economy

When the first cultured meat patty was developed and presented in 2013, the cost of production was more than US$300 000[31]. A variety of companies have begun to reduce these costs and to devise a way of scaling up reactor systems and recycling culture media for affordable large-scale production[32,33]. Obtaining conventional meat from an animal takes considerable time to raise the animal for slaughter. There is little room for cost reduction and improvement of feedstock efficiency as the model is governed by the growth of the animal. The cultured meat model allows for improvements in efficiency by directly producing muscle tissue for consumption. Significant cost reduction will occur with the removal of fetal bovine serum (FBS) from culture media with several companies already targeting serum-free media products[34-36]. Cultured meat allows for products that would otherwise not be available in the market, for example exotic meat sources or seafood in inland communities. It is hoped that countries with little land area for cattle production, which are dependent on imports, could produce cultured meat. Furthermore, compared to conventional meat, cellAg could allow more people across the globe to access high-quality meat at a cost favourable both economically and to the environment[2,3].

4 Advances in cellular agriculture

One aim of developing cellAg technology is to deliver a meat product that is safe, nutritious, sustainable, and that can meet the demand for food, which

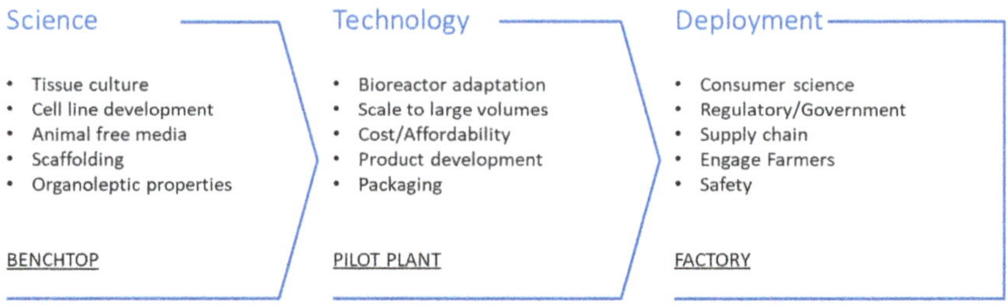

Figure 4 Technology suite and key development requirements for the successful launch of cellAg products in the market. Innovation and discovery at the laboratory scale continue to be required for the discovery and development of alternate solutions. Conceptually, in a subsequent workflow, technology is required to scale and adapt laboratory procedures to pilot plant and factory scale. Consumer and market studies are required to launch and deploy safe consumer-grade products in the context of a defined regulatory framework.

traditional livestock is increasingly struggling to provide[37]. As this is a novel field, new advancements are continually being developed by a wide range of stakeholders including both academic and private research laboratories. The field is based upon multidisciplinary research in the biological sciences, engineering, and agriculture. Key areas for development include cell line development, production of cost-effective cell media, development of cell scaffolds, product development, and engineering of novel bioreactor systems to produce quantities of biomass required at a commercial level (Fig. 4).

4.1 Discovery at laboratory scale

Cells and cell line development are at the initial core of cultured meat development. While some groups are interested in the development of liver tissue from hepatocytes or fat tissue from adipocytes, most edible meat originates from muscle tissue and therefore most R&D efforts are focused on muscle cells. Currently, the main cell line used for fundamental science and research in the muscle research field is the C2C12 myoblast cell line derived from mice. Another publicly available myoblast cell line is derived from quail (QM7). The growing cellAg field is pushing for the generation of further myoblast lines derived from commonly farmed livestock animals. The two main types of stem cells utilized are pluripotent stem cells and adult stem cells. Muscle satellite cells, mesenchymal stem cells (MSCs), and fibro (adipogenic) progenitors (FAPs) are the three main adult stem cell lines[38]. Embryonic (ESCs) and induced pluripotent stem cells (iPSCs) are the two main pluripotent stem cell line types developed. Recently, the Good Food Institute funded projects to develop cell lines from cows, pigs, shrimps, and fish. Research on their viability and myogenic potential is under way[33].

The identity of a cell usually determines its physiological behaviour. Most eukaryotic cells display an adherent phenotype, i.e. their growth and development are dependent upon being attached to a surface. For growth at scale in bioreactor vessels, this can be unfavourable and cells growing in suspension would be preferred. Interestingly, for the application of CellAg, a combination of both properties might be most desirable. In this way proliferation to high cell density might work better in suspension, whereas differentiation into edible biomass might work better with adherent cells. At this point, there is no consensus on how this impasse can be resolved best, and there have been advances in turning other cell lines like CHO into suspension[39]. Different forms of microcarriers are being explored for their potential to provide surface and buoyancy so that a bioreactor can be filled in three dimensions with growing biomass. Moreover, by using cellular aggregates, novel approaches are being explored to overcome this dilemma[40].

For decades, cell and tissue culture work has been utilizing foetal bovine serum (FBS) in conjunction with a basal medium consisting of nutrients and salts required for cell growth. In the context of cellAg, FBS is unfavourable for several reasons. It is expensive, it is animal derived, and, based upon its origin, it varies in composition and quality. Another compelling thought is the idea that FBS will become less available if cultured meat is as successful as anticipated. At present, in the cellAg field, few FBS-free media formulations are available including Essential 8[41] and TeSR[42] medium. Research is being conducted to create chemically defined media, often utilizing recombinant growth factors, which efficiently can replace serum-based medium and animal-derived growth factors. A variety of strategies are also being employed including cost reduction of recombinant growth factors or minimizing the concentrations required[35].

Preparation and optimization of cell media is the single most crucial step to developing a meaningful supply chain for the production of cultured meat. Both academic and corporate institutions drive media development such as Mosa Meat B.V., one of the early start-up companies, which managed to reduce their medium cost by 98%. This was achieved by characterizing myogenic gene expression during differentiation and by supplementing the corresponding regulators in the media to enhance its definition[36]. Other companies such as UPSIDE Foods[43] and CellMEAT[44] have claimed the development of FBS-free media as well. A range of media formulations have been developed by academic institutions, such as Tufts University developing serum-free medium for bovine cell lines[35] and differentiation media for rapid muscle cell differentiation. At present, about two dozen cell media suppliers are ready to cater their products to the cultured meat industry[33].

Animal-based meat takes shape on bones with intercalation of fatty layers and collagen-based connective tissue. Without these features muscle tissue would be pointless, and it is their properties which impart texture and

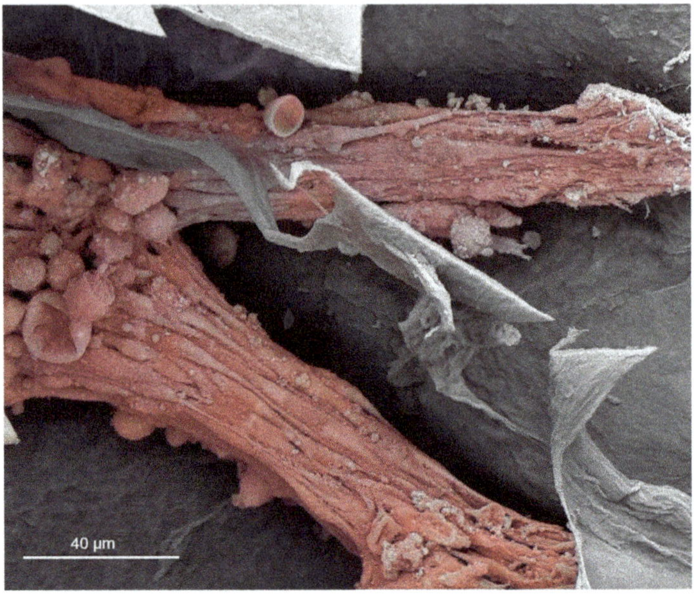

Figure 5 Scaffolding at the cellular level.[45] SEM image of differentiated myotubes (C2C12 cells) bound to a nanocellulose bioscaffold (NBS) matrix. Understanding cell-to-matrix interaction will be helpful for the successful scaling of cultured meat.

mouthfeel to a piece of animal-based meat. To create three-dimensionality, bite, texture, and mouthfeel with cultured meat, various approaches are targeted towards the development of scaffolding materials, which are combined with the growing cell mass. Suitable scaffolding is also needed for adherent muscle cells to enable a structure, which ensures efficient nutrient and waste movement to maximize the growth of biomass. Ideally, scaffolding material would be edible and potentially add nutrition to any final meat product[45]. Several types of scaffolds have been successfully developed, such as hydrogel-based, algae-based scaffolds, and nanocellulose[38] (Fig. 5).

4.2 Technology development and scaling

To deliver cultured meat products into regional and global markets, sizeable amounts need to be produced and current concepts pursue the development of bioreactors at various types and levels. In principle, a bioreactor is a vessel that provides controlled conditions regarding temperature, oxygenation, pH value, medium quality, and physical agitation. The best bioreactors - by far - are live animals. Animals provide structural support through bones, cartilage, and collagen to grow muscle tissue. Moreover, a live animal organism is equipped with blood vessels to provide for the transport of nutrients and cellular waste products. Researchers are trying to replicate the conditions for cell growth

Published by Burleigh Dodds Science Publishing Limited, 2023.

found in animals to accommodate the growth of animal cells outside their body. Credible scepticism has been expressed towards the idea of growing 'meat' in bioreactors at volumes comparable to volumes delivered by the livestock industry[46]. More on techno-economic analysis will be discussed elsewhere in this book.

Based upon the cellular system and the product of choice, i.e. muscle meat, foie gras, or seafood, etc., the size, process, and design of the bioreactor are selected. Conventional bioreactor systems fall into various categories such as stainless-steel stir tank bioreactor, single-use stir tank bioreactor, orbitally or rotationally shaken, fluidized bed, rocking wave, or hollow fibre bioreactor (HFB). Additional options include airlift and bubble column, vertical wheel, and fixed bed bioreactors. Advances have been made using several types of bioreactors for C2C12 cell proliferation, such as HFB[47] and polysulfone hollow fibres[48]. Currently, bioreactors of up to several thousand litres volume are being utilized in the cultured meat field[49]. As it has been established for the biomedical industry, for every run of a bioreactor, an optimized seed train needs to be developed consisting of several sequential steps to expand an aliquot of frozen cryoculture via shaker flasks and intermediate culture devices to the desired final main culture.

Regardless of the achievements made in scaling up of cultured meat, over the foreseeable time, pure cultured material might not be competitive in price to animal-derived meat. Therefore, with the idea to provide high-quality blends of nutrient-dense cultured meat together with fibre-rich complement, many developers pursue hybrid products together with plant-based material. Both plant and cultured approaches have their benefits and can be combined to create a range of products that cater to a spectrum of consumers. Creating a product from both will have the advantages of being nutritious, clean, and sensory equivalent to conventional meat. Interestingly, the difference in overall content at the level of the final product between a bio-scaffold used for growth of the cells and subsequently added plant material to produce a hybrid product could be exceedingly small (Fig. 6).

4.3 The emerging cellular agriculture industry

After the first 'proof of concept' hamburger had been developed and presented in 2013, the number of innovators in the new research field saw significant growth year after year. Witnessing Prof. Mark Post serve a cultured meat hamburger at a media event in London[50] has triggered a chain of events resulting in multiple cultured meat start-up companies. Currently, the number of start-ups focused on developing cultured meat products has risen to more than 100, with approximately 21 new cultured meat ventures compared to the previous year[33].

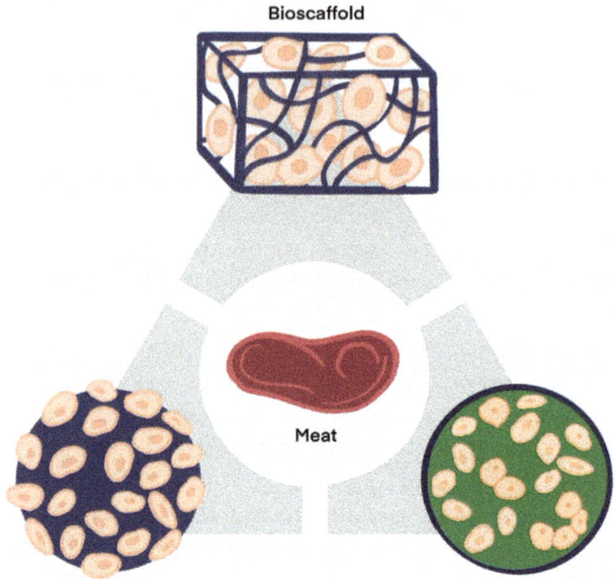

Figure 6 Successful scaling necessitates firm control over the interaction between myoblast cells and their environment at various levels of magnitude and at various time points during the production of cultured meat. At the micrometer level, it will be desirable to achieve a high ratio of cells per microcarrier. Bioscaffolds are being employed to impart dimensionality and texture to the growing biomass. Many current approaches in the emerging cultured meat industry are boosting production levels by the introduction of combined materials using cultured and plant-based materials. Thus, the edible product indicated in the centre might range from a blend of cultured meat with a high plant content to a meat without any plant material. Ongoing technology developments will provide for more seamless transitions between the various levels of interaction.

The cultured meat field has grown exceptionally and has managed to reach an investment capital of $1.3 billion in 2021[33]. Cultured meat companies such as Believer Meats (formerly Future Meat Technologies), GOOD Meat (formerly Eat Just), and UPSIDE Foods (formerly Memphis Meats) have shown evidence as to how far this field has advanced in just 8 years. To enable rollout in the market, the Singapore food authority (SFA) issued regulatory approval for the sale of cultured chicken bites by GOOD Meat[51].

CellAg aims at utility broader than replacing traditional meat products with the declared interest in other animal-sourced goods. Animal-derived products used in fashion or cosmetic industries represent another potential area for research and development, e.g. cultivating gelatine, fur, and leather[52]. Other animal food products such as milk[53], eggs[54], and caviar could also be replicated by cellAg advances. One example to note is the production of cultured human breast milk, which could represent an interesting alternative to infant formula[53].

Some companies expressed interest in using cultured meat products in the pet-food industry. Similarly, biomedical research and the biomedical industry may also benefit from the advances of cellAg with FBS representing a significant cost in cell cultivation for a range of products such as antibodies and vaccines. Growth media formulations without FBS may help to minimize any batch effects introduced by the addition of FBS.

5 Challenges for the cellular agriculture field

Cultured meat is based on the idea of producing meat and other animal products without animals. While this field has the potential to revolutionize the way we think about food, it also needs significant development and technological innovation. As mentioned, the challenge lies in replicating the muscle-growing environment of an animal inside a technology platform or a production site. Our knowledge of tissue engineering and development has been limited to medical applications and research[55]. Overall, the biggest challenge for the cellAg field is scaling at affordable cost. This challenge can be addressed by developing solutions to the speed of cell growth, to the manageable volume of cells that can be grown safely in a bioreactor, and to the details related to the procurement and production of cell media. Existing technologies will have to be altered to be applicable at a scaled-up stage. Discussed below are some of the technical challenges.

5.1 Cell line procurement and media optimization

One challenge in developing cultured meat involves maintaining the viability of cells *in vitro* and the ability to both proliferate and differentiate into the various cell types and structures[56]. Well-developed cell lines sourced from commonly consumed animals are important to cellAg as they directly impact the downstream upscaling, and therefore, the end-product[57]. Induced pluripotent stem cells (iPSCs) have been developed to mitigate the complexities of deriving cells from an embryo; however, they possess properties of embryonic stem cells[58]. Overall, it will be important to obtain better control over the differentiation/proliferation balance for different cell types used[56].

As mentioned before, FBS is a key element of traditional cell culture supplementing the media with essential nutrients, proteins, and fats. FBS is derived from the blood of a dairy cow foetus when it is being slaughtered. In addition to ethical concerns, there is a limited supply, leading to a significant price. Using serum would drive up the cost of cultured meat as FBS can cost upwards of $500 per litre[59], and approximately 50 litres are required to culture one kilogram of meat using current conditions[46]. Moreover, FBS is undefined, which can complicate the scale-up process and hamper the health of the cells at a commercial level[60]. To make cultured meat 'clean' and sustainable, it needs to be free of FBS. Thus, a renewable and accessible alternative is required that can be scaled up to accommodate the new industry.

5.2 Bioreactor adaptation

Another obstacle preventing cellAg from easily being adopted commercially is the inability to produce sufficient biomass[61]. In part this obstacle is based upon problems associated with industry-sized bioreactor design. Imperfect modelling, mass transfer, and shear stress are some of the limitations being faced[62]. Currently, modelling techniques for large-scale bioreactors are ineffective, with smaller-scale bioreactors unable to predict conditions for large-scale bioreactors[61]. Moreover, Computational Fluid Dynamics (CFD) tools used to model the operation of large-scale bioreactors are not configured for a cellAg context and are thus unreliable[63]. In lab-scale cell culture vessels, the large surface area to volume ratio means that the oxygen mass transfer occurs through the surface of the media. However, large-scale bioreactors have a much smaller surface area to volume ratio and rely on sparging and mixing as the main mechanism of oxygen mass transfer. Thus, the ability to predict and control the rate of oxygen transfer through sparging becomes a crucial design challenge in upscaling bioreactors.

Ongoing innovation and research are being dedicated to the reduction of shear stress in the bioreactors, which refers to the force experienced by cells inside the bioreactor because of velocity gradients present in the fluid following mixing or sparging[64]. The amount of shear stress is a significant factor in bioreactor design since it can cause cell death or detachment of anchorage-dependent cells from microcarriers (MCs)[62] or from bioscaffolds.

5.3 Scaffolding material

At present, methods to culture unstructured meat products are available, but not for structured products[65]. A scaffold will be required to produce structured meat products (defined meat cut analogues) such as steak. It will be easier to develop products such as minced meat[66]. Ideally, any scaffold used in cellAg would be non-animal in origin, have surface characteristics that enable cellular attachment, be affordable and easily scaled in a supply chain, and finally be able to add nutritional value to any final product.

There have been developments in using cellulose and decellularized spinach leaves as scaffolds that allow cells to grow in a defined arrangement[67,68]. This is currently on a small scale and will be required to be scaled up to produce meat in large quantities. Using 3D printing tools to impart structure and texture is being considered as an alternative, although that would add up to the overall cost of the final product[69,70].

Published by Burleigh Dodds Science Publishing Limited, 2023.

5.4 Acceptance and marketing challenges

Numerous challenges are associated with cellAg technology unrelated to scaling and the product development areas of media, bioreactor, and cost optimization. Of equal importance is the ability to market, at scale, cultured meat products, hence the significant interest in consumer expectations around cellAg products.

Even though multiple scientific publications and reports have been published around consumer expectations in the cellAg domain, there is currently no market for cultured meat[71]. Consequently, demand cannot be precisely assessed, and only forecasted.

Consumers have clear expectations towards the organoleptic properties of meat[14]. To achieve these properties with cellAg will be a challenge for everyone in this new field. It seems clear that a consumer who wants to eat meat will eat meat. Some of the published literature suggests that the proportion of participants willing to try cultured meat varies significantly, from 5% to 11%[26]. A key factor that reduces the acceptance of cultured meat is the belief that 'what is natural is good'[72]. Public and food-related risk awareness among individuals also influences the decision to accept or reject cultured meat[73]. These factors, whichever way they are mixed among populations, will affect the overall success of cultured meat products.

6 Conclusion and future trends in research

Many of our twenty-first-century food systems are suffering from parallel crises, i.e. climate change, sociocultural conflicts, large-scale global interdependencies, pandemics, and limited biodiversity – to name the most visible. CellAg and the various cultured products are being associated with numerous advantages ranging from being more sustainable, being more ethical, being nutritionally superior, to being more convenient and one day even more affordable. It will, however, require remarkable development and scientific progress to establish a robust system that can support the demand for meat. If the challenges linked with technology development, economic implementation, and consumer acceptance are being met, it is reasonable to assume that the expected advantages indeed will materialize.

For the future, it is realistic to anticipate different product offerings in the meat segment for consumers. Animal-based meat products will be co-existing with plant-based and cultured meat. It seems unrealistic to anticipate that animal-based products will be disappearing from the supermarket shelf anytime soon. Over the long run, it will be desirable to observe an alleviation of issues related to animal and livestock handling and to practices linked with factory farming.

The enthusiasm behind novel technology development is credible, and it is important to push financing, research, and implementation to the next level. At the same time, and as often, it is likely that the current and pioneering spirit in the cellAg field sooner or later will give way to a more sober and realistic view of the situation. A successful cellAg industry will have the visual appeal of stainless-steel bioreactor farms in an ultraclean production environment. Throughout its production, cultivated meat will raise the bar for hygiene and food safety as compared to conventional animal-based meat.

7 Where to look for further information

7.1 Introductory resources

- 'How it's made : the science behind cultured, clean, and cell-based meat' by Elliot Swartz (https://elliot-swartz.squarespace.com/cellbasedmeat/cleanmeat301).
- 'Is the future of meat animal-free?' by Liz Specht (https://www.ift.org/news-and-publications/food-technology-magazine/issues/2018/january/features/cultured-clean-meat).
- 'Cultured Meat and Future Food' podcast by Alex Shirazi (https://podcasters.spotify.com/pod/show/futurefoodshow).

7.2 Further reading

- Good Food Institute cultivated meat publications list (https://paperpile.com/shared/WLg4bc).
- The New Harvest OpenCellAg Repository (The New Harvest OpenCellAg Repository | Zenodo).
- Good Food Institute 'State of the industry reports' (https://gfi.org/resource/cultivated-meat-eggs-and-dairy-state-of-the-industry-report/).
- 'FDA completes first pre-market consultation for human food made using animal cell culture technology' (https://www.fda.gov/food/cfsan-constituent-updates/fda-completes-first-pre-market-consultation-human-food-made-using-animal-cell-culture-technology).
- 'Ex-ante life cycle assessment of commercial-scale cultivated meat production in 2030' (https://link.springer.com/article/10.1007/s11367-022-02128-8).

7.3 Books

- 'An introduction to cellular agriculture' by Ahmed Khan.
- 'Moo's Law: An investor's guide to the new agrarian revolution' by Jim Mellon.

Published by Burleigh Dodds Science Publishing Limited, 2023.

- 'Clean Meat: How growing meat without animals will revolutionize dinner and the world' By Paul Shapiro.
- 'Billion Dollar Burger: Inside Big Tech's Race for the Future of Food' by Chase Purdy.

8 References

1. le Coutre, J. Editorial: Cultured meat - Are we getting it right? *Front. Nutr.* 8, 675797 (2021). 10.3389/fnut.2021.675797.
2. Tubb, C. and Seba, T. *The Second Domestication of Plants and Animals, the Disruption of the Cow, and the Collapse of Industrial Livestock Farming A RethinkX Sector Disruption Report* (2019).
3. Searchinger, T. Waite, R., Hanson, C., Ranganathan, J. and Matthews, E. *Creating a Sustainable Food Future: Final Report*. Available at: www.sustainablefoodfuture.org (2019).
4. Pingali, P. L. Green revolution: Impacts, limits, and the path ahead. *Proc. Natl Acad. Sci. U. S. A.* 109(31), 12302–12308 (2012).
5. Espinosa, R., Tago, D. and Treich, N. Infectious diseases and meat production. *Environ. Resour. Econ. (Dordr)* 76(4), 1019–1044 (2020).
6. US Environmental Protection Agency. *Importance of Methane*. Available at: https://www.epa.gov/ghgemissions/overview-greenhouse-gases#methane (2022).
7. Nasir Md., N. A. N., Kamaruddin, S. A., Zakarya, I. A. and Aminul Islam, A. K. M. Sustainable alternative animal feeds: Recent advances and future perspective of using azolla as animal feed in livestock, poultry and fish nutrition. *Sustain. Chem. Pharm.* 25, 100581 (2022).
8. Rajauria, G. Seaweeds: A sustainable feed source for livestock and aquaculture. In: *Seaweed Sustainability* (Elsevier, 2015): 389–420. 10.1016/B978-0-12-418697-2.00015-5.
9. Verbeke, W., Marcu, A., Rutsaert, P., Gaspar, R., Seibt, B., Fletcher, D. and Barnett, J. 'Would you eat cultured meat?': Consumers' reactions and attitude formation in Belgium, Portugal and the United Kingdom. *Meat Sci.* 102, 49–58 (2015).
10. Dhont, K., Piazza, J. and Hodson, G. The role of meat appetite in willfully disregarding factory farming as a pandemic catalyst risk. *Appetite* 164, 105279 (2021). 10.1016/j.appet.2021.105279.
11. Wilks, M., Phillips, C. J. C., Fielding, K. and Hornsey, M. J. Testing potential psychological predictors of attitudes towards cultured meat. *Appetite* 136, 137–145 (2019).
12. van der Zee, B. What is the true cost of eating meat? *The Guardian*. Available at: https://www.theguardian.com/news/2018/may/07/true-cost-of-eating-meat-environment-health-animal-welfare (2018).
13. The Rockefeller Foundation. *True Cost of Food-Measuring What Matters to Transform the U. S. Food System*. Available at: https://www.rockefellerfoundation.org/report/true-cost-of-food-measuring-what-matters-to-transform-the-u-s-food-system/ (2021).
14. Siegrist, M. and Sütterlin, B. Importance of perceived naturalness for acceptance of food additives and cultured meat. *Appetite* 113, 320–326 (2017).

15. Bryant, C. J. Culture, meat, and cultured meat. *J. Anim. Sci.* 98(8) (2020). 10.1093/jas/skaa172.
16. Bryant, C. and Barnett, J. Consumer acceptance of cultured meat: A systematic review. *Meat Sci.* 143, 8–17 (2018).
17. Schaefer, G. O. and Savulescu, J. The ethics of producing in vitro meat. *J. Appl. Philos.* 31(2), 188–202 (2014).
18. Hollywood, J. and Madson, P. *Don't Have a Cow Man*. London: The Adam Smith Institute. Available at: https://www.adamsmith.org/research/dont-have-a-cow-man-the-prospects-for-lab-grown-meat (2018).
19. Fassler, J. Lab-grown meat is supposed to be inevitable. The science tells a different story. *The Counter*. Available at: https://thecounter.org/lab-grown-cultivated-meat-cost-at-scale/ (2021).
20. Odegard, I., Sinke, P. and Vergeer, R. *TEA of Cultivated Meat - Future Projections of Different Scenarios*. Available at: https://cedelft.eu/publications/tea-of-cultivated-meat/ (2021).
21. Bryant, C. J. and Barnett, J. C. What's in a name? Consumer perceptions of in vitro meat under different names. *Appetite* 137, 104–113 (2019).
22. Schenk, P., Rössel, J. and Scholz, M. Motivations and constraints of meat avoidance. *Sustainability* 10(11), 3858 (2018).
23. Scollan, N. D., Dannenberger, D., Nuernberg, K., Richardson, I., MacKintosh, S., Hocquette, J. F. and Moloney, A. P. Enhancing the nutritional and health value of beef lipids and their relationship with meat quality. *Meat Sci.* 97(3), 384–394 (2014).
24. Gasteratos, K. *90 Reasons to Consider Cellular Agriculture*. Available at: http://nrs.harvard.edu/urn-3:HUL.InstRepos:38573490 (2019).
25. Middleton, J., Reintjes, R. and Lopes, H. Meat plants—A new front line in the covid-19 pandemic. *BMJ* 370, m2716. 10.1136/bmj.m2716 (2020).
26. Wilks, M., Phillips, C. J. C., Fielding, K. and Hornsey, M. J. Testing potential psychological predictors of attitudes towards cultured meat. *Appetite* 136, 137–145 (2019).
27. Tuomisto, H. L. and Teixeira de Mattos, M. J. Environmental impacts of cultured meat production. *Environ. Sci. Technol.* 45(14), 6117–6123 (2011).
28. Swartz, E. *New Studies Show Cultivated Meat Can Have Massive Environmental Benefits and Be Cost-Competitive by 2030*. Available at: https://gfi.org/blog/cultivated-meat-lca-tea/ (2021).
29. Sinke, P. and Odegard, I. *LCA of Cultivated Meat: Future Projections for Different Scenarios*. Available at: https://cedelft.eu/publications/rapport-lca-of-cultivated-meat-future-projections-for-different-scenarios/ (2021).
30. Robinson, T. *Global Livestock Environmental Assessment Model (GLEAM)*. Available at: www.fao.org/gleam/en/ (2017).
31. Meat, M. *Where's the Beef? A First-Hand Assessment of Cultivated Meat Progress*. Available at: https://mosameat.com/blog/cultivated-meat-progress (2021).
32. Choudhury, D., Tseng, T. W. and Swartz, E. The business of cultured meat. *Trends Biotechnol.* 38(6), 573–577 (2020).
33. Byrne, B. and Murray, S. *State of the Industry Report-Cultivated Meat*. Available at: https://gfi.org/wp-content/uploads/2021/04/COR-SOTIR-Cultivated-Meat-2021-0429.pdf (2020).
34. Kolkmann, A. M., Post, M. J., Rutjens, M. A. M., van Essen, A. L. M. and Moutsatsou, P. Serum-free media for the growth of primary bovine myoblasts. *Cytotechnology* 72(1), 111–120 (2020).

35. Stout, A. J., Mirliani, A. B., Rittenberg, M. L., Shub, M., White, E. C., Yuen, J. S. K. and Kaplan, D. L. Simple and effective serum-free medium for sustained expansion of bovine satellite cells for cell cultured meat. *Commun. Biol.* 5(1), 466 (2022).
36. Messmer, T., Klevernic, I., Furquim, C., Ovchinnikova, E., Dogan, A., Cruz, H., Post, M. J. and Flack, J. E. A serum-free media formulation for cultured meat production supports bovine satellite cell differentiation in the absence of serum starvation. *Nat. Food* 3(1), 74–85 (2022).
37. Tubb, C. and Seba, T. *The Second Domestication of Plants and Animals, the Disruption of the Cow, and the Collapse of Industrial Livestock Farming A RethinkX Sector Disruption Report* (2019).
38. Post, M. J., Levenberg, S., Kaplan, D. L., Genovese, N., Fu, J., Bryant, C. J., Negowetti, N., Verzijden, K. and Moutsatsou, P. Scientific, sustainability and regulatory challenges of cultured meat. *Nat. Food* 1(7), 403–415 (2020).
39. Spier, R. E. and Kadouri, A. The evolution of processes for the commerical exploitation of anchorage-dependent animal cells. *Enzyme Microb. Technol.* 21(1), 2–8 (1997).
40. Eke, G., Vaysse, L., Yao, X., Escudero, M., Carrière, A., Trevisiol, E., Vieu, C., Dani, C., Casteilla, L. and Malaquin, L. Cell aggregate assembly through microengineering for functional tissue emergence. *Cells* 11(9), 1394 (2022).
41. Chen, G., Gulbranson, D. R., Hou, Z., Bolin, J. M., Ruotti, V., Probasco, M. D., Smuga-Otto, K., Howden, S. E., Diol, N. R., Propson, N. E., Wagner, R., Lee, G. O., Antosiewicz-Bourget, J., Teng, J. M. and Thomson, J. A. Chemically defined conditions for human iPSC derivation and culture. *Nat. Methods* 8(5), 424–429 (2011).
42. Sun, N., Panetta, N. J., Gupta, D. M., Wilson, K. D., Lee, A., Jia, F., Hu, S., Cherry, A. M., Robbins, R. C., Longaker, M. T. and Wu, J. C. Feeder-free derivation of induced pluripotent stem cells from adult human adipose stem cells. *Proc. Natl Acad. Sci. U. S. A.* 106(37), 15720–15725 (2009).
43. Kayser, K. *Animal Component Free: UPSIDE's Cell Feed Breakthrough Levels Up the Future of Cultivated Meat*. Available at: https://upsidefoods.com/animal-component-free-upsides-cell-feed-breakthrough-levels-up-the-future-of-cultivated-meat/ (2021).
44. Buxton, A. *South Korea's Cell MEAT Makes Fetal Bovine Serum-Free Cell Culture Media for 100% Ethical Cultivated Meat*. Available at: https://www.greenqueen.com.hk/cellmeat-fetal-bovine-serum/ (2021).
45. Rybchyn, M. S., Biazik, J. M., Charlesworth, J. and le Coutre, J. Nanocellulose from *Nata de Coco* as a bioscaffold for cell-based meat. *ACS Omega* 6(49), 33923–33931 (2021).
46. Humbird, D. Scale-up economics for cultured meat. *Biotechnol. Bioeng.* 118(8), 3239–3250 (2021).
47. Yamamoto, Y., Ito, A., Jitsunobu, H., Yamaguchi, K., Kawabe, Y., Mizumoto, H. and Kamihira, M. Hollow fiber bioreactor perfusion culture system for magnetic force-based skeletal muscle tissue engineering. *J. Chem. Eng. Jpn* 45(5), 348–354 (2012).
48. Baba, K. and Sankai, Y. Development of biomimetic system for scale up of cell spheroids - building blocks for cell transplantation. In: *39th Annual International Conference of the IEEE Engineering in Medicine and Biology Society (EMBC)* (IEEE Publications, 2017), 1611–1616. 10.1109/EMBC.2017.8037147.
49. Allan, S. J., de Bank, P. A. and Ellis, M. J. Bioprocess design considerations for cultured meat production with a focus on the expansion bioreactor. *Front. Sustain. Food Syst.* 3 (2019).

50. Jha, A. and Topham, L. *First Hamburger Made From Lab-Grown Meat to be Served at Press Conference.* Available at: https://www.theguardian.com/science/2013/aug/05/first-hamburger-lab-grown-meat-press-conference (2013).
51. Gabbert, J., Tarhuni, N., Frederick, A., Hu, S. and Midkiff. *Reinventing Meat-Lab Grown Protein is Cultivating Investor Interest.* Available at: https://pitchbook.com/newsletter/lab-grown-protein-is-cultivating-investor-interest-xlE (2021).
52. Marston, J. and Raises, J. *$2M for Its Animal-Free Collagen and Gelatin.* Available at: https://thespoon.tech/jellatech-raises-2m-for-its-animal-free-collagen-and-gelatin/ (2021).
53. Leichman, A. K. *One Step Closer to Lab-Made Mother's Milk.* Available at: https://wilkismilk.com/news/israel21c-one-step-closer-to-lab-made-mothers-milk/. ISRAEL21c (2022).
54. *Fiction Foods. Cult Food Science Portfolio Company Fiction Foods Introduces Proprietary Cell-Based Performance Scramble* (2022).
55. O'Neill, E. N., Cosenza, Z. A., Baar, K. and Block, D. E. Considerations for the development of cost-effective cell culture media for cultivated meat production. *Compr. Rev. Food Sci. Food Saf.* 20(1), 686–709 (2021).
56. Soice, E. and Johnston, J. Immortalizing cells for human consumption. *Int. J. Mol. Sci.* 22(21) (2021). 10.3390/ijms222111660.
57. Bassaganya-Riera, J., Berry, E. M., Blaak, E. E., Burlingame, B., le Coutre, J., van Eden, W., El-Sohemy, A., German, J. B., Knorr, D., Lacroix, C., Muscaritoli, M., Nieman, D. C., Rychlik, M., Scholey, A. and Serafini, M. Goals in nutrition science 2020–2025. *Front. Nutr.* 7, 606378 (2020). 10.3389/fnut.2020.606378.
58. Skrivergaard, S., Rasmussen, M. K., Therkildsen, M. and Young, J. F. Bovine satellite cells isolated after 2 and 5 days of tissue storage maintain the proliferative and myogenic capacity needed for cultured meat production. *Int. J. Mol. Sci.* 22(16) (2021).
59. Brann, M. (2022). *Foetal Calf Blood Hits Record $615 a Litre as Research Labs Secure Crucial Commodity.* Available at: https://www.abc.net.au/news/2022-06-18/foetal-calf-blood-bovine-serum-record-value-australia/101159254.
60. Djisalov, M., Knežić, T., Podunavac, I., Živojević, K., Radonic, V., Knežević, N. Ž., Bobrinetskiy, I. and Gadjanski, I. Cultivating multidisciplinarity: manufacturing and sensing challenges in cultured meat production. *Biology (Basel)* 10(3) (2021). 10.3390/biology10030204.
61. Allan, S. J., de Bank, P. A. and Ellis, M. J. Bioprocess design considerations for cultured meat production with a focus on the expansion bioreactor. *Front. Sustain. Food Syst.* 3 (2019).
62. Hanga, M. P., de la Raga, F. A., Moutsatsou, P., Hewitt, C. J., Nienow, A. W. and Wall, I. Scale-up of an intensified bioprocess for the expansion of bovine adipose-derived stem cells (bASCs) in stirred tank bioreactors. *Biotechnol. Bioeng.* 118(8), 3175–3186 (2021).
63. Verma, R., Mehan, L., Kumar, R., Kumar, A. and Srivastava, A. Computational fluid dynamic analysis of hydrodynamic shear stress generated by different impeller combinations in stirred bioreactor. *Biochem. Eng. J.* 151, 107312 (2019).
64. Li, C., Teng, X., Peng, H., Yi, X., Zhuang, Y., Zhang, S. and Xia, J. Novel scale-up strategy based on three-dimensional shear space for animal cell culture. *Chem. Eng. Sci.* 212, 115329 (2020).

65. Bellani, C. F., Ajeian, J., Duffy, L., Miotto, M., Groenewegen, L. and Connon, C. J. Scale-up technologies for the manufacture of adherent cells. *Front. Nutr.* 7, 575146 (2020).
66. Modulevsky, D. J., Lefebvre, C., Haase, K., Al-Rekabi, Z. and Pelling, A. E. Apple derived cellulose scaffolds for 3D mammalian cell culture. *PLoS ONE* 9(5), e97835 (2014).
67. Reiss, J., Robertson, S. and Suzuki, M. Cell sources for cultivated meat: Applications and considerations throughout the production workflow. *Int. J. Mol. Sci.* 22(14) (2021). 10.3390/ijms22147513.
68. Bomkamp, C., Skaalure, S. C., Fernando, G. F., Ben-Arye, T., Swartz, E. W. and Specht, E. A. Scaffolding biomaterials for 3D cultivated meat: prospects and challenges. *Adv. Sci. (Weinh)* 9(3), e2102908 (2022).
69. Kang, D. H., Louis, F., Liu, H., Shimoda, H., Nishiyama, Y., Nozawa, H., Kakitani, M., Takagi, D., Kasa, D., Nagamori, E., Irie, S., Kitano, S. and Matsusaki, M. Engineered whole cut meat-like tissue by the assembly of cell fibers using tendon-gel integrated bioprinting. *Nat. Commun.* 12(1), 5059 (2021).
70. Wang, T., Kaur, L., Furuhata, Y., Aoyama, H. and Singh, J. 3D printing of textured soft hybrid meat analogues. *Foods* 11(3), 478 (2022).
71. Rombach, M., Dean, D., Vriesekoop, F., de Koning, W., Aguiar, L. K., Anderson, M., Mongondry, P., Oppong-Gyamfi, M., Urbano, B., Gómez Luciano, C. A., Hao, W., Eastwick, E., Jiang, Z. V. and Boereboom, A. Is cultured meat a promising consumer alternative? Exploring key factors determining consumer's willingness to try, buy and pay a premium for cultured meat. *Appetite* 179, 106307 (2022). 10.1016/j.appet.2022.106307.
72. Verbeke, W., Marcu, A., Rutsaert, P., Gaspar, R., Seibt, B., Fletcher, D. and Barnett, J. 'Would you eat cultured meat?': Consumers' reactions and attitude formation in Belgium, Portugal and the United Kingdom. *Meat Sci.* 102, 49–58 (2015).
73. Bryant, C. and Barnett, J. Consumer acceptance of cultured meat: an updated review (2018–2020). *Appl. Sci.* 10(15), 5201 (2020).

Chapter 2

Talking points on the cultural politics of cultured meat

Neil Stephens, University of Birmingham, UK; and Alexandra Sexton, University of Sheffield, UK

1 Introduction

Cultured meat technologies are seeking to change the world in a significant way, and like anything that seeks to change the world - irrespective of whether it is a change we want to see or not - it requires critical inspection in terms of what that change looks like and who it will impact. In this chapter, we introduce a number of talking points on the topic of cultured meat, including issues for further consideration, and provide open questions to be explored subsequently. We have grouped the themes into four headings: the public life of cultured meat, the economic life of cultured meat, the political life of cultured meat, and cultured meat and social change. These are intended to open up key points of reflection for anyone engaged in cultured meat to think about, in order to consider what types of impact the technology might have, and where some of the uncertainties remain.

Cultured meat technology has always been articulated through a set of promissory narratives about what it will be and how it will change the world (for the better). These narratives are likely already familiar to readers of this book, and some are discussed in more depth elsewhere within this book. One key vision has been the capacity to produce meat in a way that is beneficial for the environment compared to existing production methods, with lower use of land, water, and energy, potentially resulting in lower greenhouse gas emissions.

http://dx.doi.org/10.19103/AS.2023.0130.02

A second key vision is to produce meat while killing either none, or at least far fewer animals than current systems, bringing with it moral value in terms of reducing global violence and animal exploitation. A third core vision within these promissory narratives has been producing meat that is healthier or safer than traditional production methods, with sub-themes within this pointing to meat production without antibiotic use, novel forms of engineered nutrition, and a meat free from the risks of zoonotic disease. Beyond this core three themes, there have been other narratives that become more or less visible at different moments, such as the capacity to produce meat for space travel, the economic benefits of developing new companies and sectors within national economies, and the capacity to make new eating experiences through innovative forms of meat. The narratives have also been developed over time, with more focused species or region-specific narratives developed by companies which focus on distinct markets, for example the fish-oriented companies in Asia which reference the specific marine ecology or local cultural tastes.

These promissory narratives, we have long argued (Sexton et al., 2019; Stephens, 2010, 2013; Stephens et al., 2019), do more than articulate a vision of the future. They are active parts of shaping cultured meat today, by attracting resources, being the basis of debate about how desirable or tractable the technology and its goals are, and by asserting a framework of meaning around what cultured meat actually is (as well as who will consume it and why). Cultured meat technology, and the way people talk about it, is an attempt to reshape the world of today and tomorrow; it is intended as a clear intervention into our societies, our ethics, and our planetary health.

If cultured meat is successful, it will be inherently cultural. People will eat it, identify with it, use it to further their politics, maybe even normalise it. It will become embedded within cultural and public life. But there is no guarantee that this will happen, and no clear-cut ways of knowing what form these new food cultures might take, or their broader impact, if they come into being. This chapter is written in this spirit. It raises talking points about the cultural politics of cultured meat.

2 Public life of cultured meat

A key frame used within the cultured meat community for the public life of cultured meat has been the 'consumer acceptance' model. Studies have been conducted on public and consumer acceptance issues, employing both survey and focus group methodologies (as discussed elsewhere in this book). The findings vary, with the variance connected to both the territory or demographic asked, as well as the study design and how cultured meat is introduced within it (Bryant and Barnett, 2018, 2020). Studies find publics have a variety of responses to cultured meat, with some very supportive, others very critical,

Published by Burleigh Dodds Science Publishing Limited, 2023.

and many in between. In general, the studies tend to show greater acceptance among younger, educated, male, urban individuals. They tend to show larger numbers say they would try it, as opposed to commit to eating it regularly. However, studies also tend to show that the largest cohort is still people who have not yet heard of the technology and are yet to develop a clear opinion.

Another recurring theme is that the type of information provided to people about cultured meat shapes their responses, with more positive framings leading to more positive responses (Bekker et al., 2017; Bryant and Dillard, 2019; Rolland et al., 2020). Given this, it is worth remembering that real-world public responses to cultured meat will be framed by the societal information context at the time, which will depend upon the dominant or competing narratives in popular circulation at the time. Furthermore, we suggest there has been at times an overemphasis on the role of consumer acceptance studies, as they can offer an overly reductive framing of broader publics, and under-represent the richness and nuance of people's understandings of the political, social, and identity elements of food choices. Studies that focus primarily on whether consumers will or will not purchase cultured meat, and what factors may increase their likelihood of purchasing cultured meat, are important and needed, but by limiting the core point of consideration to a binary purchasing decision, they do not allow us to fully grapple with the diversity and complexity of people's engagement with cultured meat technology, which extends far beyond whether to buy or not.

This is relevant, as meat has a long history of being associated with identity, and this has taken different forms at different times and contexts. As examples, meat has been associated with displays of masculinity, wealth, strength, and health (Adams, 2015; Rosenfeld et al., 2020). As such, the social status of meat in any given context is a variable that supporters of cultured meat need to be attentive to. Equally, it raises questions about if and how cultured meat might alter the social status of meat if it becomes adopted. As one example, the status of cultured meat within specific religious thinking has been an ongoing topic of discussion. Primarily, this focuses upon its permissibility within Islamic halal and Jewish kosher practices (Hamdan et al., 2019; Hamdan et al., 2021; Kenigsberg and Zivotofsky, 2020; Krone, 2022). Discussions of this type typically highlight specific variables that would impact upon a halal or kosher designation, with the cell source, media content, and specific process used being important in both cases. As an example, it has been suggested cultured meat would only be halal if the cells are extracted from a halal slaughtered animal, and no blood is used in the media. In terms of being defined as kosher, similarly, the importance of being from a kosher species and being kosher slaughtered has been raised. However, in the case of both halal and kosher, this discussion remains ongoing as both the technology and thinking in this area develop.

Analysis of English language media coverage of cultured meat has shown that news stories tend to be generally positive about the technology (Goodwin and Shoulders, 2013; Hopkins, 2015; Painter et al., 2020). This is because the reporting generally reflects narratives from cultured meat advocates, either the companies or supportive NGOs and individuals. These stories are often based upon company-issued press releases and feature interviews from cultured meat proponents. In our observations, the news framing has been fairly standardised for many years. Cultured meat is presented as an interesting and unusual science, which might be very beneficial for the environment, animal welfare, and health reasons, before questions are raised about whether people would choose to eat it, or more recently, how it might be regulated. Early stories were premised mostly around future-oriented claims from scientists or entrepreneurs saying the production of cultured meat could soon be possible. Increasingly, such stories now use a company announcement as a hook, for example a new image or tasting opportunity of a cultured meat prototype, or recently raised money.

In the media and elsewhere, cultured meat has been given many names (Bryant and Barnett, 2019; Stephens et al., 2019). Around 2005, *in vitro* meat was used most regularly within the community of people working to produce it. Over time this shifted to cultured meat, which was the dominant name by 2015. Then, clean meat was suggested, and became prominent for a few years, before decreasing in use again, as cell-based meat and then cultivated meat also became popular. Other names have been used outside of the community. The media frequently uses lab-grown meat, with other terms including synthetic meat, fake meat, and Frankenstein meat seen as antagonistic to the technology by those within the community (Painter et al., 2020). The choice of name is important, as it frames how cultured meat is understood by accentuating some interpretations and diminishing others. Importantly, all of these retain the word meat as a descriptor, asserting the technology's meatness.

Another important issue is the number of uncertainties about the future of cultured meat, which include the technical process it may involve, the resources needed to produce it, the price, the political and consumer response, the impact on other markets, and related impacts on the environment, landscapes, and rural communities (Stephens et al., 2018). This leads to questions about how accountable public discussion - in the media, policy, and other contexts - should best proceed to engage with that uncertainty (Ryynänen and Toivanen, 2022; Sexton et al., 2019). This, of course, is not an issue specific to cultured meat, it is also true of numerous other innovative technological projects; however, working through the specifics of this in the cultured meat context remains important.

Published by Burleigh Dodds Science Publishing Limited, 2023.

3 Economic life of cultured meat

To be successful, cultured meat needs to be an economic success. Key discussion points here include the likely market impact, as well as funding and business models being used, and how these issues relate to broader economic factors like government support and power within the food sector.

A key discussion point that shapes many other issues is what impact cultured meat might have in the market. There are many unknowns that compound the complexity of making accurate predictions. Key variables will include the quality, availability, and price of any cultured meat products (Treich, 2021; Van Loo et al., 2020). However, the question is more than just what level of market penetration cultured meat might achieve in any given time frame given these quality and price attributes. We must also consider what might happen to livestock-based meat sales, and different subcategories within these. The typical starting narrative used within the cultured meat community is that cultured meat sales will displace sales of livestock meat, as consumers switch from one product to the other. This can be characterised as a type of substitution effect, in which the total volume of meat sales remains the same, but consumers shift from livestock to cultured sources. The capacity of cultured meat to deliver environmental benefits is premised upon this reduction in livestock-based meat.

However, there are multiple ways the real-world market dynamics of cultured meat introduction may be more complex than this one-for-one substitution effect. One possibility would be an addition effect, in which livestock-based sales do not decrease as cultured meat sales increase, with the net result of more meat produced and consumed (livestock and cultured combined) (Stephens et al., 2018). The ratio of increases in cultured meat to decreases in livestock-based meat will be important, and this again has inherent complexities. Attempts could be made to predict these outcomes based upon experiences of other innovations that sought to replace established products. However, issues including the cultural significance of cultured meat and uncertainties about the technical process make such predictions difficult.

One key dynamic is 'sausages versus steaks', which asks what the different market impacts might be of cultured meat systems that can only produce processed meats (sausages, burgers etc.), as opposed to cultured meat systems that can also effectively produce more complex structures (steaks, chicken breasts, etc.). The issue here is whether the industrial production of cultured meat processed foods would impact the profitability of livestock production sufficiently to lower the number of animals farmed, or whether the market for steaks and other complex meats would be sufficient to sustain farmed animal numbers. In other words, to what extent do processed cultured meat products hit the margins of livestock farmers, and what impact might this have. It is, of course, likely that such trade-offs would play out differently in different and changing contexts.

Another question is where the market impact would be felt. For instance, we can ask whether reductions in livestock numbers would happen in the country of consumption, or whether it is more likely (or more or less beneficial) for production reductions to happen elsewhere in the world (and where that might be). This is a point we return to later in the section on social change.

Another core point to consider is the underlying funding model. The cultured meat sector has been characterised by two waves of investment models and personnel (Stephens et al., 2019). The first wave broadly covered the period from the early 2000s to 2013 and involved mostly individual university-based projects funded through small government grants - in some cases as side projects within existing research grants - and awards from third sector organisations. Early cultured meat research programmes faced multiple barriers to securing funding. Cultured meat was seen to blur existing disciplinary boundaries between biomedical and food innovation and so often failed to fit neatly into established funding bids. Many early cultured meat researchers also reported a general sense of oddness attributed to their work from their colleagues and usual sources of university support. The nascency of the technology during the first wave was also seen as too early stage, too high risk, and too expensive for many public funds.

The transition between the first and second waves of cultured meat investment in the 2010s was characterised by the entry of high net-worth philanthropic individuals (e.g. Sergey Brin, Bill Gates) and investment streams more associated with high-tech ventures, such as technology accelerator programmes and private venture capital firms. The San Francisco Bay Area in California, popularly referred to as Silicon Valley, attracted increasing cultured meat activity. The numbers of companies steadily rose from a handful in 2015–2016 to around 100 by the early 2020s, and we suspect those numbers may continue to rise in the near future. Typically, these companies begin with a founder duo, who develop the vision and business plan for the company, by looking to find a point by which they can differentiate themselves from existing companies, while still remaining sufficiently similar to the core model for the sector to attract venture capitalist support and subsequently grow their staff. There have also been a number of investments in these companies from multinational livestock companies and incumbents within the middle of the agri-food value chain (processing, food retail, and services), for example Tyson Foods and PHW Group, and globally recognised food retail brands such as KFC, which have made strategic investments and partnerships within the section, often to keep track of this emerging and potentially rival industry.

Recent years have seen increasing calls within the cultured meat industry for more government financial support, be that through public-private joint ventures or standard academic funding. Prominent corporate actors within the sector have highlighted that greater sums of capital and longer time frames of R&D than seed-level venture capital can provide may be needed to transition lab-scale prototypes to larger scale pilot and commercial facilities. Some have also highlighted frustrations with the lack of collaboration and data sharing within the industry, in part a product of the intellectual property that has been tied to the private high-tech investment streams that have to date shaped this emerging sector (Dutkiewicz, 2019). A shift to greater public funding would potentially build in more open-source models of knowledge production that could benefit the industry and possibly other sectors outside of protein food production (Newton and Blaustein-Rejto, 2021) - for example, bringing the cost of cell culture medium down could make the production of certain biopharmaceutical products cheaper. More democratic sources of funding and R&D models could also mitigate against these technologies repeating the context of genetic engineering and pharmaceuticals whereby a small number of companies reap the majority of economic benefits, and in doing so perpetuate the imbalance of power and ownership that currently drives significant inequalities in the global food system. Asking who is funding the sector, both through private and third sector means, requires continuous critical scrutiny as investment streams play a central role in shaping how technologies like cultured meat are developed and the possibilities they present in both technical and sociopolitical terms.

We also need to consider the business models that companies are using to support their work. Most of the first wave of cultured meat companies positioned themselves as largely vertically integrated companies developing their own meat products, culture media, and bioreactor technology, and a B2C business targeting sales of meat to consumers. Over time, more companies were launched that instead focused upon a specific component of the process, e.g. culture media or bioreactors, operating as B2B companies supplying others in the sector. This has seen more discussion of the early companies operating in a less vertically integrated way by working with other companies as suppliers. Recently, the sector has seen a small number of acquisitions, as larger cultured meat companies have bought smaller ones to take in-house their expertise and technology.

These business models will in part shape the impact of cultured meat on ongoing discussions and concerns about the concentration of power within the food industry (Broad, 2019; Treich, 2021). Whether cultured meat will support or challenge the increasing concentration of power among a small number of corporate actors is currently unclear, though recent developments within the sector suggest that it will likely be 'incorporated as reforms within the corporate food regime' (Broad, 2019). Continued partnerships between cultured meat ventures and multinational food corporations (aka 'Big Food') reveal that some of the most powerful agri-food incumbents are making strategic investments in cultured meat and other alternative proteins to ensure they are the disruptors, not the disrupted. It is unclear whether cultured meat will eventually displace or simply add new revenue streams alongside the conventional livestock operations of incumbents, or whether the cultured meat ventures, once acquired, will be closed down to mitigate competition. The buy-in costs of cultured meat for the near future are expected to be prohibitive to small- and medium-scale agribusinesses without some form of subsidised support. A critical question to ask, then, is not only whether cultured meat presents opportunities for conventional agribusiness, but who among incumbent supply chains has the economic power to take early advantage of this emerging sector.

Some accounts of future cultured meat production envision a form of localisation and democratisation of meat production, with many smaller-scale producers operating as a challenge to existing large-scale food production in a similar manner to previous alternative food networks (van der Weele and Driessen, 2013). Cultured meat potentially offers opportunities for making small mixed farming of heritage livestock economically viable for the producer and more affordable for the eater. This cultured meat scenario is linked more closely to current visions of a 'small farm future' (Smaje, 2020) and seeing this novel technology within a more holistic plan of farming as a suite of ecological, economic and sociocultural services.

Another key economic aspect is the role of governments, which have the capacity to support or hinder the development of cultured meat through multiple mechanisms. This includes whether governments decide to provide support for the sector in terms of research grants, or to help develop and promote training opportunities with the sector. Another is tax and subsidy regimes. This includes whether cultured meat producers should receive subsidies, and the extent to which this could relate to food production or environmental stewardship (Treich, 2021). There is also scope for discussions of whether the potential of cultured meat relates to the sometimes-mooted issue of a meat tax, or emissions trading schemes (Funke et al., 2021). Analysis is needed into how might a global agricultural subsidy models change in this context.

4 Political life of cultured meat

As we have made clear, if cultured meat is successful, and even if it is not, it will have political ramifications. This impact will be framed by issues, including power in the food system, the distribution of resources, the types of political discussions that people have, and the role of governments.

A key talking point is access and food justice. A common goal of the cultured meat sector is to create products that are cheap, tasty, and convenient. Care must be taken however not to assume that cheap food is the route to a fair food system, and that if cultured meat is cheap, it will be affordable and accessible to all. We are currently living in an age where food has never been cheaper and at the same time, food poverty and hunger have been rising in both Minority and Majority nations. As has been widely evidenced, the costs of the cheap food model are inevitably paid somewhere, and often disproportionately by the most vulnerable and marginalised populations. Depending on how cultured meat scales up as an industry, the costs of a race to the bottom could come in terms of livelihood loss, breakdown of rural communities, further consolidation of power and ownership in the agri-food system, predatory delay by incumbents to stall change, more overall production and resource use due to the addition rather than substitution effect, continued public health issues associated with cultures of over-consumption generally and of meat products specifically, and continued dependence on intensive monocrop models of agriculture (i.e. land sparing) to supply this new industry.

Achieving food justice through cultured meat is not automatic, and if unregulated, could be unlikely. More research is needed to consider what optimal modes of responsible innovation and more holistic understandings of food and farming could provide for considering how cultured meat might fit into sustainable and just agricultural models. Some ideas that have been proposed include a 'socialised' funding model (Dutkiewicz, 2019) to encourage open-source knowledge production and a 'food tech justice' approach (Broad, 2019) that works to build in established principles of food justice into the innovation processes and outcomes of cultured meat development. More research is needed to explore these kinds of alternative models for scaling cultured meat development, and it will require active commitment by the cultured meat industry to build in public goods into their processes, as well as support and incentives from the third sector.

Issues of food justice, in many nations, open up discussions about democracy and accountability. Some visions of the future of cultured meat anticipate profound social, economic, and environmental change. Indeed, these changes are for many the core reason to support the technology. However, the potential for such change leads to questions about what an optimal mode of accountability for this change might look like. It is reasonable to ask what kind of democratic mandate is necessary for change of this scale to be deemed legitimate. One possibility is to let the market adjudicate, in that if sufficient numbers of people purchase cultured meat, then that establishes a democratic mandate. However, it could be argued that this approach, first, leaves the democratic decision too late in the process to effectively shape the development pathway, and second, that food-purchasing decisions are shaped by other constraints beyond the free giving of support for or against a political project. This, then, leads to a question of what other forms of accountability may be appropriate, for example, dialogue models or government-led projects.

Clearer visions are required for what an accountable and legitimate cultured meat future could look like.

Critical commentators have also argued that cultured meat represents the familiar model of 'non-disruptive disruptions' (Goldstein, 2018) – that is, technofixes that allow businesses, politicians, and individuals to avoid taking hard decisions, such as pursuing strong policies of behaviour change and regulatory penalties for corporates engaging in ecologically damaging practices, in favour of markets and 'Science' providing solutions in the near future. In this view, cultured meat could be seen as part of a mandate for inactivity that gives licence to societies to avoid hard decisions and strong actions that are required today. Based on publicly available data, it is difficult to imagine that within the next 5 years, cultured meat will both (a) be available as an affordable mass market product in multiple countries and (b) have meaningfully replaced conventional meat production and its various associated harms. We would tentatively suggest that such outcomes are currently ambitious to expect even within the next decade, and thus provide no such mandate for inactivity on all aspects of addressing climate crisis and food justice.

Clearly, the concern is that the future promise of a set of technologies that include cultured meat could be used to absolve politicians from the moral responsibility to act now. It is widely accepted that the next decade is critical for making substantial systemic changes to global food practices to mitigate against the most extreme consequences of climate change, as well as address other looming crises such as antimicrobial resistance. Critics have argued that the quantity of funds, time, and effort being expended on technofixes like cultured meat detract from supporting agri-food solutions that already exist, such as agroecology, which have established proof of efficacy and offer more ecologically holistic and socially democratic outcomes. Where some cultured meat advocates claim there is no time to reinvent capitalism and scale lower tech approaches from the ground up, a counter-argument could see cultured meat as a form of 'predatory delay' that enables the biggest perpetuators of agri-food-related harms to continue with business as usual as they wait for technofixes to be developed. It is important to note too that there is currently no certainty that cultured meat and other cellular agriculture products will ever be technologically or economically viable as larger-scale commercial enterprises, meaning they would be unable to deliver on their promises of global-scale benefits for societies and the planet.

5 Cultured meat and social change

Changes in the market for meat products could have associated changes in related labour markets (Newton and Blaustein-Rejto, 2021). Uncertainties about the impact of cultured meat on the market thus introduce parallel uncertainties about labour market change (Stephens et al., 2018). A successful cultured meat sector would bring new employment opportunities, and even new job types, though not necessarily an increase in the number of overall food and agricultural jobs. The new jobs created by a cultured meat sector may contribute to the knowledge economy and economic growth within host nations. The impact on existing labour markets in livestock, meat processing, and associated sectors (such as feed production) depends on the substitution or addition effect of cultured meat as discussed earlier, as well as on the design and scale of cultured meat production systems. Declining livestock-based meat production would likely lead to less employment in the sector and those aligned to it. Certain technical requirements and economies of scales of cultured meat production, such as maintaining a sterile culturing environment, also favour automation over human labour. In such a context it may be relevant to discuss what 'just transition' (McCauley and Heffron, 2018) may look like in this context, as a set of social interventions to support workers whose jobs may be coming under threat to ensure workers and communities do not bear the costs of change. This given, an emergent cultured meat sector might offer new opportunities for a minority in terms of cell sourcing, working with heritage breeds, and the feed industry in producing input materials for cultured meat production, and also new possibilities in meat processing. A key discussion point is where these labour market shifts could occur, regionally and internationally.

Cultured meat presents potential opportunities and impacts for rural communities (Newton and Blaustein-Rejto, 2021). Some predict cultured meat will produce equal or greater amounts of meat than conventional livestock production on a significantly smaller land footprint. This could lead to the loss of agricultural businesses connected to conventional livestock production, which in turn could have significant sociocultural and economic impacts on rural communities. At the same time, job opportunities within conventional agriculture in places like the United States and United Kingdom have been on a downwards trend over recent decades. This is in large part due to the shift away from more labour-intensive mixed farming models to industrialised agriculture which has replaced the majority of on-farm jobs with mechanisation and chemical inputs.

Investing in rural and agricultural economies has been highlighted as a key priority in many countries. As discussed earlier, cultured meat may potentially offer new job opportunities that could benefit rural communities. Possible options for existing farmers could be to provide inputs to the cultured meat industry, such as donor cells and plant-based ingredients for the growth serum. Locating cultured meat production facilities in rural areas could provide new job opportunities; it is important that the quality, safety, and security of these jobs, as well as overall number, are assessed in comparison with those offered

Published by Burleigh Dodds Science Publishing Limited, 2023.

by existing agricultural models, ranging from industrial to agroecological. There are a number of competing factors for choosing the location of cultured meat facilities which will determine the possible opportunities for rural communities: these include land price and existing infrastructure, as well as the ecological and economic savings of being close to energy sources (e.g. wind, water), key inputs (e.g. crops for growth serum), and concentrated customer bases (e.g. urban areas).

There are certain technical requirements and economies of scale that may result in cultured meat production being significantly automated and thereby requiring a reduced labour force compared with both industrial and more extensive, low-input agricultural models. If cultured meat catalyses large-scale rewilding efforts, another and possibly greater opportunity for rural communities may come from conservation and ecotourism. The development of a 'craft' industry of cultured meat production has also been envisioned by some in the sector. While the technical and economic feasibility of smaller-scale, on-farm facilities are yet to be evidenced, it is possible that craft-cultured meat enterprises could provide rural communities and food producers with new opportunities to develop artisanal, *terroir*-based food ventures that could also support the re-introduction of small herds of native and heritage livestock breeds.

Achieving optimal social and ecological goods through land-use change is far from automatic and currently a highly contested topic, particularly concerning the role of livestock in landscape stewardship. Ongoing research suggests that small numbers of livestock farmed in agroecological ways can potentially play a key part in climate solutions, particularly in nations like the United Kingdom that have climates suited to grassland. Advocates of rewilding promote a combined ambition of 'core' rewilded areas that exist beyond human intervention and supporting farmers to practice 'wilder' and 'nature-friendly' farming. More research is needed to consider what good practice in landscape stewardship could look like if cultured meat becomes a viable large-scale industry. Strong policies and commitments by the cultured meat industry will be needed to prevent any resultant land-use change leading to worse environmental and socio-economic impacts. Consideration of what a just transition to cultured meat could look like in terms of land-use change raises a number of issues, including the interconnected issues of carbon sequestration, biodiversity and habitat restoration, air and water pollution, inequalities in land ownership, livelihood opportunities and quality of work within food production, and ensuring public access to green spaces. It should also refrain from a one-size-fits-all model and instead acknowledge the specific geographies, ecologies, economies, and cultures of different landscapes.

Connected to the theme of landscape change are questions of global food security, and where cultured meat might be made and where it will be

Published by Burleigh Dodds Science Publishing Limited, 2023.

consumed. Commentators have cautioned against the political economic forces that may see the cultured meat industry perpetuate colonial patterns of past and present food systems – i.e. where richer Minority countries have the economic and technical power to become the primary exporters of these novel foods to poorer Majority countries. This potential spatial organisation would further entrench the problematic model of global food security that has been widely criticised for placing primary emphasis on a nation's and individual's ability to *purchase* food, rather than having access to the means of producing food.

Conversely, some cultured meat proponents envision their technologies offering radically different geographical possibilities than conventional livestock given their unique uncoupling from the land. Small-scale cultured meat facilities offering localised food production and livelihood opportunities to inhabitants of refugee camps and remote rural communities around the world have been proposed (the technological and socio-economic viability of these scenarios have yet to be evidenced).

A further point to highlight is that food insecurity is an inherently geographical and geopolitical issue with a diverse range of drivers and nutritional characteristics. There has been a propensity for the narratives of the cultured meat industry to speak of food insecurity at the global scale and as an issue of insufficient overall quantity of food produced. The sector has also tended to characterise conventional animal agriculture as a homogeneous sector of practices and harms. These narratives risk neglecting the political economic factors that drive continued food insecurity in the world. Not only has food insecurity been shown to be a product of unequal distribution rather than insufficient quantity, but the role of protein in food insecurity issues is hugely context-specific – for example, it is widely established that most Minority countries consume enough, and in some cases too much, protein in their diets while protein remains deficient in a number of Majority nation populations.

6 Conclusion

If cultured meat is to achieve the most ambitious goals that have been set for it, then it will create significant change. However, exactly what that change will look like remains indeterminant. As such, at this stage, it is necessary for the collective discussion around cultured meat to remain open and reflexive about a diverse range of possibilities, and to retain a self-critical perspective. In this review of key talking points, we have outlined a number of visions for cultured meat, and discussed economic, political, and cultural aspects of these potential futures. This piece has aimed to be broad but concise, with each point introduced but not covered in depth. It is also not exhaustive, as uncertain futures open too many possibilities to make a fully comprehensive account achievable. We have also offered little in the way of solutions or resolutions for these issues. But we hope to have raised some talking points, which will aid both those who are new to thinking about the technology and those who have been involved for an extended period, in keeping their thoughts on the broader social implications of the emergent world of cultured meat.

7 Acknowledgements

Neil Stephens' work was supported by a Wellcome Trust Research Fellowship (grant number WT208198/Z/17/Z). Alexandra Sexton's work was supported by a Leverhulme Trust Early Career Research Fellowship (grant number ECF-2020-105).

8 Where to look for further information

The following articles provide a good overview of the subject:

- Broad, G. M. (2019). Plant-based and cell-based animal product alternatives: An assessment and agenda for food tech justice. *Geoforum* 107, 223–226. https://doi.org/10.1016/j.geoforum.2019.06.014.
- Newton, P. and Blaustein-Rejto, D. (2021). Social and Economic Opportunities and Challenges of Plant-Based and Cultured Meat for Rural Producers in the US. *Frontiers in Sustainable Food Systems* 5. https://www.frontiersin.org/article/10.3389/fsufs.2021.624270.
- Stephens, N., Di Silvio, L., Dunsford, I., Ellis, M., Glencross, A. and Sexton, A. (2018). Bringing cultured meat to market: Technical, socio-political, and regulatory challenges in cellular agriculture. *Trends in Food Science & Technology* 78, 155–166. https://doi.org/10.1016/j.tifs.2018.04.010.
- Stephens, N., Sexton, A. E. and Driessen, C. (2019). Making Sense of Making Meat: Key Moments in the First 20 Years of Tissue Engineering Muscle to Make Food. *Frontiers in Sustainable Food Systems* 3, 45. https://doi.org/10.3389/fsufs.2019.00045.

9 References

Adams, C. J. (2015). *The Sexual Politics of Meat - 25th Anniversary Edition: A Feminist-Vegetarian Critical Theory*. Bloomsbury Publishing.

Bekker, G. A., Fischer, A. R. H., Tobi, H. and van Trijp, H. C. M. (2017). Explicit and implicit attitude toward an emerging food technology: the case of cultured meat. *Appetite* 108, 245–254. https://doi.org/10.1016/j.appet.2016.10.002.

Broad, G. M. (2019). Plant-based and cell-based animal product alternatives: an assessment and agenda for food tech justice. *Geoforum* 107, 223–226. https://doi.org/10.1016/j.geoforum.2019.06.014.

Bryant, C. and Barnett, J. (2018). Consumer acceptance of cultured meat: A systematic review. *Meat Science* 143, 8–17. https://doi.org/10.1016/j.meatsci.2018.04.008.

Bryant, C. and Barnett, J. (2020). Consumer acceptance of cultured meat: an updated review (2018–2020). *Applied Sciences* 10(15), Article 15. https://doi.org/10.3390/app10155201.

Bryant, C. and Dillard, C. (2019). The impact of framing on acceptance of cultured meat. *Frontiers in Nutrition* 6, 103. https://doi.org/10.3389/fnut.2019.00103.

Bryant, C. J. and Barnett, J. C. (2019). What's in a name? Consumer perceptions of in vitro meat under different names. *Appetite* 137, 104–113. https://doi.org/10.1016/j.appet.2019.02.021.

Dutkiewicz, J. (2019). Socialize lab meat. *Jacobin*. Available at: https://jacobin.com/2019/08/lab-meat-socialism-green-new-deal.

Funke, F., Mattauch, L., van den Bijgaart, I., Godfray, C., Hepburn, C. J., Klenert, D., Springmann, M. and Treich, N. (2021). Is meat too cheap? Towards optimal meat taxation (SSRN Scholarly Paper No. 3801702). *SSRN Electronic Journal*. https://doi.org/10.2139/ssrn.3801702.

Goldstein, J. (2018). *Planetary Improvement: Cleantech Entrepreneurship and the Contradictions of Green Capitalism*. MIT Press.

Goodwin, J. N. and Shoulders, C. W. (2013). The future of meat: A qualitative analysis of cultured meat media coverage. *Meat Science* 95(3), 445–450. https://doi.org/10.1016/j.meatsci.2013.05.027.

Hamdan, M. N., Post, M., Ramli, M. A., Kamarudin, M. K., Md Ariffin, M. F. and Zaman Huri, N. M. F. (2021). Cultured meat: Islamic and other religious perspectives. *UMRAN – International Journal of Islamic and Civilizational Studies* 8(2), Article 2. https://doi.org/10.11113/umran2021.8n2.475.

Hamdan, M. N., Ramli, M. A. and Rahman, A. A. (2019). Penggunaan Sel Stem dalam Pengkulturan Daging: analisis Menurut Perspektif Hukum Islam: the Use of Stem Cells in Cultured Meat: analysis According to Islamic Law Perspective. *Journal of Fatwa Management and Research*, 18(1), 8–24. https://doi.org/10.33102/jfatwa.vol18no1.2.

Hamdan, M. N., Ramli, M. A., Zaman Huri, N. M. F., Abd Rahman, N. N. H. and Abdullah, A. (2021). Will Muslim consumers replace livestock slaughter with cultured meat in the market? *Trends in Food Science and Technology* 109, 729–732. https://doi.org/10.1016/j.tifs.2021.01.034.

Hopkins, P. D. (2015). Cultured meat in western media: the disproportionate coverage of vegetarian reactions, demographic realities, and implications for cultured meat marketing. *Journal of Integrative Agriculture* 14(2), 264–272. https://doi.org/10.1016/S2095-3119(14)60883-2.

Kenigsberg, J. A. and Zivotofsky, A. Z. (2020). A Jewish religious perspective on cellular agriculture. *Frontiers in Sustainable Food Systems* 3, 128. Available at: https://www.frontiersin.org/article/10.3389/fsufs.2019.00128.

Krone, A. (2022). Religion, animals, and technology. *Religions* 13(5), Article 5. https://doi.org/10.3390/rel13050456.

McCauley, D. and Heffron, R. (2018). Just transition: integrating climate, energy and environmental justice. *Energy Policy* 119, 1–7. https://doi.org/10.1016/j.enpol.2018.04.014.

Newton, P. and Blaustein-Rejto, D. (2021). Social and economic opportunities and challenges of plant-based and cultured meat for rural Producers in the US. *Frontiers in Sustainable Food Systems* 5. Available at: https://www.frontiersin.org/article/10.3389/fsufs.2021.624270.

Painter, J., Brennen, J. S. and Kristiansen, S. (2020). The coverage of cultured meat in the US and UK traditional media, 2013–2019: drivers, sources, and competing narratives. *Climatic Change* 162(4), 2379–2396. https://doi.org/10.1007/s10584-020-02813-3.

Rolland, N. C. M., Markus, C. R. and Post, M. J. (2020). The effect of information content on acceptance of cultured meat in a tasting context. *PLoS ONE* 15(4), e0231176. https://doi.org/10.1371/journal.pone.0231176.

Rosenfeld, D. L., Rothgerber, H. and Tomiyama, A. J. (2020). Mostly vegetarian, but flexible about it: investigating how meat-reducers express social identity around their diets. *Social Psychological and Personality Science* 11(3), 406–415. https://doi.org/10.1177/1948550619869619.

Ryynänen, T. and Toivanen, A. (2022). Hocus-pocus tricks and moral progressions: the emerging meanings of cultured meat in online news comments. *Food, Culture and Society* 0(0), 1–30. https://doi.org/10.1080/15528014.2022.2027688.

Sexton, A. E., Garnett, T. and Lorimer, J. (2019). Framing the future of food: the contested promises of alternative proteins. *Environment and Planning. E, Nature and Space* 2(1), 47–72.

Smaje, C. (2020). *A Small Farm Future: Making the Case for a Society Built Around Local Economies, Self-Provisioning, Agricultural Diversity and a Shared Earth*. Chelsea Green Publishing.

Stephens, N. (2010). In vitro meat: zombies on the menu. *SCRIPTed: A Journal of Law, Technology & in Society* 7, 394–401.

Stephens, N. (2013). Growing meat in laboratories: the promise, ontology, and ethical boundary-work of using muscle cells to make food. *Configurations* 21(2), 159–181. https://doi.org/10.1353/con.2013.0013.

Stephens, N., Di Silvio, L., Dunsford, I., Ellis, M., Glencross, A. and Sexton, A. (2018). Bringing cultured meat to market: technical, socio-political, and regulatory challenges in cellular agriculture. *Trends in Food Science and Technology* 78, 155–166. https://doi.org/10.1016/j.tifs.2018.04.010.

Stephens, N., Sexton, A. E. and Driessen, C. (2019). Making sense of making meat: key moments in the first 20 years of tissue engineering muscle to make food. *Frontiers in Sustainable Food Systems* 3, 45. https://doi.org/10.3389/fsufs.2019.00045.

Treich, N. (2021). Cultured meat: promises and challenges. *Environmental and Resource Economics* 79(1), 33–61. https://doi.org/10.1007/s10640-021-00551-3.

van der Weele, C. and Driessen, C. (2013). Emerging profiles for cultured meat; ethics through and as design. *Animals: an Open Access Journal from MDPI* 3(3), 647–662. https://doi.org/10.3390/ani3030647.

Van Loo, E. J., Caputo, V. and Lusk, J. L. (2020). Consumer preferences for farm-raised meat, lab-grown meat, and plant-based meat alternatives: does information or brand matter? *Food Policy* 95, 101931. https://doi.org/10.1016/j.foodpol.2020.101931.

Published by Burleigh Dodds Science Publishing Limited, 2023.

Chapter 3

Creating a regulatory framework for cultured meat products: Singapore

Johnny Yeung, Yong Quan Tan, Siew Herng Chan, Kern Rei Chng, Calvin Yeo, Jer Lin Poh, Teng Yong Low and Joanne Sheot Harn Chan, National Centre for Food Science, Singapore

1 Introduction

The Singapore Food Agency (SFA) is Singapore's national competent authority for food safety. In Singapore, all foods, including novel foods, come under the purview of SFA. The SFA was formed on 1 April 2019 under Singapore's Ministry of Sustainability and the Environment, with the vision to ensure and secure a supply of safe food. SFA consolidates all food-related resources and capabilities in the Singapore government for the holistic management of the food industry 'from farm to fork'.

2 Key food safety issues in Singapore

As a small island city-state, Singapore currently imports over 90% of its food supply to meet the demands of the population (Singapore Food Agency, 2019). As food safety challenges continue to evolve with emerging risks and an increasingly complex industry landscape, it is important for SFA to continually develop and enforce standards, guidelines and policies to keep pace with latest scientific knowledge to ensure food safety.

SFA is committed to ensuring and securing a supply of safe food for the nation through the adoption of a risk-based approach to food safety guided by science and consistent with international standards. With food safety ultimately

http://dx.doi.org/10.19103/AS.2023.0130.03

being a joint responsibility among stakeholders in the food supply chain, SFA fosters an enabling environment so that all stakeholders can each play their part. SFA's food safety system is therefore underpinned by three key considerations, as described next.

2.1 Consideration 1: farm-to-fork systems approach

SFA regulates the safety of food from import to local primary production and manufacturing and finally to food retail. This is achieved by identifying key stakeholders and control points along this farm-to-fork food supply chain. The farm-to-fork model of food safety assurance is best exemplified in SFA's import control regime which has been established with three levels of control as shown in Fig. 1. These levels are:

- Pre-import control;
- Import control; and
- Post-import control.

As with any other food, the importation of novel foods such as cultured meat will be subject to the same regime even after it has received SFA's pre-market approval.

First, before food is imported into Singapore, pre-import control measures are established to ensure safety of the food sources through a risk-differentiated approach. Measures range from the accreditation of food sources for high-risk items (such as meat and eggs), to application of licensing and import permits for food consignments or submission of export certificates for lower risk products. These measures serve to enhance food safety assurance by

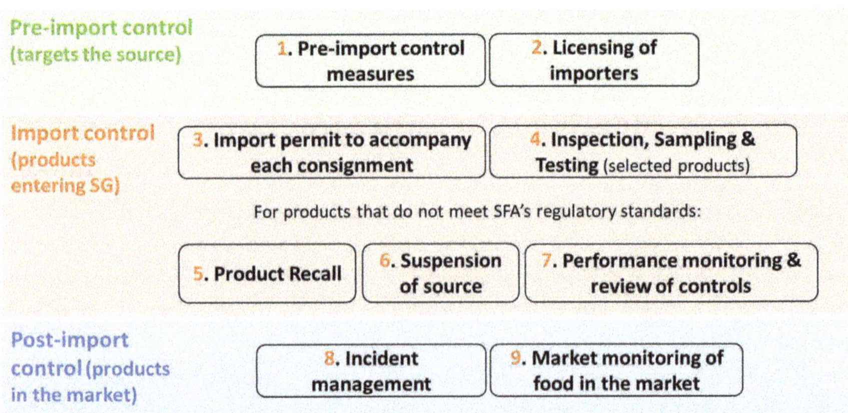

Figure 1 Summary of SFA's pre-import, import and post-import control regimes.

addressing food safety risks prior to import and protect the local food supply by preventing diseases that may affect production in local farms. These measures also help locally manufactured food products to gain overseas market access by ensuring that they are free from animal diseases of trade importance (e.g. Avian influenza, foot and mouth disease, African swine fever) (Ryan, 2016). For novel foods, pre-market safety assessments are also required to establish the safety of such food products. This is because they do not have a history of safe use and any potential safety concerns need to be identified and addressed.

Second, at the point of import, commercial shipments of food may only be imported by licensed or registered importers and must be accompanied by import permits. All importers, whether licensed or registered, need to comply with the relevant legislation. Import permits allow SFA to ensure that the relevant pre-import requirements and import conditions are met, facilitate downstream targeting of specific products for inspection as well as facilitate sampling and information collection for food traceability. Food products are also subject to SFA's inspection, sampling and testing programme whereby higher risk items are targeted and tested for commonly associated food safety hazards (e.g. *Salmonella* spp. in eggs).

Third, even beyond the point of import, food products may be recalled, and suppliers may be suspended if food does not meet regulatory standards, or when SFA is alerted by international/regional platforms (e.g. WHO INFOSAN, EU Rapid Alert System, G2G notifications) and/or horizon scanning. This is complemented by SFA's market monitoring programme which is non-regulatory in nature. Under this programme, food products are sampled and tested from the retail end to enable SFA to better understand the risk profile of imported products, identify high-risk products for more targeted checks and to refine the regulatory sampling and testing programme when necessary.

2.2 Consideration 2: science-based risk assessment and management

SFA uses science and data obtained from fieldwork (e.g. food safety incidents, non-compliances to requirements, food testing results) to assess the level of food safety risks posed at various control points along the farm-to-fork supply chain. During the assessment, SFA aligns to the risk analysis framework that has been established by the Codex Alimentarius Commission (CAC), an international food safety standards-setting body under the governance of the Food and Agriculture Organization (FAO) and United Nations (UN). The framework comprises risk assessment, risk management and risk communication (Fig. 2) (FAO, 2007).

Based on the assessed risk, SFA determines and adopts the appropriate food safety risk management measures in its regulatory regime to reduce and

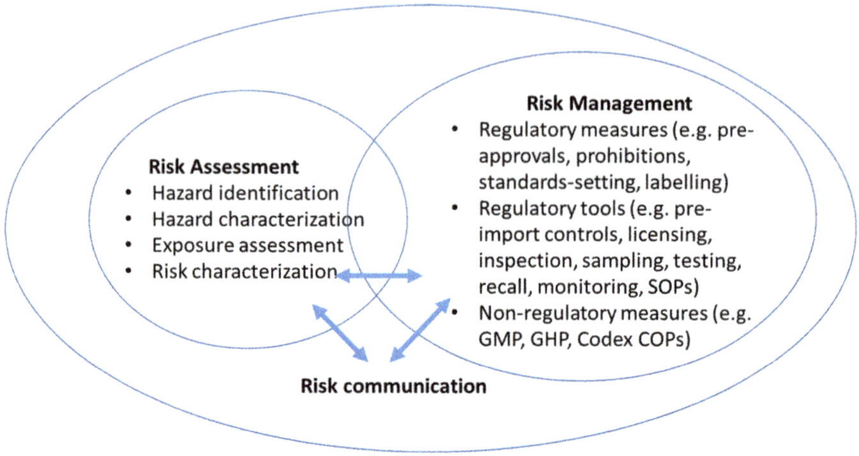

Figure 2 Codex risk analysis framework.

mitigate the foodborne risks to consumers. These risk management measures include the following:

i. *Standard setting*: Food safety standards (i.e. regulatory limits and conditions that the industry must comply with) are set to an appropriate level of protection for Singapore, based on risk assessments conducted and with reference to international standards, guidelines and recommendations such as those issued by the CAC. In the absence of international standards, reference is taken from major developed countries and jurisdictions (e.g. US, Australia, New Zealand, European Union, Japan) as well as Association of Southeast Asian Nations (ASEAN) Standards. SFA also works with stakeholders to develop industry standards that the food industry can choose to adopt to help them in food safety assurance.
ii. *Imposition and enforcement of regulatory requirements*: Regulatory levers (e.g. licensing of food operators) are established and the food safety requirements are enforced through inspection and testing.
iii. *Establishing mechanisms for food safety incident management*: Anticipatory and responsive mechanisms have been put in place. These include issuing food safety advice prior to periods of possible higher food safety risk, food recall plans by industry and suspension of affected food suppliers.

SFA recognizes that a science-based risk assessment and management approach to food safety will only be successful with robust food safety science capabilities in the ecosystem. Hence, SFA has multiple building blocks in place to support the development of such food safety science capabilities. First, SFA keeps a close watch on food safety developments through our horizon scanning programme, and reviews our standards based on developments at regional international forums and developed countries, in which SFA participates in discussions. Second, SFA houses the National Centre for Food Science (NCFS), which is recognized locally, regionally and internationally as a centre for food science excellence. NCFS has been awarded several accolades, such as the OIE Collaborating Centre for Food Safety, WHO Collaborating Centre for Food Contamination Monitoring, as well as having four ASEAN Food Reference Laboratories. Third, under SFA's Laboratory Recognition Program (LRP), a network of private laboratories has been set up, which enables NCFS' food testing capabilities to be scaled to the private industry. Furthermore, SFA has launched the Future Ready Food Safety Hub (FRESH; see later), which is a public-private partnership between SFA, Nanyang Technological University (NTU) and Agency for Science, Technology and Research (A*STAR) tasked with pursuing research and development (R&D) in food safety science.

Inevitably there are broader expertise areas that are not located within SFA. To address this limitation, SFA engages and collaborates with overseas authorities (e.g. signing of Memorandum of Understandings) with the Food Standards Australia and New Zealand (FSANZ) and French Agency for Food, Environmental and Occupational Health and Safety (ANSES), and also convenes scientific advisory committees on specific topics to tap on the domain expertise in the wider community to help fine tune scientific aspects to support SFA's regulatory decisions.

2.3 Consideration 3: joint responsibility for food safety

Food safety is a joint responsibility where the government, industry and consumers have a role to play in ensuring safe food for all. Even as SFA puts in place a regulatory system and enabling environment to ensure food safety, food business operators must play their part to maintain food safety assurance and be responsible for the safety of the food it offers to the public. To this end, SFA has embarked on the following industry-centred initiatives:

i. *Safety Assurance for Food Establishments (SAFE)*: SFA established the SAFE framework for all food establishments to incentivize them to consistently maintain high food safety standards. Food establishments that have demonstrated a good track record of food safety assurance and have put in place capabilities and systems to ensure better food safety standards will be eligible for longer licence durations and higher award tiers.

ii. *Engagements with industry*: SFA holds regular dialogues with traders, food manufacturers and retailers to raise awareness of food safety measures, encourage the implementation of food safety management systems (e.g. hazard analysis and critical control point (HACCP)), and assist them to comply with regulatory requirements.
iii. *Building food testing capability in the private sector*: SFA facilitated the development of a network of third-party laboratories that the food industry could tap on for food testing needs. SFA also introduced the LRP from April 2020 to enhance the recognition of these laboratories among the industry and foreign competent authorities.
iv. *Building capabilities in the food industry*: It is necessary for the food manufacturing and food & beverage industry workforce to be trained to support the adoption of systems and processes and ensure consistent food safety performance. Hence, SFA has rolled out an enhanced training and competency framework to upskill workers in this sector.

At the same time, consumers should be empowered to carry out good food hygiene practices and make informed food safety decisions. Through its public outreach efforts, SFA raises consumer awareness that food safety is a joint responsibility and educates consumers on foodborne hazards and steps on keeping food safe. This is done through Learning Journeys, school/workplace talks, community events as well as mainstream and social media. SFA develops communication materials such as consumer advisories on topics ahead of anticipated public concern. These are proactively pushed out on various platforms such as SFA's website ('Risk at a glance' articles), and SFA's social media platforms (e.g. Facebook and Instagram).

In the longer term, SFA's vision is to shift towards a state of food safety culture, i.e. where the food industry is progressive, self-directed and intrinsically motivated to ensure food safety along the supply chain; where consumers are informed and take proactive food safety measures; and where government enables better food safety, by creating an enabling tech- and data-based environment for industry and consumers to take on greater ownership.

3 The Singapore Food Agency's regulatory framework for cultured meat

Under SFA's novel food regulatory framework, food that has not been consumed as part of an ongoing diet by any significant human population for a period of at least 20 years is considered novel. Therefore, SFA considers cultured meat as a novel food. SFA's novel food regulatory framework has based its principal considerations in its food safety system, as described in Sections 2.1-2.3, which is applied to all foods. For cultured meat and other novel foods, SFA supplements the existing food safety system by considering and focusing on new inputs and processes that can introduce additional food safety risks. Established food safety principles and concepts pertaining to conventional foods are also adapted for novel foods. In addition, SFA utilizes its existing regulatory levers to ensure the safety of novel foods.

SFA considers cultured meat as cells from a whole animal that have been cultured in a bioreactor and supplemented with a growth medium for cell proliferation and differentiation. These cultured cells differentiate into muscles or fat cells and propagate into protein-rich biomass or complex tissue structures, which are ultimately converted into meat products. Under SFA's regulatory framework for novel foods, companies are required to conduct and submit pre-market safety assessments for SFA's review and obtain SFA's approval before they are allowed to produce/manufacture, import, distribute or sell their cultured meat product in Singapore.

There is currently no internationally accepted term to describe products obtained from animal cell culture intended for consumption. Several terms have been proposed to describe cultured meats, including clean, *in vitro*, synthetic, cellular, lab grown, cultivated and cultured. For example, the Good Food Institute (GFI) and FSANZ have adopted the terms 'cultivated meat' and 'cell-based meat', respectively. In Singapore, SFA has adopted the working term 'cultured meat' to describe meat developed from animal cell cultures in bioreactors supplemented by culture media. Other government agencies such as Enterprise Singapore and Singapore Economic Development Board were also consulted on the use of 'cultured meat' terminology to synchronize communications between government bodies as well as the public. Regardless of the terminology, it is mandatory for companies to label their products with qualifying terms such as 'cultured', 'cultivated' or 'cell-based', to accurately convey the true nature of their food product.

SFA's regulatory framework for novel foods is primarily focused on food safety to ensure that novel foods are safe for consumption. SFA understands that new food sources and production systems prompt diverse discussions among numerous stakeholders. Apart from food safety risks, these diverse discussions could include topics such as nutritional, sustainability, socioeconomic, animal welfare and cultural implications of new food sources and production systems. SFA's regulatory framework for novel food is focused on food safety risks and is not intended to capture considerations not related to food safety. SFA is of the view that food safety considerations of novel foods should not be intermingled with other considerations that are not related to food safety. This is to ensure scientific robustness in the safety assessment of novel foods.

SFA's safety assessment framework for cultured meat captures the food safety considerations during the cultured meat manufacturing process and it covers three levels. First, the safety of individual inputs in the production process (e.g. cell lines, culture media and reagents) are assessed to ensure that potential safety risks have been addressed. Second, the production process and quality controls are guided by detailed risk assessments and risk management plans to ensure that there is no introduction of additional food safety risks. As cultured meat typically involves cell culture processes, the framework for cultured meat takes guidance from good practices in biopharmaceutical manufacturing, cell culture, microbiology and molecular biology that ensure stability and traceability of the cell line throughout the manufacturing process. The production process must be properly controlled and adhere to good food safety and hygiene practices. Last, the final product must comply with the Singapore's Food Regulations. This includes compliance to the regulatory limits for additives and contaminants.

As each cultured meat manufacturing process is unique and certain techniques and methodologies are new to food manufacturing, SFA's safety assessment framework utilizes food safety principles and guidelines that can be generally applied to all cultured meat manufacturing processes that are currently conceivable. SFA posits that a safety assessment framework that is based on broad scientific and food safety principles, and that which is not unnecessarily prescriptive, facilitates current and future innovation in food technology while still ensuring food safety. Details of food safety considerations in SFA's safety assessment framework for cultured meat are depicted in Fig. 3 and further elaborated next.

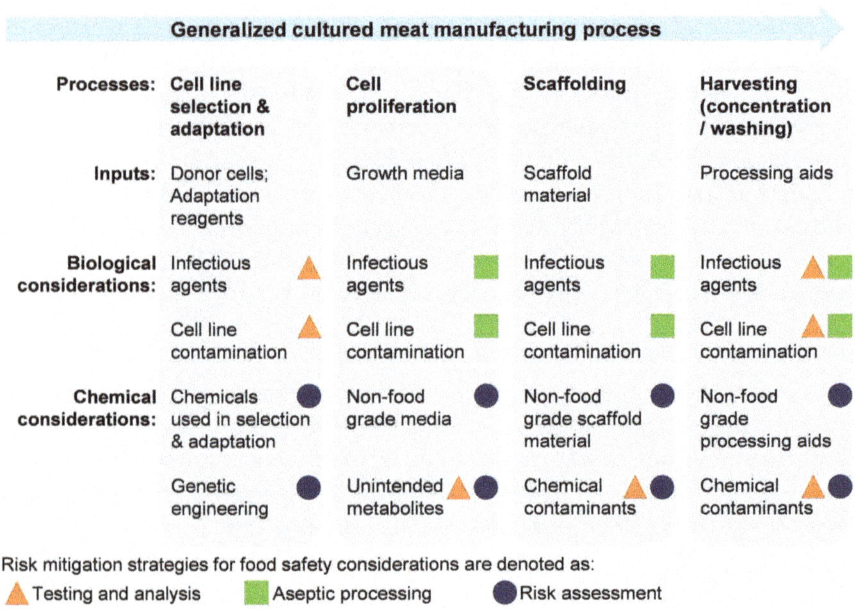

Figure 3 Food safety assessment framework for cultured meat.

i. *Infectious agents*: Foodborne pathogens, which include viruses, bacteria, fungi and other parasites, can be introduced at any stage of the manufacturing process. In the presence of nutrient-rich cell culture growth media, these pathogens can easily proliferate to levels that could infect and cause disease for the consumer. Arguably, the most important control point for removing foodborne pathogens in the manufacturing process is at the cell isolation step. Here, attention is focused on the presence of pathogens, particularly those known to be present in the particular animal type from which the cells are derived. If cells are isolated from whole animals, whether live or slaughtered, it is typically necessary to use antibiotics and/or other drugs that inhibit the growth of pathogens naturally present in the whole animal. Absence of these pathogens must be confirmed through standard immunological, microbiological or molecular biology methods. After ensuring the sterility of the donor cells, the highest standards of cleanliness must be maintained during cell proliferation and processing, and routine analysis of foodborne pathogens is necessary to provide assurance on the absence of infectious agents. In recent years, metagenomics tools, such as 16S rRNA sequencing and shotgun sequencing, have been increasingly recognized for its usefulness in rapidly identifying a wide range of pathogens, including those that are less well characterized (Kovac et al., 2017). However, metagenomics tools currently suffer from lower sensitivity and species/sub-species resolution, and should only be used to complement standard detection methods at this point (Snipen et al., 2021). If bovine products such as foetal bovine serum are introduced into the process, companies should ensure that these products are sourced from regions with no indigenous cases of bovine spongiform encephalopathy (BSE). This is because the presence of prions in the bovine products, which are the infectious agent causing BSE, can lead to the formation and accumulation of prions in cells (Hedlin et al., 2012). Consumption of meat infected by prions has been linked to development of Creutzfeldt-Jakob disease in humans, which is incurable and invariably fatal (Roma and Prayson, 2005).

ii. *Cell line contamination by other cell lines and adventitious agents*: SFA's safety assessment for a cultured meat product is specific to the cell line from which the product is derived from. As such, measures that minimize the possibility of cell line contamination are important considerations in the safety assessment framework. During the isolation of the target primary cell line from a donor animal, scientifically validated cell sorting methodologies such as flow cytometry, magnetic-activated cell sorting and microfluidic cell sorting should be used (Wyatt Shields et al., 2015). Primary cell lines that are directly sourced from a cell line supplier should ensure that the supplier is reputable, and the cell line has been authenticated. It is widely recognized that continual subculturing of cells exerts selective pressure on these cells and this can cause significant genetic drifts, resulting in cells that are not genotypically and/or phenotypically representative of the initial cell line (Hughes et al., 2007). Good cell banking protocols that are widely practised in biopharmaceutical manufacturing, such as establishment of a Master Cell Bank and Working Cell Bank, as well as adhering to good cell subculture protocols, can mitigate the effects of genetic drifts (Zhu et al., 2017). Routine monitoring of cell line stability via genomics (e.g. whole-genome sequencing, single-nucleotide polymorphism microarray) and transcriptomics (e.g. RNA-seq) approaches, as well as phenotypic approaches (e.g. cell doubling time, oxygen consumption rate), contributes to a comprehensive depiction of cell line stability that supports food safety assessment. Where significant differences in gene expression are observed between the initial cell line and the cells in the cultured meat product, companies should investigate if the differences would result in food safety risks. Examples of such risks include increased expression of allergenic proteins or production of unintended metabolites.

iii. *Genetic engineering*: Some companies utilize genetic engineering to endow cell lines with useful traits, predominantly immortalization of cell lines beyond their natural replication limit (termed Hayflick limit) (Shay and Wright, 2000; Wikandari et al., 2021). Genetic engineering may entail the integration of transgenes into the genome of the cell line, or to make precise alterations or deletions of existing genes using more recent genome editing tools (Soice and Johnston, 2021). The process of genetic engineering can introduce food safety risks that must be assessed. If the cell line has been engineered to produce proteins that are novel to the cell line (i.e. foreign proteins), such proteins must be assessed for safety to ensure that they do not present toxicity, allergenicity or digestibility issues when consumed. All intentional genetic alterations, be it gene insertion, alteration or deletion, must also be proven to be stable through consecutive subcultures. In addition, companies must verify that all nucleotide sequences used during the genetic engineering process, but that are not intended to remain in the cells (e.g. plasmid backbone, nuclease gene) are indeed absent from the cell line used in production. Verification must be performed using established molecular biology methodologies, such as Southern blotting and next-generation sequencing. Currently, prevailing methods of cell line immortalization via genetic engineering involve the expression of foreign oncogenes or inactivating the expression of endogenous tumour suppressor genes (Soice and Johnston, 2021), which may increase the oncoprotein content of the cells. Further research is warranted to determine if such genetic engineering methods would introduce further food safety risks to the cultured meat product.

iv. *Chemical contaminants*: The cultured meat manufacturing process can introduce chemical contaminants into the final food product. Some of these contaminants may present food safety risks. For example, chitin, which is used in some cell culture scaffolds, may trigger allergic reactions in some individuals. (Lee, 2009; Campuzano and Pelling, 2019) Certain antibiotics can cause allergic reactions, liver damage and other adverse health effects in sensitive individuals, while the overuse of antibiotics in food production can contribute to antimicrobial resistance, a public health concern (Tang et al., 2017; Bacanlı and Başaran, 2019). Safety assessments of chemical contaminants in the cultured meat product are necessary to ensure food safety. Testing and analysis of chemical contaminants should be done using protocols and guidelines that are established in the scientific literature or supported by OECD, AOAC and other reputable trade or scientific organizations. Acceptable maximum levels of chemical contaminants in all foods, including cultured meat, are stipulated in Singapore's Food Regulations. Food containing chemical contaminants that exceed the stipulated levels will not be allowed for sale in Singapore.

v. *Non-food grade chemicals*: The cultured meat manufacturing process requires the use of a wide range of chemicals that, while routinely used in cell culture, are novel to food processing. Thus, it is not feasible for only food-grade chemicals and reagents to be used in the process. Some of the non-food-grade chemicals used may be found at low levels in the final cultured meat product and may present health risks. Notwithstanding the absence of food-grade certification, chemicals sourced from reputable suppliers come with specification documents that support food safety risk assessment. Some chemicals that are not traditionally used in food processing or manufacturing have associated toxicological data, such as a no-observed-adverse-effect level (NOAEL) and benchmark dose level (BMDL), that are available in the scientific literature. Both NOAEL and BMDL serve as departure points to estimate health-based guidance values such as acceptable daily intake (ADI) of the chemical in the cultured meat product (Hardy et al., 2017). However, many other chemicals do not have associated toxicological data. In the absence of established toxicological data, companies may adopt the threshold of toxicological concern as a pragmatic approach towards food safety assessment of these chemicals (More et al., 2019), but only if the chemical is suitable for assessment by such an approach (one of which is that it must be demonstrated to be non-genotoxic). A number of decision-tree approaches to estimate toxicological effects of small molecules, such as Toxtree, have been developed to support both food and environmental risk assessments (Kroes et al., 2004; Patlewicz et al., 2010; Williams et al., 2017).

SFA's safety assessment framework for cultured meat also takes into consideration the hazard control plan (HCP) for the manufacturing process, regardless of whether the cultured meat product is imported or manufactured locally. Companies should provide information pertaining to food safety management systems that are put in place, and to provide relevant certification or documentation pertaining to HACCP, good manufacturing practices and good cell culture practices. Process flow diagrams may also be supplied to better identify critical control points that contribute to food safety of the product. Companies that intend to manufacture cultured meat products in Singapore are required to first obtain a licence from SFA for a food processing establishment.

On top of existing licensing requirements for food processing establishments that manufacture traditionally consumed food, cultured meat companies may also need to put in place additional control measures to address specific risks associated with cultured meat production. Examples of these control measures include the following:

- a designated storage facility for sterile instruments and tools;
- provision of a designated media preparation room;
- cold storage facility for cell banks and the cultured meat product; and
- provision of a designated cell culture room that is kept at a positive pressure relative to other rooms to minimize entry of infectious agents outside of the room.

4 Strengthening food safety capabilities in cultured meat products through partnerships

SFA's emphasis on regulatory agility (Soumyanarayanan et al., 2021) is key to enabling regulatory compliance to ensure food safety while facilitating new food innovations including cultured meats. Our regulatory framework on novel food aims to bring the industry and the regulator in close coordination based on the firm belief that food safety is a joint responsibility. To build trust between the regulator and the industry, SFA encourages and facilitates early engagement opportunities with the industry through the organization of novel food virtual clinics and pre-submission consultations. Such engagements are beneficial to both parties: the industry can gain access to the regulator's guidance and incorporate safety assurance measures as they develop their food innovations; meanwhile the regulator can obtain early insights from the industry and make timely risk assessments to formulate regulatory considerations in preparation for upcoming food innovations.

Apart from the industry, the academic sector is another key stakeholder that SFA partners to build up the wide-ranging and diverse domain expertise necessary for the effective assessment and regulation of food innovations. This is particularly essential since regulators are unlikely to have the breadth and depth of technical expertise and knowledge in-house to comprehensively evaluate food innovations that are constantly and rapidly evolving. Therefore, SFA has implemented various strategies to tap on the academic sector to partner in shaping research and development directions through targeted grant calls and funding to align academic research to the needs of regulatory science, establish public-private partnership platforms to facilitate synergy and integrate of capabilities within the food science ecosystem, and tap on multidisciplinary experts to contribute scientific advice on deep technical issues for regulatory assessments and considerations. Details on initiatives that SFA has participated in pursuit of these strategies are outlined as follows:

i. *Future REady Food Safety Hub (FRESH)*: The Singapore Food Story (SFS) was initiated to support Singapore's national agenda of strengthening food security. The R&D programme under the SFS focuses serves as a platform to drive the creation of new biotech-based foods and ingredients that are underpinned by a robust future-ready food safety system. One means of securing Singapore's food supply and economic value proposition involves building local food safety capabilities to support growing food innovations in Singapore, including in alternative proteins.

To this end, a public-private partnership platform known as FRESH was set up to drive food safety research to support food innovations. FRESH, with Agency for Science, Technology and Research (A*STAR) as its implementing agency, aims to foster synergistic collaborations between industry, academia and regulators, and aggregate relevant existing scientific expertise in Singapore's food science ecosystem to address emerging challenges in food safety, particularly in the development of alternative proteins and new processing technologies. In addition, FRESH plays an important role in facilitating novel food product development, by providing scientific technical support on safety risk assessment to the local industry. At the same time, FRESH provides a medium for regulatory scientists to collaborate and be exposed to upstream food safety issues in industry research and development. Finally, FRESH also provides relevant scientific and risk assessment training to build up a pipeline of food safety talent pool in Singapore over the longer term.

ii. *Novel Food Safety Expert Working Group*: SFA recognizes the deep and complex science driving new food innovations and is cognizant that is important to work with experts in various specialized scientific domains to ensure that food safety assessments are scientifically robust and carried out in a timely fashion. Therefore, SFA taps onto the expertise outside of the agency by forming a novel food safety expert working group to provide scientific advice to SFA for safety risk assessment reviews of novel foods. The expertise residing in the workgroup is diverse and has high relevance to cutting-edge food technologies and food safety sciences such as food toxicology, nutrition, bioinformatics, epidemiology and public health.

In addition to engaging the academic sector, SFA actively engages other regulatory counterparts to foster consensus-building in novel food regulations. Inaugurated in 2019 as the Regulator's Forum for Novel Foods, the Roundtable for Novel Food Regulations, as it is now known as, is organized under the umbrella of the Singapore International Agri-Food Week. This Roundtable provides an opportunity for the industry, research community, as well as local and overseas regulators to pursue discussions on the rapidly developing novel food sector, including the challenges and possible solutions in the safety assessment of novel food. The third Roundtable, organized in November 2022 (most recent iteration at time of writing), brought together 160 participants from the industry, academia, and governments across 15 jurisdictions to share and discuss about the rapidly developing novel food sector. This included members at the forefront of innovative food, including international organizations such as the CAC, the FAO of the UN and the World Health Organization. The Roundtable also included a panel discussion in which developments on the safety assessment, regulatory approvals and risk communications around novel foods were discussed.

5 Building consumer confidence in the safety of cultured meat products

While food safety risk assessment and management programmes are pivotal to ensuring that food is safe for consumption, it is also key for regulators to conduct risk communication effectively, to facilitate consumers' understanding of food safety risks and to build confidence in the food regulatory system.

SFA anticipated that its regulatory approval, the first in the world, of a cultured chicken product was likely to generate high levels of interest among consumers and industry experts, despite its limited commercial availability at launch. SFA therefore adopted a responsive risk communication strategy to assure the public of the rigorous nature of our safety assessment approach using a range of tools and channels targeting different audiences. These tools and channels included as follows:

i. *Articles for consumers*: Short articles and infographics are regularly published on SFA's social media platforms, which are written in accessible language. Articles covering various alternative proteins including cultured meat were initiated in the lead-up to SFA's announcement of our regulatory approval of the cultured chicken product. The articles explained how SFA assesses and monitors the safety of these foods.
ii. *Articles for industry players*: SFA also published a detailed explanation of our regulatory framework and safety assessment approach for novel foods on the SFA website to provide more information to industry experts and interested members of the public.
iii. *Sharing to stakeholders in workshop/meeting*: In preparation for publicity on Singapore's regulatory approval for cultured chicken, SFA conducted a local media briefing to share on its novel food regulatory framework with local media outlets and safety assessment approach, while A*STAR shared on the technology used in the creation of alternative proteins, Singapore's research and development efforts in future foods, as well as its potential contribution to sustainability and food supply. Meanwhile, other government agencies such as Singapore's Economic Development Board (EDB) also regularly roll out messaging on topics such as the potential of alternative proteins to contribute towards Singapore's economy.

Singapore's regulatory approval for cultured chicken garnered high traction on social media platforms and was widely publicized in international and local news. Online sentiments primarily focused on the 'artificial' nature of cultured meat, and some members of the public expressed concern over potential side effects of consuming cultured meat. Others reacted positively and viewed cultured chicken as a more humane alternative to conventionally farmed meat, and applauded SFA's regulatory agility and rigorous food safety assessment processes. This observation corroborates with published studies. For example, a survey conducted by researchers from NTU showed that a third of respondents found genetically modified food unappealing and that these negative feelings extended to their feelings for 'nano-enabled food' (Ho et al., 2020). On the other hand, a more recent study by the Singapore Management University surveyed Singaporeans and Americans and found that Singaporean participants had greater acceptance of cultured meat as compared to their American counterparts (Chong et al., 2022).

SFA continues to build its risk communication capabilities to raise understanding, acceptance and trust among consumers towards new food production technologies that could play an important role in meeting Singapore's food supply needs in the years ahead. For example, SFA has co-funded research projects on consumer science and risk communication on novel foods through the Singapore Food Story Research and Development Programme. Research outcomes will be translated into risk communication messages that are easily accessible to consumers. Nonetheless, even as SFA continues its risk communications efforts, the cultured meat industry will need to play its part to address public perception and consumer acceptance of cultured meat.

6 Conclusion

As a single agency overseeing all food-related matters in Singapore, SFA has leveraged on its unique position to ensure food safety through an integrated food safety system. To ensure the safety of food arising from innovation in Singapore, SFA has developed a regulatory framework for novel foods based on science and risk. Such policy thinking enabled SFA to ensure adequate space for the industry and research community to innovate while protecting consumers' health.

Nevertheless, it is important for SFA to remain agile in regulating novel foods, especially when the novel food ecosystem is developing rapidly. To this end, SFA actively engages with companies through Novel Food Virtual Clinics and pre-submission consultations. SFA also actively pursues advancements in food safety science through collaborative research with the academic sector and establishing a local ecosystem for novel foods (e.g. SFA's Novel Food Expert Panel and FRESH), as well as maintaining close alliances with SFA's regulatory counterparts outside of Singapore through bilateral and multilateral engagements (e.g. annual Roundtable for Novel Food Regulations).

At the same time, it is important to build up consumer confidence in new food innovations and address any concerns that they might have through an effective communications approach. Leveraging on both mainstream and social media platforms, a communications strategy has been crafted and implemented related to cultured meat, and media and consumer sentiments on this topic are also being monitored. This is an ongoing endeavour with active research efforts being conducted and the research outcomes will help to fine tune the risk communication strategy surrounding cultured meat.

7 References

Bacanlı, M. and Başaran, N. (2019). Importance of antibiotic residues in animal food, *Food and Chemical Toxicology* 125, 462. doi: 10.1016/J.FCT.2019.01.033.

Campuzano, S. and Pelling, A. E. (2019). Scaffolds for 3D cell culture and cellular agriculture applications derived from non-animal sources, *Frontiers in Sustainable Food Systems* 3, 38. doi: 10.3389/FSUFS.2019.00038/BIBTEX.

Chong, M., Leung, A. K. Y. and Lua, V. (2022). A cross-country investigation of social image motivation and acceptance of lab-grown meat in Singapore and the United States, *Appetite* 173, 105990.

EFSA Scientific Committee, Hardy, A., Benford, D., Halldorsson, T., Jeger, M. J., Knutsen, K. H., More, S., Mortensen, A., Naegeli, H., Noteborn, H. and Ockleford, C. (2017). Update: Use of the benchmark dose approach in risk assessment, *EFSA Journal* 1, 15, e04658. doi: 10.2903/J.EFSA.2017.4658.

EFSA Scientific Committee, More, S. J., Bampidis, V., Benford, D., Bragard, C., Halldorsson, T. I., Hernández-Jerez, A. F., Hougaard Bennekou, S., Koutsoumanis, K. P., Machera, K. and Naegeli, H. (2019). Guidance on the use of the threshold of Toxicological Concern approach in food safety assessment, *EFSA Journal* 6(17), e05708. doi: 10.2903/J.EFSA.2019.5708.

Food safety risk analysis. Available at: https://www.fao.org/3/a0822e/a0822e00.htm.

Hedlin, P., Taschuk, R., Potter, A., Griebel, P. and Napper, S. (2012). Detection and control of prion diseases in food animals, *ISRN Veterinary Science*. Hindawi Publishing Limited 2012, 254739. doi: 10.5402/2012/254739.

Ho, S. S., Goh, T. J., Chuah, A. S. F., Leung, Y. W., Bekalu, M. A. and Viswanath, K. (2020). Past debates, fresh impact on nano-enabled food: A multigroup comparison of presumed media influence model based on spillover effects of attitude toward genetically modified food, *Journal of Communication* 70(4), 598.

Hughes, P., Marshall, D., Reid, Y., Parkes, H. and Gelber, C. (2007). The costs of using unauthenticated, over-passaged cell lines: How much more data do we need?, *BioTechniques*. Eaton Publishing Company 43(5), 575. doi: 10.2144/000112598.

Kovac, J., Bakker, H. D., Carroll, L. M. and Wiedmann, M. (2017). Precision food safety: A systems approach to food safety facilitated by genomics tools, *TrAC Trends in Analytical Chemistry*. Elsevier 96, 52. doi: 10.1016/J.TRAC.2017.06.001.

Kroes, R., Renwick, A. G., Cheeseman, M., Kleiner, J., Mangelsdorf, I., Piersma, A., Schilter, B., Schlatter, J., van Schothorst, F., Vos, J. G., Würtzen, G. and European branch of the International Life Sciences Institute. (2004). Structure-based thresholds of toxicological concern (TTC): Guidance for application to substances present at low levels in the diet, *Food and Chemical Toxicology* 42(1), 65. doi: 10.1016/J.FCT.2003.08.006.

Lee, C. G. (2009). Chitin, chitinases and chitinase-like proteins in allergic inflammation and tissue remodeling, *Yonsei Medical Journal*. Yonsei University College of Medicine 50(1), 22. doi: 10.3349/YMJ.2009.50.1.22.

Patlewicz, G., Jeliazkova, N., Safford, R. J., Worth, A. P. and Aleksiev, B. (2008). An evaluation of the implementation of the Cramer classification scheme in the Toxtree software, *SAR and QSAR in Environmental Research*. Taylor & Francis Group 19(5–6), 495. doi: 10.1080/10629360802083871.

Roma, A. A. and Prayson, R. A. (2005). Bovine spongiform encephalopathy and variant Creutzfeldt-Jakob disease: How safe is eating beef?, *Cleveland Clinic Journal of Medicine* 72(3), 185. doi: 10.3949/CCJM.72.3.185.

Ryan, J. (2016). *Biosecurity and Bioterrorism: Containing and Preventing Biological Threats* (2nd edn.). Oxford: Elsevier.

Shay, J. W. and Wright, W. E. (2000). Hayflick, his limit, and cellular ageing, *Nature Reviews: Molecular Cell Biology*. Nature Publishing Group 1(1), 72. doi: 10.1038/35036093.

Singapore Food Agency. (2019). *SFA|Our Singapore Food Story – The 3 Food Baskets, Singapore Food Agency Website*. Available at: https://www.sfa.gov.sg/food-farming

/sgfoodstory/our-singapore-food-story-the-3-food-baskets (Accessed: 18 May 2022).

Snipen, L., Angell, I. L., Rognes, T. and Rudi, K. (2021). Reduced metagenome sequencing for strain-resolution taxonomic profiles, *Microbiome*. BioMed Central Ltd 9(1), 1–19. doi: 10.1186/S40168-021-01019-8/TABLES/1.

Soice, E. and Johnston, J. (2021). Immortalizing cells for human consumption, *International Journal of Molecular Sciences*. Multidisciplinary Digital Publishing Institute 22(21), 11660. doi: 10.3390/IJMS222111660.

Soumyanarayanan, U., Choong, M., Leong, J., Lumpkin, M. M., Rasi, G., Skerritt, J. H., Vogel, S. and Lim, J. C. W. (2021). The COVID-19 crisis as an opportunity to strengthen global regulatory coordination for sustained enhanced access to diagnostics and therapeutics, *Clinical and Translational Science* 14(3), 777. doi: 10.1111/CTS.12954.

Tang, K. L., Caffrey, N. P., Nóbrega, D. B., Cork, S. C., Ronksley, P. E., Barkema, H. W., Polachek, A. J., Ganshorn, H., Sharma, N., Kellner, J. D. and Ghali, W. A. (2017). Restricting the use of antibiotics in food-producing animals and its associations with antibiotic resistance in food-producing animals and human beings: A systematic review and meta-analysis, *Lancet. Planetary Health* 1(8), e316. doi: 10.1016/S2542-5196(17)30141-9.

Wikandari, R., Manikharda, Baldermann, S., Ningrum, A. and Taherzadeh, M. J. (2021). Application of cell culture technology and genetic engineering for production of future foods and crop improvement to strengthen food security, *Bioengineered*. Taylor & Francis 12(2), 11305. doi: 10.1080/21655979.2021.2003665.

Williams, A. J., Grulke, C. M., Edwards, J., McEachran, A. D., Mansouri, K., Baker, N. C., Patlewicz, G., Shah, I., Wambaugh, J. F., Judson, R. S. and Richard, A. M. (2017). The CompTox chemistry dashboard: A community data resource for environmental chemistry, *Journal of Cheminformatics*. BioMed Central 9(1), 61. doi: 10.1186/S13321-017-0247-6.

Wyatt Shields, C., Reyes, C. D. and López, G. P. (2015). Microfluidic cell sorting: A review of the advances in the separation of cells from debulking to rare cell isolation, *Lab on a Chip* 15(5), 1230. doi: 10.1039/C4LC01246A.

Zhu, M. M., Mollet, M., Hubert, R. S., Kyung, Y. S. and Zhang, G. G. (2017). Industrial production of therapeutic proteins: Cell lines, cell culture, and purification, In *Handbook of Industrial Chemistry and Biotechnology*. Nature Publishing Group, p. 1639. doi: 10.1007/978-3-319-52287-6_29.

Chapter 4

Creating an infrastructure for cultured meat

Irfan Tahir, The University of Vermont, USA; Miranda Stahn, Independent Consultant, Canada; N. Stephanie Kawecki, University of California Los Angeles, USA; Vicky Andriessen, Mosa Meat, The Netherlands; Bianca Datta, Massachusetts Institute of Technology, USA; Dwayne Holmes, Stichting New Harvest, The Netherlands; Lejjy Gafour, CULT Food Science Corp, Canada; and Isha Datar, New Harvest Inc., Canada

1 Introduction

While industrial animal agriculture feeds billions globally, its growth is creating an industry that propagates the inhumane treatment of animals while accelerating the adverse effects of climate change.[1-7] In addition, the production of animals at a high density increases the risk of spreading epidemic viruses, foodborne illness, and antibiotic resistance, deeply threatening human and animal health as well as food security.[8-10] For example, the mass slaughtering of animals is a common technique used to mitigate the spread of disease, resulting in food supply chain disruptions, increased cost of consumer products, and reduced accessibility to nutritious food. Thus, the practice of factory farming greatly reduces the resilience of food systems, which is particularly concerning in a climate-changed world. To address these issues, one proposed alternative is the use of cellular agriculture technology to create animal products such as meat, milk, and eggs from cell cultures rather than whole animals.[11]

Cellular agriculture is a new, interdisciplinary field of study and burgeoning industry with the potential to end the global dependence on animal agriculture. Over the past decade, US$4 billion has been privately invested in over 100 global companies, aiming to develop agriculture products from cell cultures rather than whole plants or animals.[12] While some are developing products such as leather, milk, and eggs from cell cultures, the

http://dx.doi.org/10.19103/AS.2023.0130.06

majority of companies are focused on creating cultured meat or advancing aspects of cultured meat supply chains. By contrast, in the same time period, less than US$100 million has been invested into public sector research and development via philanthropy and government grants.[13] In all, public investment is approximately 2.5% of private investment as of the time of this chapter's writing.

To assess the state of public and private funding within the field of cultured meat, it is helpful to compare it to an existing funding trajectory seen within an adjacent field. For example, complex biotechnology such as biopharma would follow a typical pathway from discovery to market as described (Fig. 1).

The first third of the pathway is publicly funded by government grants and other public sources with the typical intention of building education and knowledge infrastructure for the field. The middle, or the notorious 'valley of death,' is the region where most translational research efforts aim to intervene.[14] The last third is funded by venture capital (VC) with the intention of building intellectual property (IP) and driving the translation of developed technologies. This multi-stakeholder, multi-funder discovery pipeline is typical for highly complex technologies. Due to a series of compounding factors, such as a lack of dedicated funding streams through government funding agencies and the boundless availability of private investment capital, cultured meat sees a disproportionate amount of private funding for the translation and scale-up of technologies (Fig. 2). This scenario in favor of private entities raises concerns, given that the concurrent focus on IP often leads to a lack of available information for policymakers and consumers, as well as shortages of independent and unbiased experts who can develop, evaluate, and optimize

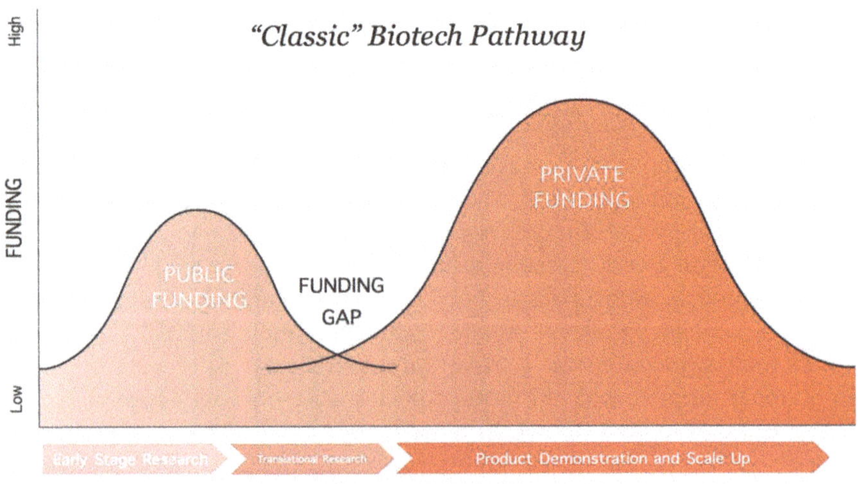

Figure 1 Classic biotechnology product development pathway.

Published by Burleigh Dodds Science Publishing Limited, 2023.

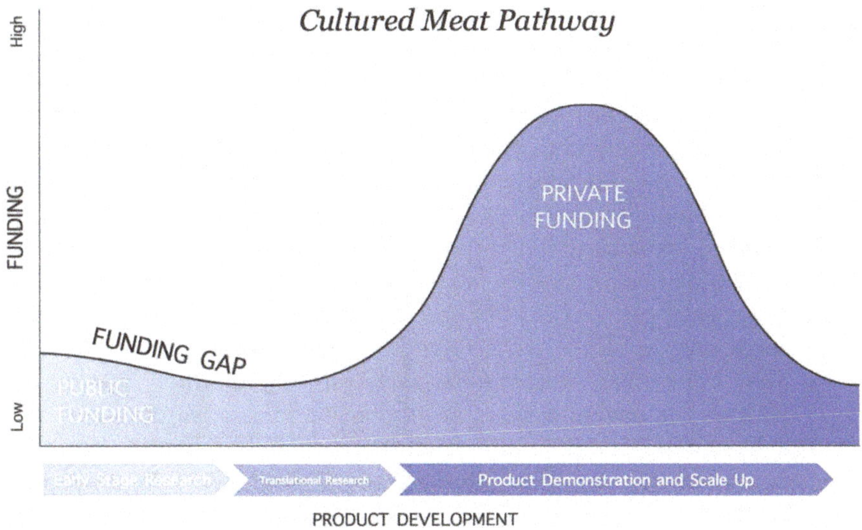

Figure 2 Cultured meat product development pathway.

cultured meat technologies. In short, cultured meat is beginning to follow an atypical funding trajectory that has traditionally formed the foundation for scientific innovation and industrial growth.[15]

If we think of cultured meat technologies as a city, it is currently one without public infrastructure. Privately owned 'houses' are under development, but the roads, bridges, sewers, and public squares - all of which are crucial to a thriving ecosystem - remain unbuilt. Thus, switching focus toward developing foundational knowledge, frameworks, tools, and techniques that are openly accessible would accelerate and strengthen the field of cultured meat by filling in the current gaps in infrastructure.

On the surface, it might seem that simply pouring public funding into classic institutions would address this problem. However, as cultured meat is an atypical field, the required infrastructure may go beyond traditional institutional support. There seems little chance that traditional avenues for publicly funded research could keep pace in a meaningful way with innovations in the highly funded private sector.

Right now, most research in cultured meat is performed by private companies, creating several obstacles to efficient progress in the field, such as:

1. The desire to protect potential IP drives an intentional lack of data sharing, blocking the common scientific practice of peer review. This, in turn, generates information 'silos' that may result in the duplication of research efforts, including lines of research that will fail, and slows down the overall pace of innovation. Additionally, it leads to a lack of public

information critical for evidence-based decision-making, delaying the education of regulators and of safety approvals.

2. The need to attract investors may create incentives for companies to obscure the actual state of their technology, which can lead to the misallocation of funding when successful marketing strategies are rewarded by investors over otherwise promising lines of scientific research.
3. Wanting to appear farther along than competitors encourages the misallocation of resources into a premature race to the market; funding projects to gain access before others, with immature 'products' that serve only publicity goals, regardless that it takes money and time away from actual research. But such 'publicity products' that do not address the animal welfare, environmental impact, sustainability, and safety goals offered by the technology are not desirable, are contrary to arguments made by private developers themselves to fund research, and in the end, are unlikely to achieve long term success if they are not sustainable. This makes such diversions away from research an inherent misallocation that, unfortunately, attracts further misallocation by others to appear competitive.
4. Both over-hyped marketing practices and premature access to markets can create false expectations and confusion for the public and investors, possibly hindering the proper development or roll-out of the technology. For example, beyond the potential for both private investors and the public at large to lose interest in cultured meat if claims are discovered to be empty, stakeholders may not realize where and to what degree public research might be useful to speed development. Should private interest in cultured meat collapse, it is possible that academic interest may rise and sustain the field until viable technologies are established. With proper infrastructure, such a collapse can be avoided in the first place.

If academic research largely avoids these obstacles, the need to follow the academic incentive system presents its own challenges. Time and resource requirements for publication are added burdens not essential for commercial decision-making, making publication prohibitive to much of the industry as well as a diversion for their research teams. Additionally, the lack of, or high cost for, open access in some high-end journals can present another unintentional block to data sharing.

Proposing research infrastructure which is purpose-built for cultured meat as it exists today would create opportunities to overcome these obstacles by avoiding the obligations that divert industry and academia. For example, without marketing requirements, an independent institute could

Published by Burleigh Dodds Science Publishing Limited, 2023.

be fully transparent with its research in a way that for-profit, IP-centered companies seeking investment in a competitive field simply cannot. Without the publication requirements of academic labs, an institute would be capable of operating at the speed of industry to develop the data and evidence stakeholders need, potentially even working directly with stakeholders to develop solutions. This unique niche could provide benefits to both industry and academia: offering potentially faster insights, tools, or products to the former and data spurring more extensive research opportunities for the latter. It can also innovate mechanisms to share data outside slower-paced and often closed publication platforms. Finally, being informed and designed from the outset along principles of responsible research and innovation, the proposed institute could work to ensure benefits reach those who will actually build or adopt the technologies (e.g. consumers, farmers, and others within the current animal product value chains) and highlight the issues these underrepresented stakeholders face.

This chapter aims to illustrate potential pitfalls of the existing infrastructure ecosystem for cultured meat, highlight what's missing, and propose potential avenues for remedying these pitfalls as the field evolves. Outlined are also some key points on how cultured meat technology has the potential to advance faster when stakeholders work together as compared to working in isolation.

2 Gaps in existing infrastructure

2.1 The talent gap

An unprecedented and growing talent gap exists within Science, Technology, Engineering, and Math (STEM) sectors. Demands for highly qualified personnel (HQP) within STEM sectors have increased by over 80% within the last 30 years.[16] Cellular agriculture, and specifically the subsector of cultured meat, is unlikely to be immune from these talent gaps. Due to the particular and cross-disciplinary skills required for alternative protein and cultured meat development, talent acquisition and retention could be significant limitations for the sector's growth if not immediately addressed.

Other emerging sectors such as synthetic biology, bioengineering, and clean tech can serve as examples of how to scale the global talent pool to quickly meet ecosystem needs. Additionally, several approaches can be adapted from these sectors to fill the cultured meat talent gap, including expanding the curriculum of traditional academic programs, creating fellowship programs, or launching upskilling programs for the established workforce. This subsection will explore specifically what these programs would look like for cultured meat and how their role out can be streamlined to align with sector growth.

The intuitive approach to solving this gap is introducing cultured meat courses and degree offerings into established university and college faculties. These programs have been quite successful in the brewing industry, whereby specific programs at small institutions (Olds College of Agriculture & Technology, Alberta, Canada) and dedicated Masters programs (Masters of Brewing and Distilling, Heriot-Watt University, Edinburgh, Canada) have created shovel-ready HQP that have the skills to take over established protocols with minimal mentorship. However, academic opportunities for students to pursue formal training in cultured meat and alternative protein development are limited, even at the graduate level; many that are available in the sector are often side or passion projects of established academics in affiliated faculties such as biomedical engineering, genetics, agriculture, or medicine.

Presently, students interested in pursuing coursework related to cultured meat (either out of personal interest or to increase their hireability within the sector) must piece together an array of specialized courses across institutions to supplement their standard biomedical/science education. Courses offered at Harvard University, Stanford University, Singapore's Nanyang Technological University, Technical University of Munich, and Tufts University feature broad overviews of meat alternatives. Although these courses typically feature a generalized curriculum as opposed to specific topics on cultured meat, they are dually beneficial to increasing students' hireability.[17]

The widespread implementation of specific cultured meat curricula is expected to be slow. For comparison, the emergence of synthetic biology as a sector can be traced back to the 1961 discovery of the lac operon by Francois Jacob and Jacques Modod.[18] Yet, despite significant discoveries in the field, including the pioneering of CRISPR in 2010 by Jennifer Doudna and Emmanuelle Charpentier, and the generation of 'synthetic life,' synthetic biology would not become a mainstream and impactful part of formal academic curricula until the mid-2010s.[19] Instead, universities have been able to spin out pools of highly qualified talent in synthetic biology and engineering biology by introducing students to the sector through professional development opportunities and extracurricular engagement. An example of such programs in this space is the International Genetically Engineered Machine (iGEM) Program.[18] The iGEM has defined itself as 'an independent non-profit organization dedicated to the advancement of synthetic biology, education and competition, and the development of an open, collaborative, and cooperative community.'[20] The organization is most notably known for its annual jamboree, where promising students compete as teams to develop innovative omics-enabled solutions to some of the world's most pressing issues. Most teams materialize as university student clubs with faculty sponsorship. In addition to proposing a technological solution, many of these clubs actively engage in campus and community outreach to promote the impact of synthetic biology on society.

Although implementing a global training network as extensive as iGEM might be difficult in the short term for the growing cultured meat sector, it demonstrates that student extracurricular engagement can drive emerging sectors forward. For example, recruiting campus leaders to create campus clubs focused on awareness and outreach could effectively educate local ecosystems about cultured meat (as well as cellular agriculture as a whole sector). An example of this is independent student clubs such as the Good Food Institute's Alt Protein Project, which includes more than 20 student groups around the globe dedicated to turning universities into engines for alternative protein education, research, and innovation.[21] On a smaller scale, universities and their respective innovation hubs/and commercialization offices could support competitions such as cultured meat hackathons to promote sector innovation and thought leadership.

An example of this is the Cultivate Tomorrow Hackathon, an annual student-organized hackathon focused on providing students with the opportunity to explore interests and apply their skills and knowledge to address challenges present in the field of sustainable food technology.[18] Students partner with industry mentors from leading cultured meat and precision fermentation companies to gain insights into the real-life hurdles these companies were facing. As a result, dozens of participants continued to pursue job opportunities in the sustainable food industry.

Not only have initiatives like iGEM and local technology hackathons like Cultivate Tomorrow and Thought for Food helped students supplement their curriculum knowledge of emerging sectors, but they also have the potential to create several spin-off companies from universities by bringing like-minded individuals together. iGEM, in particular, is credited with generating several major companies, including Gingko Bioworks, Hyasynth Bio, and PvP Biologics.[22] Therefore, replicating this model within the cultured meat space could produce multiple benefits: first, generating a steady pipeline of HQP ready to take on the demands of the sector, and second, support the development of innovative and impactful cellular agriculture start-ups capable of solving sector problems and addressing critical issues in the global agri-food landscape.

Smaller institutions, including colleges and polytechnics, do provide training opportunities that are easily adaptable to cultured meat research programs and companies. Technicians skilled in fermentation and general lab techniques can be trained at local technological institutions. These roles will be critical for scaling cultured meat technologies and raise an important point that emerging industries require a range of individuals across a spectrum of educational backgrounds. It is important to ensure that academic institutions are not only producing enough creative thinkers to lead research and development teams but also skilled manual workers to support product scale-up and production.

Nevertheless, all the options mentioned previously create a financial burden on institutions. For example, if initial enrollment in cellular agriculture and cultured meat-based curriculum is low, it could be challenging for universities to justify these investments. Institutions could partner with funding agencies to recruit and fund curricula in this space, or it could be an option to explore the establishment of federally and VC-funded, both private and public funding, maker spaces or centers of excellence, allowing students to benefit from onsite extracurricular professional development programs. These could include tech accelerator programs and entrepreneurship and communications fellowships, providing the opportunity to develop technology outside standard university research programs.

The maker-space model has become a cornerstone of most academic institutions, specifically within science and engineering faculties. Offering cultured meat agriculture-focused maker spaces could provide a unique revenue stream for academic institutions with already established agriculture faculties or strong biomedical engineering programs. For example, Concordia University in Montreal, Quebec, Canada, has created The Concordia Biofoundry, which not only services student innovation but also draws in revenue from across Canada to the institution for academics looking to generate and scale up the fermentation of custom microorganisms.[23] Another successful example is the Synbridge maker space at the University of Lethbridge in Alberta, Canada.[24] Although these facilities are unlikely to generate massive returns on government investment, they serve as centralized training grounds for students looking to gain specific skill sets - like those which would be necessary for successful careers in cultured meat, which, in turn, can be viewed as a larger (more altruistic) investment in the overall ecosystem

Finally, another avenue for highly skilled talent in the cultured meat sector is nestled within the already established workforce. Individuals with graduate and undergraduate training in degrees such as bioengineering, biochemistry, and genetics can easily transition between major biopharmaceutical and cellular agriculture technologies throughout their careers. Training a post-doc adept in human cell culture to culture, maintain, and genetically engineer animal cells would be straightforward, although incentivizing them to leave competitor industries may be challenging due to a perceived lack of job security and financial disincentives. Other opportunities lay in the field of adjacent sectors that have become less lucrative over time, such as general biology, where recent graduates or established professionals could be able to translate their skills to the cultured meat field with minimal training. Given the multidisciplinary workforce needed for cultured meat product development, it is important to note that expertise from established industries can be directly translated to this sector. Some examples include meat and food scientists, where existing skill sets and training can be applied to applications within cultured meat.

Published by Burleigh Dodds Science Publishing Limited, 2023.

A relevant example of this can be found in the global energy sector which has effectively begun re-training its workforce to be ready to assume new roles in the growing cleantech sector. Examples of this include online micro-credential programs, professional accelerators, and paid fellowship opportunities. Amidst the hundreds of programs available, a notable case study in this space would be the Avatar Accelerator Program. Based in Calgary, Alberta, and funded by energy sector giants such as Cenovus, Shell, etc., this program gives established energy professionals the opportunity to explore the cleantech sector and develop skills that would enable them to lead their respective organizations in the energy transition. Translating this line of thinking to cultured meat, the traditional farming sector can glean a tremendous amount of highly skilled geneticists, and veterinary scientists, who could lend their expertise to the sector. Programs like this have begun to emerge in the cultured meat space, including the Cellular Agricultural Society Fellowship, Good Food Institute Fellowship, New Harvest Fellowship, and the Rockey FFAR Fellows Program (created in conjunction with North Carolina State University). However, increased accessibility to and funding for these programs will be critical to serving the upcoming talent gap within the sector.

Ideally, the aforementioned solutions will not be carried out in isolation. To successfully grow, the cultured meat industry will need to simultaneously develop multiple talent pipelines to fill the mass of expected jobs as the field grows. As cultured meat has the potential to reduce national greenhouse gas emissions and diversify local economies significantly, there is a strong case for government investment in these training opportunities to enable the field's growth as a whole.[25]

2.2 The funding gap

Generally, two notable funding gaps exist within the innovation ecosystem. The first is during the idea-to-prototype phase, where founders struggle to access capital.[26] Their technology and enterprises are usually too early to provide sufficient evidence of viability for investors. The second occurs later in the commercialization phase, where it can be difficult for founders to raise sufficient funding to pilot and scale their technology to commercial viability.[27,28] Achieving these milestones can be even more difficult for companies rooted in interdisciplinary sectors, such as cultured meat.

A key challenge in finding public funding for the cultured meat space is that the sector is often considered too biotech-focused for traditional agricultural departments and simultaneously too food-focused for biomedical departments. Beyond overarching themes, like reducing greenhouse gas emissions and segueing to net-zero economies, it can be difficult to contextualize cultured meat projects in the broader scope of most funding programs. This accentuates the need for funds specifically allocated to cultured meat. Governments have been incentivized to facilitate this and in 2022, the world's largest public funding for cultured meat to date was announced by the Dutch government, which earmarked €60 million (roughly US$65 million) to support education, research, and upscaling of the field.[29-31] Another important public fund was announced in the United States of America, where the U.S. Department of Agriculture granted US$10 million to create a National Institute for Cellular Agriculture in Massachusetts.[32] Led by Dr David Kaplan of Tufts University, the institute will join researchers from Virginia Tech, Virginia State, University of California-Davis, Massachusetts Institute of Technology, and University of Massachusetts-Boston. Public funds like these are important, as it requires all involved parties to discuss how to allocate resources, resulting in an optimal infrastructure, and thereby supporting the development of the cultured meat ecosystem further.

A different example of political involvement is funding provided for highly specific projects. A collaboration between cultured meat company Mosa Meat and food processing company Nutreco called the 'Feed for Meat' project received US$2 million in funding from the European Union to study the possibilities of using waste from the animal food chain for cultured meat production.[33] A parallel example is a collaboration between cultured meat company Meatable, and the Technical University of Delft, The Netherlands, which received around €1 million (which is equal to roughly US$1 million) from the Dutch Research Council to study the use of protein-based biopolymers for potential food and healthcare applications.[34] Although funds like these help further development of cultured meat, it is noteworthy that they are sometimes limited to specific government priorities or require substantial amounts of in-kind and cash matching contributions from established industry partners. Internal capital might therefore be required, so the aforementioned funds cannot always be utilized to bridge funding gaps.

In addition to public funding, VC investments are a resource for cultured meat companies. VC is a form of early-stage investing and finances a company's move from small to large-scale operations.[35] Generally, VC investors favor traditional sectors or established industries like energy, technology, and software, as the associated history of these sectors generates a sense of security within the investment community. The contrary is observed in the field of environmental innovations: the lack of historical data and absence of stable and competitive markets facilitates uncertainties and risks related to investments.[36] These inclinations are amplified in sub-sectors like cultured meat, due to the high technical risk, long development pipelines, and large uncertainty of the outcome.[36,37]

Despite these risks, VC investment has become increasingly consistent within the cultured meat space, as companies have matured. As of mid-2022, over US$4 billion was raised for cellular agriculture companies.[12] Although the high amount of VC investment is appreciated, it poses a large pressure on individual companies. Investment of private capital typically comes with expectations of progress within relatively constrained timeframes to generate investor returns.[38] If misaligned with realistic timelines for company growth and development, enterprises can be urged to accelerate toward the next stage of company development prematurely. While this may make the investment more liquid, it can disrupt efficient research and development, as expectations are ahead of reality. This adverse impact of VC investment is not isolated to cultured meat and is observed more often in novel industries when investor interest peaks.[39]

Introducing VC funding into a sector may also introduce more competition, especially pertaining to IP management. Although this competition can motivate companies to engage in a 'race to market' and accelerate technology development, the monetization of technology comes at the expense of openly available foundational knowledge and communal sector development of reliable business development patterns.[40] This lack of openness can detriment emerging sectors from both technical and capital perspectives. Companies on the verge of commercialization may be resolving problems already addressed by competitors, resulting in capital being spent reiteratively. Unfortunately, there is no solution to this yet. However, in a more positive light, VC does unlock the potential to increase the popularity of investment in the cultured meat space, especially with the emerging trend of celebrities investing in startups as angel investors.[28,41]

What remains unclear is if VC as an industry can adapt to the timelines of scaling the biotechnology industry. As cultured meat production technology is still immature, substantial investments are required to sustain long-term research projects.[42,43] This highlights the need for proper due diligence and investor education, to mitigate expectations on challenges and timelines for returns on investment. For instance, significant portions of investment will be allocated to scaling up the production process, and the financial requirements are expected to be comparable to scaling up in the pharmaceutical industry, where scaling up happens in steps and requires jumps in capital investment.[44]

The funding coming from VC investments is currently of great importance for the development of the cultured meat field, as the amount of public funding is not sufficient. However, as mentioned earlier, substantial investments are required to sustain long-term research projects.[43] Therefore, financial systems, like banks and institutional investors, play an important role, as they can invest large amounts of funds.[28,43] A research institute with the sole purpose of doing research and development for a technical aspect of cultured meat technologies could be an approach to close the funding gap without being dependent on VC investments. This type of organization, also known as a Focused Research

Organization (FRO) would provide effective means of learning about the latest research and sharing knowledge.[17] For example, an FRO could create the first open cell bank and share protocols for research and production, avoiding the risk of capital being spent reiteratively.

3 Possible avenues for remedy

3.1 Toward a more collaborative ecosystem

Pre-competitive collaboration occurs when two or more researchers, groups, or companies work to address common challenges within the same field to advance technology and set standards. The goal is to engage in collaborations that do not compete with or affect the success of the individual parties involved, ultimately providing mutual benefits. Due to the encouragement by VC firms, technology is often developed separately and privately within companies, resulting in zero transparency and little information sharing. Cultured meat companies are not immune to this phenomenon, with the majority of funding currently being given to IP-driven research.[12,13] Concerns have arisen on whether or not the approach is effective to accelerate the field of cultured meat. Thus, the question is, can pre-competitive collaborations assist in building a robust infrastructure to help advance the field of cultured meat more quickly?

Pre-competitive collaborations have the potential to advance the field of cultured meat through the contribution of foundational research and knowledge that is accessible to all. In doing so, there would be a reduction in public and private entities 'reinventing the wheel' and duplicating failed lines of research. However, building a more collaborative ecosystem between public and private organizations is often challenged due to conflicting goals as well as time and resource restrictions. A more unique challenge specific to the field of cultured meat arises due to the variety of R&D goals. Specifically, entirely different technology will likely be needed to develop different products. For example, consider the technology needed to produce ground versus structured cuts of meat, using genetically modified organisms or not, culturing cells from different animal species, and even the different cell types within the same animal species. In short, there is no such thing as a single cultured meat technology identifying where pre-competitive collaborations can exist. Thus, it will be important to identify key commonalities that every stakeholder can benefit from within the field of cultured meat and strategize how collaborations can be built to fill in the identified pre-competitive gaps.

Setting standards in the field of cultured meat is one potential area that can benefit from pre-competitive collaborations. Preliminary efforts to establish product and safety standards are currently underway. In October 2022, three industry associations including the Alliance for Meat, Poultry, and Seafood Innovation, the APAC (Asia-Pacific) Society for Cellular Agriculture and

Cellular Agriculture Europe established an alliance with the goal of facilitating information sharing for increased transparency and harmonizing regulatory requirements to promote consistency as products reach the market.[45] Given that the alliance has 30 industry companies and organizations in the field, one strategy could be for companies to contribute data that would inform regulatory agencies to establish safety guidelines and product standards. Furthermore, forming additional collaborations with food safety and environmental and consumer advocacy organizations could provide additional perspectives on areas for improvement within existing regulatory and safety frameworks.[46-48] Essentially, building a new and transformative food system will likely require multiple collaborations and perspectives to not only ensure its success but to also improve existing food systems that will be utilized within the industry.

Though it seems that pre-competitive collaborations would be a rare occurrence in industry, many examples can be referenced to determine how the concept may be utilized within the field of cultured meat. In 2013, 17 salmon aquaculture companies contributed resources to establish the Global Salmon Initiative (GSI), a pre-competitive collaboration that has since worked to have 40% of salmon companies globally receive Aquaculture Stewardship Council (ASC) certification.[49] By doing so, each company is trained to standardize salmon feeding to increase efficiency and practice strategies to mitigate disease. This standardization ultimately leads to a more efficiently produced and safer product.

While the ASC provides an example of companies working together, public-private pre-competitive collaborations also exist. The Biomarkers Consortium was launched by the National Institute of Health and the Food and Drug Administration (FDA) in collaboration with companies and nonprofits within the field of pharmaceuticals.[50] The goal of the collaboration is to collectively develop a library of biomarkers to diagnose and treat diseases. As a result, the biomarkers developed are foundational knowledge that benefits everyone in the field and no one company or organization is responsible for generating this knowledge. Today, the consortium has upwards of 70 members globally and has launched over 30 projects, resulting in dozens of publications, 9 tools available to the pharmaceutical industry to make clinical trial decisions, and has assisted with advancing 12 FDA approvals for drug therapies. A more recent example is the Center for Environmental Sustainability through Insect Farming (CEIF), which was established in 2021 by Mississippi State University, Texas A&M University, and Indiana University-Purdue University Indianapolis after being awarded funding through the National Science Foundation (NSF).[51] The goal of CEIF is to produce high-quality research in collaboration with academia and industry to advance the field of insect protein and inform regulatory agencies.

We are already seeing examples of pre-competitive collaborations forming in the field of cultured meat such as the Tufts University Center for

Published by Burleigh Dodds Science Publishing Limited, 2023.

Cellular Agriculture which can support pre-competitive collaborations to address fundamental technical challenges in the field.[52] In 2022, RESPECTfarms was launched with the goal of bridging the current gap between farmers and scientists working on cultured meat.[53] Their plan is to establish cultured meat farms in both the Netherlands and Germany that will work on developing a platform for scientists and farmers to utilize their expertise and collaborate on cultured meat projects. RESPECTfarms additionally has plans to work toward streamlining technological implementation through the design of decentralized models that can be utilized by any farmer on their own land at a smaller production scale. Thus, in this example, we see how pre-competitive collaborations could also be beneficial in contributing social innovation to the field of cultured meat.

So, what would allow for pre-competitive collaborations to exist in the field of cultured meat? To determine areas for partnership, common problems and research questions must be identified. For example, standardizing best practices for food product safety is an area of research that will benefit all relevant stakeholders. This could be broken down into subsets that would involve parties working on a specific product type, such as ground meat versus specific cuts of meat, where the focus can be given to specific bioreactor designs and production scale-up according to the goals of each subset. Another group can focus on advancing genetically modified technology to alter cells to react identically to the same physio-chemical cues to create a universal cell media formulation. One could even imagine these efforts being overseen by one umbrella organization that is subdivided into different areas of focus.

Pre-competitive collaboration has the potential to advance the field of cultured meat and benefit all stakeholders involved. By doing so, the benefit of collaboration can be multifold and has the potential to result in significant strides toward improving our food system for the benefit of all.

3.2 Community engagement and education

Beyond research disciplines, cultured meat demands the collaboration of industry stakeholders. As this field is still in its early stages, potential opportunities exist to collectively and equitably address ecosystem needs by challenging the dominant practices, technologies, and discourse in existing innovation spaces.[54] Those in the field have the opportunity to explore alternatives to traditional technological cycles of development, pulling in social communities and relevant industries to evolve existing establishments.[55] Building a new field creates the opportunity to also include marginalized and underrepresented stakeholders at the forefront of the process to ensure that the newly developed practices benefit all who adopt these technologies. For cultured meat, some examples include those involved in

Published by Burleigh Dodds Science Publishing Limited, 2023.

traditional agricultural production, such as animal care and husbandry, and food production laborers. Policymakers, grant institutions, and departmental agencies that determine funding and standards should also be included, with representatives from diverse cultures and varying relationships to land and animals. Additionally, economic factors such as companies that produce and market new food items and consumers who choose these new technologies are also important stakeholders. In this section, we discuss the potential for community engagement through (1) bridging the gap between technological development and broader acceptance, (2) developing shared resources and knowledge, and (3) considering the location and regional context.

3.2.1 Bridging technological developments and broader acceptance

In tackling challenges around sustainability, nutrition, animal welfare, and biotechnology, grassroots and local efforts provide many valuable insights that can ground larger industrial transitions. However, bridging these efforts with more established formal scientific agencies and other institutions is complex.[54,56] Expanding beyond grassroots innovation requires a careful balance between representing local community interests and attempting to disseminate broader solutions, with the added complexity that these solutions offer answers to issues perpetuated by larger socioeconomic forces.[57] Historically, this has often meant that only some elements of a local or niche practice reach widespread adoption. For example, the push toward organic food in the UK began in the 1920s, centered around soil, plant, and animal synergies, cyclical local systems, and decentralized consumption as a critical alternative to industrial agriculture.[56] The movement was dismissed for decades, and ultimately a narrower scope of sustainability-minded organic practices gained traction in the UK in the 1980s and 1990s, focused primarily on farming without chemicals and significantly less on decentralized food systems of symbiotic animal, soil, and plant relationships.[56,58-61] Concerns remain around implications of long-distance transportation of goods, energy demands of processing and storage, packaging, and farm biodiversity, and the typical consumer will now choose between both organic and conventional foods.[56,58-61] Thus, this initially niche concept has achieved a much broader reach, with a narrower scope, reliant upon the reinterpretation and restructuring of interests by those involved in its growth.[49]

3.2.2 Building shared resources and knowledge

As cultured meat technologies continue to advance, the development of standards, oversight, and accountability within the field and externally will become increasingly critical. Open innovation networks and collaborations across industry, academia, and policy, such as Open Cell Ag, can help address this need.[62] Researchers must identify strategies for supporting open innovation and addressing translational challenges, including a focus on modularization and reproducibility, implementation of standards, policies to increase data sharing, investment in innovation hubs that bring together various stakeholders, promoting trust in the field (both externally and internally), and open communication and feedback with potential users.[62]

In addition to developing a shared language and providing open information, proponents of cultured meat can also promote more interdisciplinary STEM education to prepare future researchers and developers. Researchers have identified an 'engagement gap' in traditional engineering programs, which often do not prepare students to work with the communities for whom they will be designing solutions.[63,64] Integrating socially-directed approaches early on in the research and education process can have profound long-term impacts. The Braided River model for STEM education provides a framework for diversity of experience in science and technology careers, emphasizing variability and responsiveness, and promoting partnerships with nonscientific experts and industries, policymakers, and communities.[65] While in some ways the development of cultured meat has engaged nontechnical communities through conferences, courses, and hackathons, further scaling of the field will require sustained outreach and engagement of this type. By broadening our understanding of what scientific training encompasses, we can encourage more individuals with nontraditional experience to enter the field and emphasize collaborations, community partnerships, and broader impacts.[65] Incorporating more perspectives in STEM education at all levels can also prepare students for the complexity and nuance of community engagement.[63,64]

In recent years, more programs focused on incorporating societal perspective in engineering education have emerged, including Station1, a nonprofit higher education institution that integrates science and technology with humanistic fields and the social sciences in order to interrogate, understand, and shape technologically-driven societal impact toward more equitable and sustainable outcomes.[66] In person, cross-disciplinary workshops provide opportunities for individuals of different groups to share perspectives and better understand varied interests and concerns surrounding transitioning fields. As an example, a recent gathering on 'Social Implications of Cellular Agriculture' which brought together industry representatives, nonprofit advocates, academic researchers, members of local indigenous communities, and dairy farmers allowed participants to collectively imagine desirable food futures and 'good' transitions to cultured salmon and dairy.[67,68]

3.2.3 Physical location and regional context

The location and physical context of research and development have profound implications beyond the realm of scientific discovery, shaped by regulations, available funding models, access to resources, regional politics, and public opinion. Physical research and development spaces have consequences on multiple levels: from the labs in which this work is initially done, to the larger-scale manufacturing plants and supply chains necessary to bring cultured meat products to market. Traditional biotechnology work occurs in specialized laboratories, where the specificity of research conducted necessitates training and preparation to enter the physical spaces. As private enterprises in this field are generating prototypes, some have created opportunities for the public to visit, tour facilities, and try the products. Recent examples include cultured meat company like Upside Foods live streaming the opening of their facility in 2021, or Wildtype providing a map of their pilot plant on their website and hosting tastings. So far, these spaces remain heavily curated and niche.

As cultured meat products gain momentum, scaling the industry will require the development of manufacturing facilities. Historically, large companies have had little engagement with local communities around the impact of such facilities on local ecosystems. The development of manufacturing facilities can have a profound impact on local communities, and the degree of entanglement of companies with the community can significantly impact outcomes for relevant populations. The presence and authority of local institutions focused on civic engagement, environmental impact, and corporate oversight are very influential on emission and pollution rates for plants located far from company headquarters.[69] Life cycle assessment studies to evaluate the potential environmental impacts and broader social context around novel food products, and specifically cultured meat products, can provide useful tools for evaluating the consequences of production on local environments throughout the process.[70-73]

Recent techno-economic analyses have shed light on the challenges, costs, and viability of scaling cultured meat, considering factors such as bioreactor design, energy costs, contamination safeguards, and growth factor costs.[74,75] Together, these tools can help those in the field understand the impact of infrastructure and geographical impacts on communities, examine the cultural impacts on existing food systems, and allow them to learn from existing just transition efforts.

As people become increasingly intentional about their food choices and sources, having open access to information and to the physical spaces where their consumables are made provides an opportunity to make consumer perspectives a central motivation. Developers could even look to other specialized fields that have recently created more spaces for education, outreach, and public engagement, reducing the barrier to specialized training before entry. The DIY Bio movement has the potential to facilitate new practices and collaborations in unconventional spaces, leading to some cheaper or simpler technologies and broader public engagement.[76] This movement has been met with funding and maker space creation in a way that has allowed it to proliferate and adapt, as a potential parallel of how cultured meat could mobilize resources for public engagement. Finally, food plays a significant role in our daily lives, making it crucial to prioritize transparency, inclusivity, and openness in our approach to it. The socioeconomic and cultural factors influencing food production and consumption vary widely by region and heritage. Our engagement with animals, the products we use, and the animals we consume differ across cultures. It is important to note that cultured meat technologies are currently being developed with a focus on the needs and preferences of developed economies. However, this approach may create discrepancies because the rising demand for meat will also be needed to feed undeveloped economies. As we consider the proliferation of cultured meat products, we should take these factors into account to avoid unintentionally excluding certain populations. For example, many alternative protein companies focus on chicken, beef, or pork replacement products, which may not align with the preferences of cultures that do not traditionally consume these meats. Cultural preferences around animal consumption are influenced by complex socioeconomic factors, with certain cuts of meat becoming central to regional dishes due to their affordability and availability. It remains unclear how this will translate to cultured meat products. Therefore, it is important to consider the implications of cultured meat production beyond just the economic benefits and to explore ways to ensure that the technology can benefit all populations, regardless of their level of economic development.

4 Conclusion

The current lack of infrastructure in cultured meat technologies critically limits education, knowledge and resource sharing, and foundation building which are crucial for ongoing innovation and industry growth. A beneficial infrastructure can be established by creating more opportunities for talent generation, closing the funding gap, promoting cooperation over competition, and mobilizing the ecosystem.

To develop into an established sector, the cultured meat industry needs to develop multiple talent pipelines to avoid talent gaps and fill the mass of expected jobs as the field grows. As the cultured meat field is interdisciplinary and specific skill sets are required for the jobs that are necessary to successfully develop this emerging field, talent acquisition and retention could be a significant limitation in the sector's growth if not immediately addressed.

An intuitive approach to solving this talent gap is introducing cultured meat courses and degree offerings into established university and college faculties.

At the time of writing, academic opportunities in cultured meat are limited, and widespread implementation of cultured meat into curricula is not expected in the short term. There could be an opportunity to explore the establishment of federally and VC-funded maker spaces or centers of excellence, allowing students to benefit from onsite extracurricular professional development programs. Offering cultured meat agriculture-focused maker spaces could provide a unique revenue stream for academic institutions with already established agriculture faculties or strong biomedical engineering programs. Another approach to generating highly skilled talent is by transitioning people from already established workforces, like biopharma, agriculture technologies, or post-doc experiences with cell culture, to the cultured meat field. Expertise and skill sets from established industries like meat and food science can be directly translated and applied to the cultured meat industry.

Navigating funding gaps can be challenging for companies in any field, while it can be even more difficult for companies rooted in interdisciplinary sectors, like cultured meat. Due to its interdisciplinary origin, cultured meat cannot be easily categorized, as the sector is often considered to be too biotech-focused for traditional agricultural tranches, while simultaneously too food-focused for biomedical tranches. This causes problems when applying for funding, accentuating the need for the sector to have its own allocated resources.

So far, most investments in the cultured meat sector are obtained from VC sources with limited public investment. Private investment creates a competitive space in which developments are urged to be as fast as possible. This can be beneficial for research developments, but it can also limit the progress of the field. It causes an intentional absence of data sharing, it can lead to misallocation of funding and resources, and it can create false expectations for the public and investors. Although academic research does not have this effect, another, unintentional, lack of data sharing is presented by the lack of, or high cost of, open-access publishing in some high-impact journals.

As the cultured meat field is not a competitive industry yet, there are many pre-competitive challenges that can be solved by pre-competitive collaborations. These collaborations not only advance research faster but result in a more sustainable approach as resources can be shared, and it provides the opportunity to set standards early on in the field. In terms of infrastructure, there are going to be commonalities that every stakeholder can benefit from within the field of cultured meat, providing benefits to both public and private parties involved.

Cultured meat demands the collaboration of industry stakeholders. The relevance of food to daily life encourages the consideration of transparency, inclusivity, and openness. The socioeconomic and cultural factors influencing food production and consumption vary widely by region and heritage. Providing opportunities for engagement throughout the development and implementation process would allow individuals from diverse food cultures and those who will build or adopt emerging cultured meat technologies to have ownership and collaborate on the process with developers.

5 References

1. Kolkmann, A. M., Post, M. J., Rutjens, M. A. M., Van Essen, A. L. M. and Moutsatsou, P. Serum-free media for the growth of primary bovine myoblasts. *Cytotechnology* 2020 72(1), 111-120.
2. Poore, J. and Nemecek, T. Reducing food's environmental impacts through producers and consumers. *Science* 2018 360(6392), 987-992.
3. Ritchie, H. and Roser, M. Meat and dairy production. *Our World in Data* 2017.
4. Gerbens-Leenes, P. W., Mekonnen, M. M. and Hoekstra, A. Y. The water footprint of poultry, pork and beef: A comparative study in different countries and production systems. *Water Resources and Industry* 2013 1-2, 25-36.
5. Parker, R. W. R., Blanchard, J. L., Gardner, C., Green, B. S., Hartmann, K., Tyedmers, P. H. and Watson, R. A. Fuel use and greenhouse gas emissions of world fisheries. *Nature Climate Change* 2018 8(4), 333-337.
6. Rojas-Downing, M. M., Nejadhashemi, A. P., Harrigan, T. and Woznicki, S. A. Climate change and livestock: Impacts, adaptation, and mitigation. *Climate Risk Management* 2017 16, 145-163.
7. Victor, K. and Barnard, A. Slaughtering for a living: A hermeneutic phenomenological perspective on the well-being of slaughterhouse employees. *International Journal of Qualitative Studies on Health and Well-Being* 2016 11(1), 30266.
8. Aijuka, M. and Buys, E. M. Persistence of foodborne diarrheagenic *Escherichia coli* in the agricultural and food production environment: Implications for food safety and public health. *Food Microbiology* 2019 82, 363-370.
9. Dhont, K., Piazza, J. and Hodson, G. The role of meat appetite in willfully disregarding factory farming as a pandemic catalyst risk. *Appetite* 2021 164, 105279.
10. Van Boeckel, T. P., Brower, C., Gilbert, M., Grenfell, B. T., Levin, S. A., Robinson, T. P., Teillant, A. and Laxminarayan, R. Global trends in antimicrobial use in food animals. *Proceedings of the National Academy of Sciences of the United States of America* 2015 112(18), 5649-5654.
11. Ong, K. J., Johnston, J., Datar, I., Sewalt, V., Holmes, D. and Shatkin, J. A. Food safety considerations and research priorities for the cultured meat and seafood industry. *Comprehensive Reviews in Food Science and Food Safety* 2021 20(6), 5421-5448.
12. Khan, A. *The Cellular Agriculture Investment Landscape Report* (Vol. 2021) 2021.
13. Inc, N. H. *New Harvest Strategic Plan 2020-2025* 2020.
14. Auerswald, P. E. and Branscomb, L. M. Valleys of death and Darwinian seas: Financing the invention to innovation transition in the United States. *The Journal of Technology Transfer* 2003 28(3), 227-239.
15. Bashi, Z., McCullough, R., Ong, L. and Ramirez, M. *Alternative Proteins: The Race for Market Share Is On*. Denver, CO: McKinsey & Company, 1-11 2019.

16. O'Rourke, B. Growing gap in STEM supply and demand. Available at: https://news.harvard.edu/gazette/story/2021/11/increasing-access-and-opportunity-in-stem-crucial-say-experts/ (accessed 30th August).
17. Martin, B. R. and Tang, P. *The Benefits From Publicly Funded Research*. Sussex: Science Policy Research Unit, University of Sussex, 2007.
18. Cameron, D. E., Bashor, C. J. and Collins, J. J. A brief history of synthetic biology. *Nature Reviews. Microbiology* 2014 12(5), 381–390.
19. Kozovska, Z., Rajcaniova, S., Munteanu, P., Dzacovska, S., Demkova, L. and CRISPR. CRISPR: History and perspectives to the future. *Biomedicine and Pharmacotherapy* 2021 141, 111917.
20. iGEM iGEM Homepage. Available at: https://igem.org/.
21. GFI the Alt Protein Project GFI. Available at: https://gfi.org/the-alt-protein-project/.
22. Warmbrod, K. L., Trotochaud, M. and Gronvall, G. K. iGEM and the biotechnology workforce of the future. *Health Security* 2020 18(4), 303–309.
23. University, C. Concordia genome foundry. Available at: https://www.concordia.ca/research/genome-foundry.html.
24. Lethbridge, U. and SynBridge, O. Synthetic biology maker space. Available at: https://www.ulethbridge.ca/research/centres-institutes/alberta-rna-research-and-training-institute/synbridge-synthetic-biology-maker-space.
25. Tuomisto, H. L., Allan, S. J. and Ellis, M. J. Prospective life cycle assessment of a bioprocess design for cultured meat production in hollow fiber bioreactors. *Science of the Total Environment* 2022 851(1), 158051.
26. Gulbranson, C. A. and Audretsch, D. B. Proof of concept centers: Accelerating the commercialization of university innovation. *The Journal of Technology Transfer* 2008 33(3), 249–258.
27. Murphy, L. M. and Edwards, P. L. *Bridging the Valley of Death: Transitioning From Public to Private Sector Financing*. Golden, CO: National Renewable Energy Laboratory, 2003.
28. Bonini, S. and Capizzi, V. The role of venture capital in the emerging entrepreneurial finance ecosystem: Future threats and opportunities. *Venture Capital* 2019 21(2–3), 137–175.
29. Pointing, C. and Dutch Government. Backs cell-based meat with €60 million investment. Available at: https://plantbasednews.org/lifestyle/dutch-government-cell-based-meat-investment/.
30. Groeifonds, N. Cellulaire agricultuur. Available at: https://www.nationaalgroeifonds.nl/projecten-ronde-2/voorstellen-toegangspoort/cellulaire-agricultuur.
31. Knežić, T., Janjušević, L., Djisalov, M., Yodmuang, S. and Gadjanski, I. Using vertebrate stem and progenitor cells for cellular agriculture, state-of-the-art, challenges, and future perspectives. *Biomolecules* 2022 12(5), 699.
32. Waltz, E. Cellular meat gets first government investment. *Nature Biotechnology* 2021 39(12), 1484.
33. Nutreco Nutreco and Mosa Meat receive grant taking cellular agriculture a step closer to commercial viability. Available at: https://www.nutreco.com/en/news/nutreco-and-mosa-meat-receive-grant-taking-cellular-agriculture-a-step-closer-to-commercial-viability/#:~:text='Feed%20for%20Meat'%20aims%20to,the%20process%20of%20cultivating%20beef.
34. Askew, K. Protein-based biopolymers as cultured meat scaffold: 'The material is affordable, animal-free, scalable and edible'. Available at: https://www.foodnavigator

.com/Article/2022/12/21/Protein-based-biopolymers-as-cultured-meat-scaffold-The-material-is-affordable-animal-free-scalable-and-edible.

35. Samila, S. and Sorenson, O. Venture capital as a catalyst to commercialization. *Research Policy* 2010 39(10), 1348–1360.
36. Ghisetti, C., Mancinelli, S., Mazzanti, M. and Zoli, M. Financial barriers and environmental innovations: Evidence from EU manufacturing firms. *Climate Policy* 2017 17 (Suppl 1), S131–S147.
37. Hall, B. H., Moncada-Paternò-Castello, P., Montresor, S. and Vezzani, A. Financing constraints, R&D investments and innovative performances: New empirical evidence at the firm level for Europe. *Economics of Innovation and New Technology* 2016 25(3), 183–196.
38. Metrick, A. and Yasuda, A. Venture capital and other private equity: A survey. *European Financial Management* 2011 17(4), 619–654.
39. Dodson, B. P. and Levine, A. D. Challenges in the translation and commercialization of cell therapies. *BMC Biotechnology* 2015 15(1), 70.
40. Holmes, D., Humbird, D., Dutkiewicz, J., Tejeda-Saldana, Y., Duffy, B. and Datar, I. Cultured meat needs a race to mission not a race to market. *Nature Food* 2022, 1.
41. Sexton, A. E., Garnett, T. and Lorimer, J. Framing the future of food: The contested promises of alternative proteins. *Environment and Planning. E, Nature and Space* 2019 2(1), 47–72.
42. Guan, X., Lei, Q., Yan, Q., Li, X., Zhou, J., Du, G. and Chen, J. Trends and ideas in technology, regulation and public acceptance of cultured meat. *Future Foods* 2021 3, 100032.
43. Cecere, G., Corrocher, N. and Mancusi, M. L. Financial constraints and public funding of eco-innovation: Empirical evidence from European SMEs. *Small Business Economics* 2020 54(1), 285–302.
44. Werner, R. G. Economic aspects of commercial manufacture of biopharmaceuticals. *Journal of Biotechnology* 2004 113(1–3), 171–182.
45. https://www.apac-sca.org/post/apac-sca-forms-a-global-alliance-with-amps-innovation-and-cellular-agriculture-europe.
46. Safety, W. F. Available at: https://worldfoodsafety.org/.
47. World Health Organization. Available at: https://www.who.int/health-topics/food-safety#tab=tab_1.
48. Safety, C. F. F.
49. Devine, K. *The Business Case for Pre-Competitive Collaboration: The Global Salmon Initiative (GSI).*
50. Menetski, J. P., Hoffmann, S. C., Cush, S. S., Kamphaus, T. N., Austin, C. P., Herrling, P. L. and Wagner, J. A. The Foundation for the National Institutes of Health Biomarkers Consortium: Past accomplishments and new strategic direction. *Clinical Pharmacology and Therapeutics* 2019 105(4), 829–843.
51. *CEIF.*
52. Nicholas, I. and Tufts, M. S. Receives $10 million grant to help develop cultivated meat. Available at: https://now.tufts.edu/2021/10/15/tufts-receives-10-million-grant-help-develop-cultivated-meat.
53. The RESPECT Farms Team. Available at: https://www.respectfarms.com/about-us.
54. Fressoli, M., Arond, E., Abrol, D., Smith, A., Ely, A. and Dias, R. When grassroots innovation movements encounter mainstream institutions: Implications for models of inclusive innovation. *Innovation and Development* 2014 4(2), 277–292.

55. Hess, D. J. and Sovacool, B. K. Sociotechnical matters: Reviewing and integrating science and technology studies with energy social science. *Energy Research and Social Science* 2020 65, 101462.
56. Smith, A. Translating sustainabilities between green niches and socio-technical regimes. *Technology Analysis and Strategic Management* 2007 19(4), 427–450.
57. Smith, A., Fressoli, M. and Thomas, H. Grassroots innovation movements: Challenges and contributions. *Journal of Cleaner Production* 2014 63, 114–124.
58. Buller, H. J. *Agricultural Change and the Environment in Western Europe* 1992.
59. Foreman, S. Loaves and fishes: An illustrated history of the Ministry of Agriculture. In: *Fisheries and Food*. London: HMSO, 1989.
60. Cox, G., Lowe, P. and Winter, M. From state direction to self regulation: The historical development of corporatism in British agriculture. *Policy and Politics* 1986 14(4), 475–490.
61. Tate, W. B. The development of the organic industry and market: An international perspective. *The Economics of Organic Farming: An International Perspective* 1994 11, 25.
62. Yuen, K. K.-C. *New Sustainable Models of Open Innovation to Accelerate Technology Development in Cellular Agriculture*. Cambridge: Massachusetts Institute of Technology 2017.
63. Harsh, M., Bernstein, M. J., Wetmore, J., Cozzens, S., Woodson, T. and Castillo, R. Preparing engineers for the challenges of community engagement. *European Journal of Engineering Education* 2017 42(6), 1154–1173.
64. Woodson, T. S., Harsh, M., Bernstein, M. J., Cozzens, S., Wetmore, J. and Castillo, R. Teaching community engagement to engineers via a workshop approach. *Journal of Professional Issues in Engineering Education and Practice* 2019 145(4), 06019001.
65. Batchelor, R., Ali, H., Gardner-Vandy, K., Gold, A., MacKinnon, J. and Asher, P. Reimagining STEM workforce development as a braided river. *Eos* 2021 102, (10.1029).
66. *Station1*. https://www.station1.org/.
67. Institute, F. a. A. Social implications of cellular agriculture 'learning exchange'. Available at: https://www.ufv.ca/food-agriculture-institute/the-research/cellular-agriculture/learning-exchange/.
68. Rao, V. V., Datta, B. and Steinmetz, K. The role of natural scientists in navigating the social implications of cellular agriculture: insights from an interdisciplinary workshop. *Front. Sustain. Food Syst.* 2023 7, 1134100. doi: 10.3389/fsufs.2023.1134100.
69. Grant, D., Trautner, M. N. and Jones, A. W. Do facilities with distant headquarters pollute more? How civic engagement conditions the environmental performance of absentee managed plants. *Social Forces* 2004 83(1), 189–214.
70. Stephens, N., Di Silvio, L., Dunsford, I., Ellis, M., Glencross, A. and Sexton, A. Bringing cultured meat to market: Technical, socio-political, and regulatory challenges in cellular agriculture. *Trends in Food Science and Technology* 2018 78, 155–166.
71. Ülkü, M. A., Skinner, D. M. and Yıldırım, G. Toward sustainability: A review of analytical models for circular supply chains. In: *Circular Economy Supply Chains: From Chains to Systems* 2022.
72. Mattick, C. S., Landis, A. E., Allenby, B. R. and Genovese, N. J. Anticipatory life cycle analysis of in vitro biomass cultivation for cultured meat production in the United States. *Environmental Science and Technology* 2015 49(19), 11941–11949.

73. Scharf, A., Breitmayer, E. and Carus, M. Review and gap-analysis of LCA-studies of cultured meat. *Institude Ecology Innovations* 2019(v).
74. Humbird, D. Scale-up economics for cultured meat. *Biotechnology and Bioengineering* 2021 118(8), 3239–3250.
75. Risner, D., Li, F., Fell, J. S., Pace, S. A., Siegel, J. B., Tagkopoulos, I. and Spang, E. S. Preliminary techno-economic assessment of animal cell-based meat. *Foods* 2020 10(1), 3.
76. Landrain, T., Meyer, M., Perez, A. M. and Sussan, R. Do-it-yourself biology: Challenges and promises for an open science and technology movement. *Systems and Synthetic Biology* 2013 7(3), 115–126. Available at: https://www.apac-sca.org/post/apac-sca-forms-a-global-alliance-with-amps-innovation-and-cellular-agriculture-europe.

Published by Burleigh Dodds Science Publishing Limited, 2023.

Part 2

Technologies

Chapter 5

Cell line sourcing and characterization for cultured meat product development

Leonard Nelson and Richard Siller, Occam Biosciences, UK; and Gareth Sullivan, Occam Biosciences, UK and Oslo University Hospital, Norway

1 Introduction

There is a global food science revolution underway to address the growing demand for meat. Current intensive animal agriculture methods are seen as incompatible with environmental sustainability goals as the world population is projected to grow to 10 billion by 2050 (Fig. 1). Population growth coupled with increasing economic prosperity and social development, particularly in developing nations, will require an estimated doubling of protein production in the coming decades.

Cultivated meat is an alternative method to conventional livestock farming for meat production. Also known as lab-grown meat, clean meat, or cell-based meat, cultivated meat is meat (beef, chicken, pork, etc.) that is produced using a few cells isolated from an animal, rather than slaughtering and using an entire animal. In simple terms, animal stem cells are grown in a bioreactor using tissue engineering approaches to produce a final meat product such as a burger (see Fig. 2). The sector is part of an emerging 'cellular agriculture' biotechnology

http://dx.doi.org/10.19103/AS.2023.0130.07

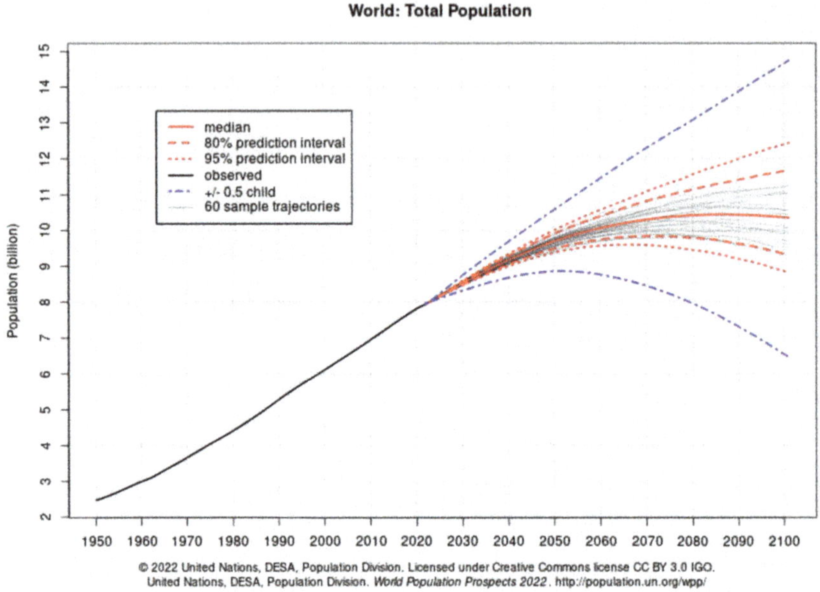

Figure 1 UN projections for world population growth this century. (Source: https://population.un.org/wpp/Graphs/Probabilistic/POP/TOT/900).

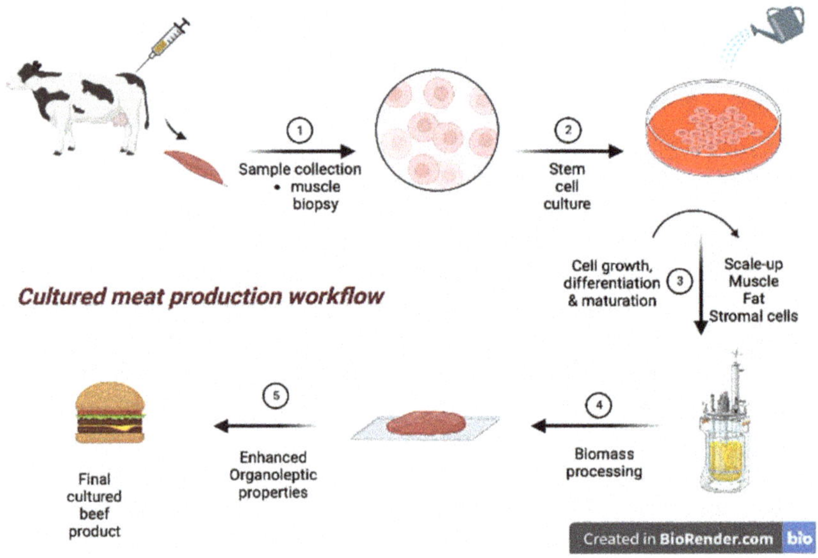

Figure 2 Key steps in cultured meat production: A small biopsy of skeletal muscle is taken from a living animal. From this, stem cells are isolated and proliferated in a bioreactor with specialized nutrient-enriched cell culture medium containing growth factor (GF) proteins. The cells are then differentiated into several cell types including muscle and fat cells. This biomass is then processed to form the edible final product.

industry that has the potential to revolutionize the way we produce a wide range of products.

Whilst there are many benefits of cultivated meat production, there remain significant technological hurdles to overcome towards achieving appropriate texture/structure, at sufficient scale and price parity to rival conventional meat production. In this chapter, we discuss the rationale underpinning cultivated meat and outline the principal benefits and technical challenges associated with producing 'lab-to-fork' cultivated meat products able to provide a nutritious, affordable, and appetizing source of proteins for consumers.

2 Why cultured meat?

Intensive livestock farming has evolved over millennia across many human cultures to provide a source of animal protein. Population growth coupled with increasing economic prosperity is driving a surge in global meat consumption. The world's resources are finite, and current intensive industrial-scale farming methods are unsustainable. They are associated with undesirable environmental impacts, animal welfare issues, and public health concerns - such as food-borne pathogens and drug residues. Cost-effective, disruptive innovations in meat-derived protein production technology are therefore required to meet UN Sustainable Development Goals (SDGs). Shifting to cultivated meat production would replace some of the animals used in the food value chain with cell culture technology, disrupting an estimated $1 trillion global market in livestock production (with beef accounting for $467 billion) (https://www.statista.com/statistics/502286/global-meat-and-seafood-market-value). Cell-cultivated meat has the potential to provide substantial benefits over conventional animal-based meat production methods.

2.1 Environmental impact and sustainability

Cultivated meat production requires less land and produces fewer greenhouse gases (GHGs) (carbon dioxide (CO_2), nitrous oxide (N_2O), and methane (CH_4)) than conventional livestock production. Recent techno-economic analysis and life-cycle assessment (LCA) modelling of theoretical cultivated meat production facilities demonstrate their potential feasibility in terms of environmental impact and cost-competitiveness (https://cedelft.eu/publications/rapport-lca-of-cultivated-meat-future-projections-for-different-scenarios/) (Risner et al., 2020). LCA studies also suggest cultured meat would involve lower consumption of water and energy. Cultivated meat could therefore have a significantly reduced carbon footprint (Fig. 3). These estimates are essentially best-case scenarios based, e.g., on optimal use of green energy in industrial-scale plants, overcoming current issues in cell culture technologies, efficient

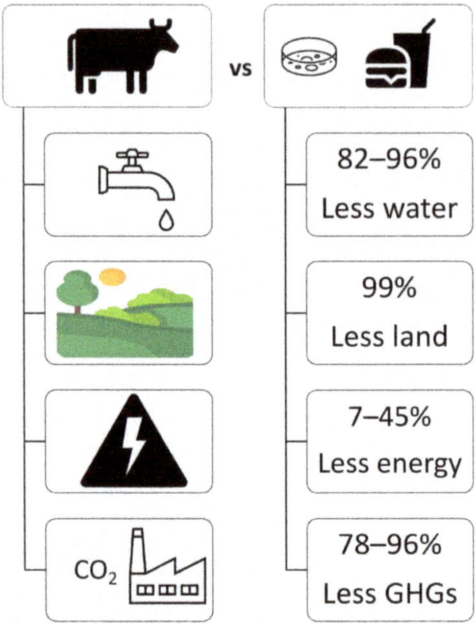

Figure 3 Estimated environmental impact reductions comparing conventional livestock and cultured meat production. (Source: GFI_sustainability_cultivated_meat.pdf = https://gfi.org/wp-content/uploads/2021/01/sustainability_cultivated_meat.pdf).

energy conversion, and recycling of spent media. A recent LCA by Risner et al. (2023) suggests that the environmental impact of animal-cell-based meat production (as they term cultivated meat) could be greater than conventional beef production systems if a highly refined pharmaceutical grade growth medium is used. The industry is currently investigating greener production materials and methods to address this challenge.

The development of more efficient, cost-effective cultivated meat production could also help ensure global food security and access to an affordable nutritious diet. Products are produced in a controlled environment that is not subject to the same risks as traditional livestock farming, such as exposure to pathogenic and chemical contaminants or environmental risks to production such as drought or flooding, disease outbreaks, and natural disasters.

2.2 Ecology

Livestock production chains account for more anthropogenic GHG emissions than all global transport combined and require an estimated 70% of all arable land to be used for grazing livestock or growing livestock feed, contributing to deforestation (https://ourworldindata.org/global-land-for-agriculture). In

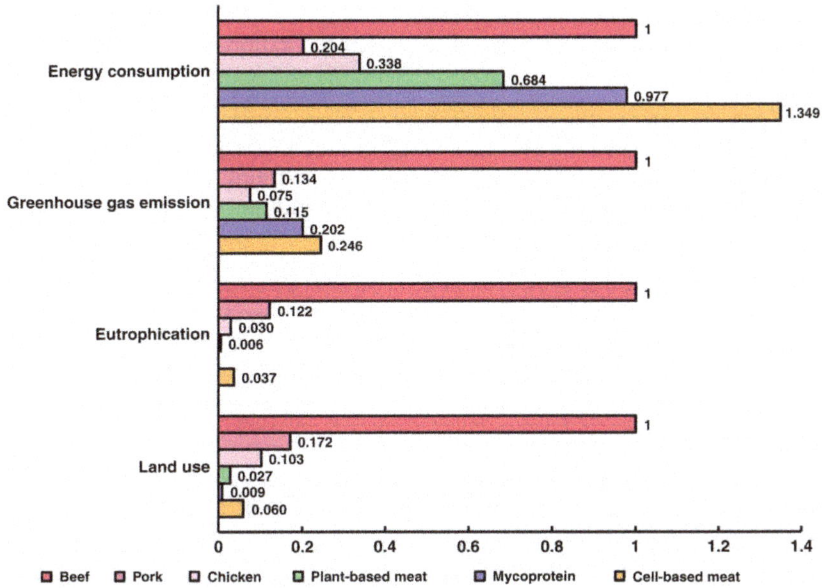

Figure 4 Comparison of the environmental impact of meat and meat analogues.

addition, intensive livestock farming (often of a single large animal species) can impact biodiversity through habitat displacement of other species and can also lead to water pollution through the release of animal waste into waterways, leading to eutrophication (Fig. 4).

2.3 Ethical concerns

In addition to environmental issues, the large-scale production of animals, such as cows, pigs, and chickens for food, presents significant social challenges including potentially cruel animal husbandry practices causing pain and suffering. Overuse of prophylactic antibiotics may also contribute to increasing antimicrobial resistance (AMR) in humans. Cultured meat could thus substantially improve animal welfare as production does not involve the slaughter of animals and only a few animals would be used as donors to obtain cell and tissue biopsies. In addition, commercialization of cultivated meat could positively impact public health in areas such as reducing zoonotic foodborne illness (e.g. *Salmonella*, *E.coli*, and *Listeria* contamination in conventional meat) given commercially cultivated meat requires aseptic conditions (Post et al., 2020; Godfray et al., 2018).

2.4 Economic benefits

A recent report by McKinsey & Co suggests that the market for cultivated meat could reach $25 billion by 2030 (0.5% of global meat supply) (https://www

.mckinsey.com/industries/agriculture/our-insights/cultivated-meat-out-of-the-lab-into-the-frying-pan). This new industry has the potential to create new jobs, as it will require a large number of skilled workers and could stimulate economic growth across sectors, including cell culture technology, facilities engineering, chemical engineering, software development, and retail - to name a few. In addition, it could eventually, with growing economies of scale, reduce the cost of meat production, which could make it more affordable for consumers. However, economic success depends heavily on many factors coming into alignment including consumer acceptance, economic bottlenecks (to achieve price parity with conventional livestock production), and overcoming specific scientific and technological pain points.

Global investment in start-up cultivated meat companies continues apace, focussing on popular meat varieties: beef, pork, chicken, and fish (Fig. 5). Potentially, economic returns are high compared with livestock farming, given that industrial-scale cell-cultivated meat production is feasible from cell/tissue samples taken from only a few animals. The value of the beef industry to the UK is £2.93 billion p.a. Even a 1% shift in this market represents a potential market value for cultured meat of £29.3 m in the UK and £3.88 billion globally.

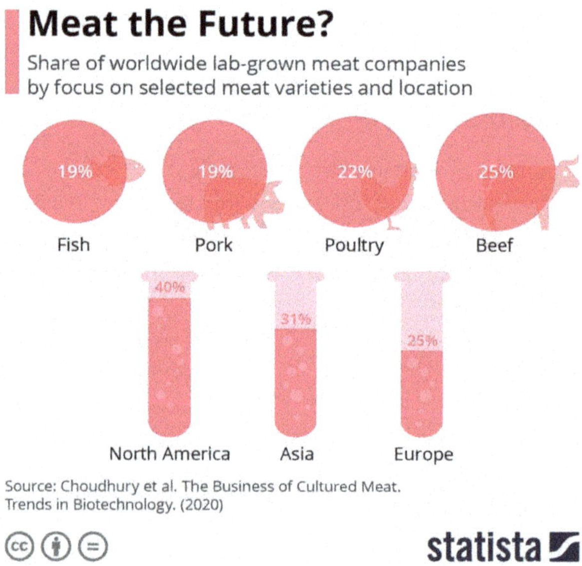

Figure 5 Share of global cell-cultivated meat companies by focus on selected meats varieties and location (as of Feb 2020). Wiki Commons: [https://www.statista.com/chart/15983/food-sustainability/].

2.5 Public acceptance

Public perception and acceptance of cultivated meat is also a key driver of success for the industry. There are complex socio-economic and psychological issues involving individual consumer preferences as well as cultural and religious (e.g. halal or kosher meat preparation) issues. Human preferences are of course multifaceted, for example, some consumers may favour white breast versus brown cultivated chicken meat, while others are influenced by the description of cultivated meat including the name (nomenclature). A recent study on nomenclature revealed 'cell-cultured' and 'cell-cultivated' were the best all-round labels in terms of clarity and appeal. There is some evidence of public acceptance of cell-cultivated meat, driven by consumers' preference to adopt a more sustainable, protein-rich diet. Indeed, plant-based 'meat analogues' such as Quorn are gaining traction in the alternative protein market space (Godfray et al., 2018; Malerich and Bryant, 2022).

In summary, the development of economically viable cell-cultivated meat production methods has the potential to be a sustainable, humane, and healthy alternative to conventional livestock farming - by significantly reducing the environmental impact of the meat industry, with benefits for animal welfare, public health, food security, and economic growth.

3 Challenges in cultivated meat production

As noted, cell-cultivated meat is meat grown from animal stem cells in a bioreactor using tissue engineering approaches. Briefly, cell culture technology involves taking a tissue biopsy and isolating cells (fibroblasts, blood, muscle, or fat) from the sample. The isolated cells can be grown to produce a potentially limitless supply of cells via reprogramming or immortalization (a process allowing cells to be cultured indefinitely), providing the potential for large industrial-scale production of meat from a single biopsy.

Cell lines can be differentiated in specialized cell culture medium formulations [containing protein growth factors (GFs) and essential nutrients] that drive maturation into either muscle cells/ tissue (myocytes) or adipose (fat) cells, as well as other tissue types including the liver. Advanced tissue engineering approaches and cell culture parameter optimization will be required to scale up stem cell production by improving cell productivity; i.e. cell doubling time, density, and biomass. Subsequent growth and maturation in bioreactors to yield the final product utilize bioprocessing technologies similar to those used extensively in industrial fermentation processes, such as for the production of monoclonal antibodies or vaccines. The final cell/tissue constructs can then be combined with scaffolds (to provide structure) with flavouring and organoleptic enhancement to form edible meat products.

However, as noted, there are critical techno-economic bottlenecks for this nascent industry and thus areas of focus for further research and development. Development of the sector will require a steep change related to cost, efficiency, and scalability to reach price parity with conventional livestock production. Indeed, some analysts suggest it could take over a decade to address the key technical challenges to achieve commercial scale and even a 1% share of the global meat market at current commodity prices (https://www.foodnavigator-usa.com/Article/2023/01/20/Cultivated-meat-in-focus-at-Tufts-cellular-ag-innovation-day-From-bioreactor-bottlenecks-to-a-global-battle-cry-for-governmentfunding?utm_source=copyright&utm_medium=OnSite&utm_campaign=copyright).

A key cost challenge is the use of critical GFs. Achieving a full price-competitive and commercial cultured meat industry will require disruptive innovations in protein production and cell culture technologies. This will require expertise and resources to solve knowledge gaps in cultivated meat R&D production stages. In this section, we focus on these critical technologies and technical 'pain points' that must be overcome to achieve commercialization of cultivated meat products.

3.1 Tissue engineering

For over two decades, tissue engineering approaches have been developed mainly in academic labs to generate *in vivo*-like 3D organotypic tissue. These approaches typically use organ-specific cells, protein scaffolds, and a cell culture nutrient medium for trophic support (containing animal-derived serum), contained within a bioreactor to maintain an optimal environment (temperature/pH and humidity) to promote growth and maturation of tissue. Cultivated meat production often follows the same basic processes, incorporating developmental biology protocols to encourage proliferation/differentiation of stem cells such as myogenic satellite cells, fibroblasts, and MSCs (Hong et al., 2021).

The industry must now go beyond this 'proof of concept' phase of laboratory-scale cultivated meat production. It is also important to address the currently fragmented research and development pipeline in moving from lab to pilot to industrial scale. Indeed, a recent analysis highlights that the sector should adopt a 'race to mission' strategy, i.e. with industry players working together (rather than the present fragmented approach involving competing players) in tackling unresolved pan-industry issues (Holmes et al., 2022).

3.2 Cell culture media

Cell culture medium is essentially a blood analogue that supports the growth of cells grown *in vitro*. Cell culture media design and optimization is a significant barrier towards commercialization of cultivated meat. Developed and refined over 50 years, a cell culture media is a complex matrix that provides trophic support, including essential nutrients (e.g. glucose, amino acids, vitamins, and minerals) and GF proteins together with small molecules such as insulin and glucocorticoids which allow cell growth and survival.

Traditional cell culture protocols use 'undefined' serum sourced from foetal calves (FBS: foetal bovine serum) to supply a milieu of morphogens, lipids, and transfer proteins. However, this is not feasible for cultivated meat production for economic, food safety, and ethical reasons. Worldwide, 1 million 3-9-month-old bovine foetuses are harvested to supply 500 000 L of FBS annually for biotech cell culture applications. Apart from safety and welfare concerns, scaling requirements for cultivated meat production renders FBS incompatible for commercial-scale bioreactors, at a cost of $1000 per litre. Intense efforts are underway to produce optimized serum and animal-free medium formulations in order to make cell culture media significantly more affordable and address ethical and other concerns (O'Neill et al., 2022).

3.3 Growth factors

The high costs of GFs remain a significant bottleneck to commercialization. Cell culture media contains protein GFs used in stem cell growth and subsequent differentiation/maturation for bioprocessing into final products (muscle, fat, or connective tissue cells). Recombinant GFs used in academia and industrial labs are readily available from many vendors. These are high-grade proteins (USP-cGMP) requiring expensive bio-separation/purification protocols. This critical step becomes extremely expensive when transitioning from laboratory to production scale. Modelling studies show current GF protein production and separation methods cost up to $400 000 per kg using the GF FGF-2 at a cost of $5000/mg in a 20 000-L bioreactor. This hurdle must be overcome to achieve sustainable 'green' GFs in sufficient (kilogram) quantities at a competitive cost. Only by reducing the cost and/or the amount of FGF-2 can we approach commodity-level prices of $2 per kg.

Several innovative solutions are in the R&D pipeline. These include the development of recombinant GF bulk-expression platforms and supplements that lower the effective concentrations of GFs, combined with cost-effective and sustainable protein bio-separation and purification methodologies. Such advances will enable the supply of low-cost highly purified GFs. Additionally, synthetic biology approaches could also be harnessed to eventually develop cheap, thermostable GFs, e.g. FGF2-G3 (https://qkine.com/thermostable-fgf2/).

3.4 Other technical considerations

Other areas for development include optimization of cell-type-specific media, improvement of cell lines, and bioreactor design. Current bioprocesses used in fermentation and vaccine production could be repurposed for cultivated meat but will require massive capital investment in industrial-scale facilities. Indeed, it has been estimated that to achieve only a 1% global market share, the cultivated meat sector will require exceeding the combined capacity of the pharma industry's current cell-culture capacity by ten-fold (from 10 million to >100 million litres - equivalent to about 100 Olympic-size swimming pools!). Finally, a reduction in energy costs will need to be achieved across the production chain including electricity used to heat media - through recycling of media and implementation of heat exchangers (https://www.mckinsey.com/industries/agriculture/our-insights/cultivated-meat-out-of-the-lab-into-the-frying-pan; https://www.foodnavigator-usa.com/Article/2021/03/15/CellRx-on-a-mission-to-slash-cost-of-growth-factors-used-in-cell-cultured-meat-production).

3.5 Regulatory acceptance

The Singapore Food Agency (SFA) was the first national regulatory body to approve a cultured meat product (chicken nuggets), whilst the U.S. Food and Drug Administration (FDA) and U.S. Department of Agriculture (USDA) have recently approved cultivated chicken meat for human consumption. Nonetheless, food safety concerns remain and risk analysis of cell-cultured products is still very much in its infancy (Ong et al., 2021). In the US, the FDA and USDA are formulating a joint regulatory framework for cultivated meat production. The FDA will oversee compliance of cell culture facilities for food safety requirements, risk/hazard analysis, and current good manufacturing practices (GMPs), whilst the USDA will ensure food safety at the post-bioreactor cell culture phase.

Two industry hurdles in achieving this goal are cost and scale, key cost drivers being recombinant GFs as components of specialized cell culture media. R&D improvements are necessary to compete on both price and quality with conventional meat. First and foremost, the cost of media (the broth in which cells grow) will need to come down. Ways to lower media costs include minimizing the use of GFs, optimizing filtration processes, and recycling media. Improving cell productivity is another way to increase output per batch. This is achieved by increasing cell density in media through cell-line optimization and improving cell-culture feeding processes to maximize density and minimize media consumption and batch time (https://www.mckinsey.com/industries/agriculture/our-insights/cultivated-meat-out-of-the-lab-into-the-frying-pan).

4 Cell line sourcing: overview

A key challenge that must be overcome for developing livestock cell sources is the ability to develop/source cell lines with the proliferative capacity and differentiation potential to allow large-scale commercial production. This requires cell lines that can be integrated into current production pipelines. These can come in different forms from primary to immortalized lines. This and the following sections discuss the sourcing of different cell types, starting with adult stem cells.

Source material can be acquired from a tissue biopsy from either a live animal such as an ear punch or using post-mortem tissues (abattoir) from the animal of interest. This leads to the production of a primary cell source, for example, muscle-resident progenitor cells (adult stem cells), which can be isolated from skeletal muscle tissues collected from the animal. The primary material can then be used to establish immortalized lines through spontaneous immortalization, genetic immortalization, or reprogramming to an induced pluripotent stem cells (iPSCs). PSCs (described in Section 7, which are derived from primary material (fibroblasts, blood, etc.), need to be differentiated into muscle cells. The following sections review different cell types. In each case, the discussion reviews the relative pros and cons.

5 Adult stem cells

Cell-based meat requires the ability of sourced cells to self-renew and proliferate to the numbers required, combined with the ability to differentiate into the cell types relating to components of meat such as muscle fibre (myocytes), fat (adipocytes), connective tissue (collagen/chondrocytes/fibroblasts), and blood cells (Listrat et al., 2016).

Adult stem cells provide a viable cell source, in the form of undifferentiated progenitor cells that reside in different tissues (tissue-specific progenitors). Adult stem cells have a number of favourable properties including multipotency: the ability to differentiate into a restricted number of cell types and are a relatively easily attainable source of cells.

The best progenitor/stem cell populations are muscle satellite cells (myoblasts) and mesenchymal stem cells (MSCs)/adipose-derived stem cells. These progenitor populations can be differentiated into mature cell types present in meat, i.e. skeletal myocytes, adipocytes, chondrocytes, and fibroblasts.

Myoblasts or muscle satellite cells are quiescent mononucleated myogenic cells found between the sarcolemma and basement membrane of muscle fibres. These cells are very abundant. One person has approximately 1×10^{10}–2×10^{10} satellite cells, making these an attractive cell type (Morgan and Partridge, 2003). Their isolation in animal systems (including porcine, bovine, ovine, and avian systems) is well established (Kim et al., 2020; Choi et al., 2020; Baquero-Perez et al., 2012; Lee et al., 2021). However, the isolation, maintenance, expansion, and differentiation of muscle satellite cells have a strong dependency on FBS (Li et al., 2015; Wan et al., 2022; Zhu et al., 2022), which raises regulatory and ethical concerns.

There have therefore been concerted efforts to do away with serum and develop serum-free media alternatives. Post and colleagues, e.g., have developed a chemically defined media for bovine myoblast culture (Kolkmann et al., 2022). This study replaced animal products with recombinant human products, e.g., by replacing bovine serum albumin with human serum albumin. A recent report from the Kaplan lab describes the development of simple and effective media for sustained expansion (Stout et al., 2022). However, there have been concerns about using recombinant human proteins in the food chain. Where possible, species-specific GFs are preferable. More work is needed to generate species-specific recombinant surrogates.

Another source of primary cells are MSCs and adipose-derived stem cells. These can provide an important component to the final product, fat, which is an essential dietary component, also providing the associated flavour, juiciness, and tenderness that consumers expect (Zagury et al., 2022). MSCs are found in several areas within the body including the bone marrow, cord blood, circulating blood, and placenta. However, MSCs are very limited with regard to number of stem cells. For example, bone-marrow-derived MSCs account for ~ 0.001–0.01% of bone marrow mononuclear cells in humans (Bhat et al., 2021).

In contrast, adipose-derived stem cells can be derived from easily accessible fat tissue deposits at multiple sites/depots in large quantities from slaughterhouses. Cells reside in subcutaneous white adipose tissue in larger animals and in visceral white adipose tissue (abdominal regions) in avian and fish species. Adipose-derived stem cells account for 1–10% of the adipose tissue population (Sharun et al., 2021). A feature that supports the use of adipose-derived stem cells is their lifespan. MSCs demonstrate senescence after just 6–7 passages, while adipose-derived stem cells do not (Kern et al., 2006). Both MSCs and adipose-derived stem cells share many features such as surface markers, morphology, transcriptome, and differentiation (Mohamed-Ahmed et al., 2018; Musina et al., 2005; Liu et al., 2007).

The process of producing fat, or adipose tissue is called adipogenesis. In this process, multipotent adipose-derived stem cells become pre-adipocytes and differentiate along the adipogenic lineage to become mature adipocytes containing lipid droplets. Adipocyte biology is well studied in humans and mice though less so in the context of cultured meat production. However, a recent review assesses a range of studies on the isolation, growth, differentiation, and/or analysis of adipogenic cells from farm animals and fish (Sugii et al.,

2022). This provides a resource for identifying potential isolation procedures and culture conditions for expansion/differentiation to adipocytes. Culture conditions for the isolation and expansion of adipose-derived stem cells are heavily dependent on FBS. However, there are reports of serum-free approaches from the regenerative medicine field that could be translated into cultured meat (Zanicotti and Coates, 2017; Bindroos et al., 2009). Primary adult stem cells currently remain the most common source for the production of cultivated meat. However, they have their limitations. They require regular visits to a slaughterhouse and have a limited proliferative capacity. Alternative cellular sources are discussed below.

6 Immortal cell lines

A key challenge that must be overcome is that differentiated (primary) cells such as muscle (myoblasts) and fat (adipose-derived stem cells) have a limited ability to replicate before senescencing. A way to overcome this is to generate immortalized cell lines. Cell lines that can be expanded indefinitely (proliferative) while maintaining the attributes (identity) of the tissue or can be differentiated effectively into muscle and fat are very attractive candidates for cell lines.

6.1 Cancer-derived cell lines

Most of the commonly used cell lines *in vitro* are immortalized. These can be derived from tumours, such as the ubiquitously utilized HeLa cell line that was essential for the development of the modern field of *in vitro* cell culture (Almeida and Korch, 2023). There are immortalized cancer cell lines that are representative of diverse tissues, including those of interest for cultivation for meat production including the liver, muscle, and fat, but these are predominantly human and mouse lines due to their use in disease modelling. While these cell lines have expanded our knowledge in the medical field, they will have limited utility for meat production as they are potentially tumorigenic and not suitable for human consumption.

6.2 Spontaneous immortalization of cell lines

Spontaneously immortalized cells are not considered to be genetically modified (GM), making them an attractive cell type for the European market which has strict regulations on GM foods (ECJ, 2018. https://curia.europa.eu/jcms/upload/docs/application/pdf/2018-07/cp180111en.pdf). However, other jurisdictions regard spontaneously immortalized cells as equivalent to cancerous cells.

Spontaneous immortalization is a very low probability event (a rate in the range of 10^{-6} in mice) (Fridman and Tainsky, 2008). Cell lines become immortal when they lose their cell cycle checkpoint pathways and evade senescence. Different cell types have different propensities to spontaneously immortalize and can acquire features such as serum-free growth and high-density cultures (Nachtigal et al., 1989). The mechanism behind spontaneous immortalization is poorly understood. But what is known is that there is a requirement for the dysregulation of key pathways, e.g. inhibition of the proto-oncogenes p53 and Rb or ARF-p53 pathway (Delia et al., 2003). Additionally, one can select clones for high telomerase reverse transcriptase (TERT) expression or low p15/p16/Rb expression.

Spontaneous immortalization has now been demonstrated in multiple species including fish, which have a particularly high propensity for spontaneous immortalization (Saad et al., 2023; Liu et al., 2022; Soice and Johnston, 2021). The Kaplan lab has produced two immortal cell lines from mackerel (Saad et al., 2023). They coaxed cells to crisis (spontaneous immortalization) through repeated passaging (the process of transferring cultured cells to a fresh growth medium). The crisis was characterized by overt apoptosis/senescence, leaving clusters of quiescent cells that then grow out. These mackerel-derived myoblast cultures demonstrated increased proliferation as well as maintenance of myogenic and adipogenic potential (Saad et al., 2023). Other species that are amenable to spontaneous immortalization include birds (Pasitka et al., 2023), cows (Zhao et al., 2010), and sheep (Suleiman et al., 2022).

6.3 Artificially immortalized (engineered) cell lines

Terminally differentiated cell lines can be artificially immortalized using several approaches including, exposure to chemical carcinogens or radiation, viral transformation or genetically. Exposure to radioactive substances or carcinogenic chemicals can immortalize a subpopulation of the cells by increasing the formation of cancer cells (Namba et al., 1988; Marczynska et al., 1991). This is thought to be a multistep process involving clonal evolution with repeated 'selection and succession' so that cells that manifest a growth advantage will outcompete the remaining population. This is a result of acquired mutations that confer a selective advantage. However, these lines are effectively considered cancer lines and have little utility for cultured meat.

Another approach to immortalize primary cells is via viral infection. Miller reported that Epstein-Barr virus (EBV) could immortalize bone-marrow-derived lymphocytes (Miller, 1982). Other DNA viruses have been reported to immortalize cells including the polyomavirus simian virus 40 (SV40) through the SV40 large T antigen (Bryan and Reddel, 1994). The human papillomavirus has been demonstrated to immortalize kerantinocytes (Blanton et al., 1991), resulting in the upregulation of TERT. However, this approach is not appropriate as these are viruses that can potentially cause cancer.

Researchers have investigated viral-induced immortalization, identifying the genetic drivers of immortalization. Commonly employed viral genes for immortalization include SV40 T antigen whose action is via binding and inactivation of p53 and Rb proteins and activation of TERT (Bryan and Reddel, 1994). By introducing both SV40 T antigen and human TERT, more effective immortalization was observed in human primary cells (Matsumura et al., 2004). hTERT alone has been demonstrated to immortalize various cells including fibroblasts, retinal pigmented epithelial cells, and porcine pancreas cells (Rambthala et al., 2002; Cao et al., 2011; Noble et al., 2004). Another approach is the production of conditionally immortalized cell lines using a temperature-sensitive T-antigen, with activity conditionally controlled by changing the temperature between permissive and non-permissive temperatures. Growth and differentiation of cell lines can be conditionally controlled by changing the temperature (Obinata, 2007).

The Kaplan lab published a study describing the generation of immortalized bovine myoblasts, targeted for cultured meat applications. They used a combination of TERT and cyclin-dependent kinase 4, delivered using a transposon-based approach. They achieved a high doubling rate and maintained myogenic differentiation potential (Stout et al., 2023). A key issue to address is what vector system to use. Integrative viral-based systems commonly employed in the research environment (i.e. Lenti- and Retro-viral based systems) are not acceptable on safety/regulatory grounds. A potential solution is gene editing with CRISPR, where the immortalizing gene(s) can be targeted to a safe harbour site such as the ROSA26 locus and, as shown in pigs, a muscle-specific safe harbour site, the *CKM* gene (Xiong et al., 2022). This ensures that a single copy is introduced into a designated site. Additional safety measures to regulate immortalization include FLOXing, which allows the removal or inversion of the gene, or the PiggyBac system which is a bidirectional transposon system (Chen et al., 2020). However, genetic modification of cells remains problematic in countries where the use of these techniques in food production is restricted or consumer acceptance is low. However, Upside Foods in the US has recently achieved FDA approval for an immortalized chicken meat product.

7 Pluripotent stem cells

There has been a focus on the use of PSCs, initially developed to study diseases *in vitro*. PSCs have two key characteristics which make them an attractive platform for the production of cultured meat:

- The capacity to self-renew indefinitely; and
- the ability to differentiate into all cell types of an organism (pluripotency).

Having unlimited capacity for cell division facilitates the cultivation of massive numbers of cells. Cells are stable and maintain their stemness across many passages, enabling reproducibility of protocols with the same cells over many years. Pluripotency makes it possible to differentiate cells to a myriad of cell types present in the body, giving PSCs a clear advantage over adult stem cells and immortalized cell lines which are restricted in their cellular identity (i.e. an immortalized liver cell line will never become a muscle cell). PSCs can provide the foundations for the generation of all the cell types required for the production of cultivated meat, i.e. muscle, fat, and blood. To that end robust protocols for differentiation to these cell types have been developed (Hafner and Dani, 2014; Mohsen-Kanson et al., 2013; Iberite et al., 2022). PSCs come in different forms which will be discussed in the following sections.

7.1 Embryonic stem cells

Embryonic stem cells (ESCs) are derived by mechanical dissection of the inner cell mass from the blastocysts and cultured under defined conditions. ESCs were first derived in mice (Evans and Kaufman, 1981). The first human ESCs were derived in the 1990s (Thompson et al., 1998), human ESCs have revolutionized the study of human disease, regenerative medicine, and tissue engineering. However, the derivation of human ESCs requires the destruction of an embryo which raises ethical concerns. The development of animal-derived ESCs has been slow and challenging, due to a lack of understanding of reproductive biology and differences in embryonic development between species. 'Embryonic-like cells' have been derived from pigs, cattle, buffalo, goat, and horses (reviewed in Nowak-Imialek and Niemann, 2012). ESCs can be more easily derived in chickens and other avian species via *in vitro* fertilization of eggs (van de Lavoir and Mather-Love, 2006).

7.2 Induced pluripotent stem cells

Ethical concerns about ESCs have led to the development of alternative strategies to source PSCs, notably the production of iPSCs which has been transformative to the field of stem cell biology. iPSCs have the same key features as ESCs (limitless self-renewal, differentiation potential) with the advantage of no embryos being sacrificed. They also represent the genetic characteristics of the donor from which they were derived. iPSCs have been derived from a myriad of species including mice, humans, ruminants, pigs, horses, dogs, cats, rabbits, and endangered species such as the rhinoceros (de Figuiredo Pessôa et al., 2019).

Pioneering work on cloning demonstrated that the entire genome of a cell could be 're-wound' to an undifferentiated embryonic state - with the remarkable capacity to develop into a clone of the donor. While cloning has proven useful for the derivation of ESCs, more recent advances have demonstrated that only a few key genes are required to convert a cell such as a fibroblast to an embryonic cell state. In the 1980s, Weintraub and colleagues showed that one master regulator gene (MyoD) was capable of converting fibroblasts into muscle cells (Davis et al., 1987). This concept was applied by Yamanaka and colleagues for the derivation of mouse and human-induced pluripotent stem cells from fibroblasts. Instead of MyoD, genes involved in the maintenance of the pluripotency network were used. These included the octamer-binding transcription factor 4 (OCT4/POU5F1); SRY-Box 2 (SOX2); Kruppel-like factor 4 (KLF4); and proto-oncogene MYC (c-MYC) (Takahasi and Yamanaka, 2006). Other genes have also been deployed to reprogram, improve efficiency, and enable reprogramming of recalcitrant species (Yu et al., 2007; Verma et al., 2012; Verma et al., 2013).

A prerequisite for reprogramming is the efficient delivery of the transcription factors into cells. Early protocols used integrative viral approaches such as retro- and lenti virus (Takahasi and Yamanaka, 2006). However, viral integration can result in reprogramming factors being reactivated, i.e. c-MYC (Ben-David and Benvenisty, 2011). This can be addressed by transposon-based approaches such as PiggyBac or Sleeping Beauty. These are integrative but can be seamlessly excised from the genome post-reprogramming (Woltjen et al., 2009). Alternative non-integrative approaches include Sendai-virus (Chen et al., 2013), episomal vectors (Bang et al., 2018), modified mRNA (Wang, 2021), self-replicating mRNA (Yoshioka et al., 2013), and protein delivery via cell-penetrating peptides (Seo et al., 2017). Another approach that does not require genetic or protein factors is the use of small molecule/chemical surrogates that activate the respective pluripotency networks, but this has only been demonstrated in mice and humans (Hou et al., 2013; Guan et al., 2022).

Another approach to reprogramming is the employment of small molecules in combination with reprogramming factors to enhance the efficiency iPSC derivation. For example, the small molecules PD0325901 (MEK/ERK inhibitor), Thiazovivin (ROCK inhibitor), and SB431542 (selective inhibitor of the transforming BF-β) used in combination significantly increased reprogramming efficiency (Lin et al., 2009). Epigenetic modifiers such as Histone deacetylase (HDAC) inhibitors such as valproic acidand trichostatin A or DNA methyl transferase inhibitor, 5-Azacytidine resulted in enhanced reprogramming (Federation et al., 2014). In fact, small molecules have been heavily employed in the generation of farmyard animal iPSCs for an overview see Ogorevc et al. (2016).

7.3 Cell types used for induced pluripotent stem cell reprogramming

A consideration in the derivation of iPSCs is what initial cell type to reprogramme. There has been much speculation around retention of epigenetic memory with respect to the cell type of origin (Williams et al., 2022). To that end, iPSCs for cultured meat production could be generated from the target cell type being generated such as myoblasts or adipocytes. But it is also possible to reset/erase the epigenetic memory of the donor tissue through exposure to chromatin-modifying compounds (Kim et al., 2010).

iPSCs have been reprogrammed from many cell types such as adult, foetal, and embryonic cell types, including fibroblasts, peripheral blood mononuclear cells (PBMCs), and myoblasts (Takahasi and Yamanaka, 2006; Vlahos et al., 2019; Ezashi et al., 2009; Winder and Trokovic, 2021). In the case of farmyard animals, the vast majority have been reprogrammed from fibroblasts (adult, foetal, and embryonic) with a handful of cases using keratinocytes, liver, stomach, and MSCs (Ogorevc et al., 2016; Scarfone et al., 2020). Fibroblasts are the most common starter population in animals for reprogramming. These can be obtained from a skin biopsy, e.g. when ear tagging is performed. Prior to reprogramming, cells are isolated by chopping up the biopsy and culturing in a tissue culture plate supplemented with growth media for several weeks to establish the fibroblast line. After the fibroblasts are established, they are banked and reprogrammed.

7.4 Types of pluripotent stem cell and media

Another consideration is the type of pluripotent stem cell to be derived in terms of development characteristics. During mammalian embryonic development, the zygote develops through multiple stages to form the blastocyst. During this process, the cells are considered totipotent up to the development of two-cell blastomeres (in mice)/eight-cell-stage blastomeres (in humans). Later in development, the epiblast is the source of PSCs, which are classed by potency. These are:

- PSCs that can generate embryonic and extra-embryonic cell lineages; or
- epiblast-derived PSCs which only generate the embryonic lineages.

Epiblast-derived PSCs show different states including naïve and primed (Nichols and Smith, 2009; Pera and Rossant, 2021).

A recent report has described a population of mouse two-cell-like PSCs (Genet and Torres-Padilla, 2020). Human eight-cell-stage-like cells capable of producing both embryonic and extra-embryonic lineages have also been described (Mazid et al., 2022). Expanded potential stem cells (EPSCs) from mouse eight-cell-stage blastomeres have also been established, which have properties of the blastomere and produce both embryonic and extra-embryonic lineages (Yang et al., 2017). Porcine EPSCs and human EPSCs have now also been reported (Gao et al., 2019). The majority of established animal lines to date are either naïve or primed, with one example of EPSCs in pigs (Gao et al., 2019). These PSCs have been generated either through reprogramming or via isolation from an embryo.

However, dictating the identity of the desired PSC subtypes requires the correct culture conditions. This is achieved through applying different combinations of small molecules and GFs to the culture medium. For example, in the establishment of porcine ESPSCs, over 400 combinations of 20 small-molecule inhibitors and cytokines were assessed (Gao et al., 2019). For naïve conditions, the classic mouse conditions of 2i (MEKi;GSKi)/LIF have been employed for porcine, bovine, ovine, and other applications (Ogorevc et al., 2016). FGF2 has also been used for the culture and maintenance of PSCs. There have also been efforts to enhance the efficiency and stability of PSCs by utilizing additional small molecules (discussed in Section 7.2). Since derived lines will potentially be used in the food chain, the potential toxic effects of utilized factors (small molecules) need to be assessed.

8 Regulatory issues

Regulatory compliance for derived cell lines for cultured meat production in UK, EU, and US, e.g. requires supplying certain key data, as discussed in the following paragraphs.

First, the origin of cells should be documented. The following data should be recorded: the donor animal details (species, age, sex, etc.) will need to be registered as well as the health history of the animal from the farmer/owner and vet reviewed/filed. This will include health certificates, medication(s), vaccination records, and tissue(s) biopsied, biopsy method, and storage conditions. Importantly, the consent of the farmer or owner of the animal should be collected and documented to ensure freedom to operate. This is especially important when it comes to collecting samples from abattoirs as in many cases the abattoirs do not own the animals and are performing a service to slaughter the animal. This is the case even when considering buying material from an abattoir as this will be for consumption and not commercialization. Additional information on the health status of the farm and information on other species kept on the farm is also of use.

For cell lines established from organ/tissue biopsies and eggs, traceability and veterinary requirements are laid down under Regulation (EC) No 853/2004 (European Parliament and of the Council of 29 April 2004), laying down specific hygiene rules for food of animal origin.

8.1 Identity and authenticity

Both identity and authenticity of the cell line are critical. During the derivation stages, the phenotypic characteristics of the cells (e.g. morphology, enzyme activity, cell surface antigens, transcription factors, etc.) should be established and monitored on a regular basis.

A major risk is cross-contamination with other cell lines, and this is a relatively common occurrence in the research environment (Markovic and Markovic, 1998). This can be addressed by introducing quality control (QC) procedures, based on collated data from initial identification and isolation. QC should be based on a number of parameters including karyotyping and next-generation sequencing - whole genome sequencing (WGS). Short tandem repeat (STR) and single-nucleotide polymorphism (SNP) assays are widely used to authenticate and detect contamination but suffer from insensitivity (Rubio et al., 2020). Employing deep WGS will increase sensitivity significantly and allow line authentication as well as assessing/identifying potential contamination.

8.2 Sterility and safety

Another key consideration is demonstrating that a cell line is sterile and free of adventitious agents. We recommend using metagenomics for detecting potential adventitious agents which is both rapid and sensitive (MacDonald et al., 2021). Extra assessments may be required for some species, e.g. prion risk for bovines. The types of biopsy taken should also be considered, with high-risk material such as the brain or spinal cord being avoided. Other potential hazards include animal-derived components such as feeder cells and FBS. These materials will also require the aforementioned testing to minimize the introduction of adventitious agents. The best policy is to minimize or exclude animal-derived components.

The entire manufacturing process should be considered since microbiological contamination can occur throughout. Mycoplasma are the smallest replicating bacteria, can be pathogenic to humans, able to infect a wide range of hosts, hard to detect, and are resistant to antibiotics (Gautier-Bouchardon, 2018). Viruses and other pathogens can also be introduced from animal-derived components used in cell isolation, expansion, and adaptation, and by operators during routine passage and handling of cells. While there are no specific guidelines, as a minimum, cell lines should be tested for adventitious viruses (species-specific viruses depending on the origin of the cells and contact with other animal-derived components), mycoplasma, bacteria, yeasts, and moulds. Ideally, methods should be internationally recognized, and international standards should be followed where appropriate.

8.3 Genetic stability

Demonstrating genetic stability over time is important, as phenotypic heterogeneity may occur during passaging, leading to the emergence of genetic and epigenetic variants during the derivation or expansion phases (Pamies et al., 2022). Genetic variants include point mutations, deletions, insertions, rearrangements, duplications, inversions, gene amplifications, and chromosome loss or aneuploidy. Datasets to demonstrate genetic stability during the manufacturing process could include WGS at different time points (passages). It is important to assess if any genetic differences are present in cells from the master cell bank versus the final product (cultured meat). If cells are GM, the stability of the modification/insertion, copy number, or the presence of vector components should also be determined.

8.4 Cell storage

Some cryoprotectants used during cell storage may be toxic if they end up in the final food. Cryoprotectants such as sugars (trehalose, sucrose, and sorbitol) are used in the food manufacture and have been deemed safe at certain concentrations. During the thawing process, cryoprotectants would be washed away or diluted to very low concentrations in the final cell biomass. It is important to check no residues are present final cell biomass, i.e. glycation end products or acrylamide (Shi et al., 2021).

A critical issue in cell storage is clear labelling, including establishing a master cell bank and a working bank, along with maintenance and control of storage conditions (e.g. monitoring temperature and nitrogen levels). Procedures should be in line with the FDA's code of federal regulations, title 21 CFR §610.18(a) (https://www.accessdata.fda.gov/scripts/cdrh/cfdocs/cfcfr/CFRSearch.cfm?fr=610.18). This should also be supported by 'Good Cell Culture Practice (GCCP)' to assure reproducibility and quality. As the regulatory framework with respect to cell line production in the cultured meat space is a grey area, it is also advisable to follow International Conference on Harmonization (ICH) Guidelines on cell line safety for pharmaceuticals and the FDA Guidance (Point to Consider – 1993, CFR 9, 113) along with clear procedures for documentation and traceability.

8.5 Cell culture media

Cell culture media is critical for the isolation, maintenance, expansion, and differentiation of cells. Components of the culture medium can potentially become part of the final food product. Novel ingredients, components used at higher concentrations, downstream by-products (residues), and other sources of contamination, raise additional safety issues (Ong et al., 2021).

All regulatory agencies require details of the composition of the culture media being employed. However, many providers/companies of culture media do not disclose formulations/compositions of media on intellectual property (IP) grounds, providing a significant challenge when assessing safety/risk. In an ideal world, components should be food grade (approved for use in food production). If not, the components should be food safe, based on toxicology studies. To date, the current landscape does not prevent the use of pharmaceutical grade/research grade components, but the possible presence of impurities at levels that are not safe for use in food needs to be considered. Where components are not deemed food grade, a risk assessment and an understanding of how they are used by the cells should be undertaken. This should include information about the purity of the components and if they are expected to remain in the final product or will be completely removed. It is recommended that culture media components conform to the purity specifications outlined by the British Pharmacopoeia, European Pharmacopoeia, Joint FAO/WHO Expert Committee on Food Additives (JECFA), or Food Chemical Codex.

8.6 Antibiotics/antimycotics

Initial cell isolation requires the acquisition of tissue/cells from either live or slaughtered animals. These tend to be under nonsterile conditions which increases the risk of microbial contamination. To combat this risk antibiotics/antimycotics are employed to control microbial growth during isolation. However, the use of antibiotics should be minimized as there should be no antibiotic residues in the final product. Guidelines for maximum residue limits of animal drugs are available from the EMA, UK and FDA, etc. (https://www.ema.europa.eu/en/veterinary-regulatory/research-development/maximum-residue-limits-mrl; https://www.gov.uk/guidance/maximum-residues-limits-mrl; https://www.fda.gov/media/102492/download). In addition, use of antibiotics/antimycotics should be recorded.

8.7 Growth factors

GFs/morphogens are a critical component of cell growth, maintenance and differentiation to different tissue lineages. However, the use of GFs can have potential safety issues. These include the source of the GFs, e.g. animal-derived or recombinant which may e.g. introduce adventitious agents. GFs produced in plants are an attractive alternative source as this process is deemed GRAS compliant (Lee et al., 2022; Schillberg and Finnern, 2021). The use of human recombinant proteins is a concern since the use of human proteins in the food chain should be avoided and species-specific native sequence GFs employed. The presence of growth factors in the final cell product is a concern due to unwanted off-target effects. The EU has strict rules on the use of growth hormones which are restricted or banned (Kuiper et al., 1998).

8.8 Animal serum and other animal-derived components

There are ethical considerations around the use of FBS and adult serum, as well as from a practical standpoint. These include lability, potential viral contamination, and batch-to-batch variability (Pamies et al., 2022) FBS is often required for initial cell isolation/adaptation. However, there is risk of viral and prion contamination. This risk can be reduced if the FBS is sourced from prion-free (bovine spongiform encephalopathy (BSE)) or low-risk countries such as New Zealand. However, rare BSE mutations have popped up in countries deemed low risk (Ferguson-Smith and Richt, 2009). There are concerted efforts to develop serum-free approaches that utilize recombinant proteins or preparations supplemented with purified growth factors/hormones in a chemically defined base media.

8.9 Culture media components

Raw materials utilized in media production (amino acids, vitamins, and minerals) are generally research or pharmaceutical grade. However, in the case of cultured meat, they should ideally be food grade as outlined in JECFA or the Food Chemical Codex. Another consideration is the source of these components, in terms of purity, batch variability, and allergenicity. Allergen risk can be introduced with the use of soy and wheat hydrolysates. There is also the risk of mycotoxin contamination from cereals (Leslie et al., 2021). Pesticide residues are also a potential problem. Guidelines for the maximum residue limits are available from the EMA (https://www.ema.europa.eu/en/veterinary-regulatory/research-development/maximum-residue-limits-mrl), UK government (https://www.gov.uk/guidance/maximum-residues-limits-mrl), and FDA (https://www.fda.gov/media/102492/download). In addition, there is growing regulatory concern about the presence of nanomaterials in food. The European Food Safety Authority has produced guidance on potential risks of nanomaterials (particles below 100 nm in size), setting up a supplier approval program, testing and traceability procedures.

9 Conclusion

The cultured meat space is still in its infancy, and there are many hurdles to jump before we reach the position where we attain price parity with live stock generated meat and mainstream usage. The ability to generate cell lines, scale production of growth factors/additives and media are technical challenges that can be relatively easily overcome. However, the major hurdles to overcome are the regulatory requirements and guidelines for cell line, media, growth factor development, and implementation into the food space. This is an area of concern as regulatory matters are always playing catch-up with the technology. Thus a more proactive / progressive arena that develops these in parallel and/or predicts emerging technologies is required, lest we heavely invest in the development of highly cost-effective methods and approaches that fail to pass regulatory compliance.

10 Where to look for further information

The following articles provide a good overview of the subject:

- FORMAT- Mason, T. 2017. Maize breeding: A current challenges and opportunities. *Crop breeding J* 43(2): 1–20.
- CultMeat_Technical-Advances_FoodSciAnimalRes_2021.pdf.
- Vickers, J. 2016. From marker-assisted selection to gene editing: emerging technologies in maize breeding. *Crop Science* 50: 42–64.
- CultMeat_Challenges_NatureFood_2020.pdf.

Key research in this area can be found at the following organisations:

- FORMAT - International Maize and Wheat Improvement Center (www.cimmyt.org).
- https://www.mckinsey.com/industries/agriculture/our-insights/cultivated-meat-out-of-the-lab-into-the-frying-pan.
- https://www.fda.gov/food/food-ingredients-packaging/human-food-made-cultured-animal-cells.
- United States Department of Agriculture (USDA) (www.usda.gov).
- International Institute of Tropical Africa (IITA) (www.iita.org).

11 References

Almeida, J. and Korch, C. 2023. Authentication of human and cell mouse lines by short tandem repeat (STR) DNA genotype analysis. In: Markossian, S., Grossman, A., Brimacombe, K., Arkin, M., Auld, D., Austin, C., Baell, J., Chung, T. D. Y., Coussens, N. P., Dahlin, J. L., Devanarayan, V., Foley, T. L., Glicksman, M., Gorshkov, K., Haas, J. V., Hall, M. D., Hoare, S., Inglese, J., Iversen, P. W., Kales, S. C., Lal-Nag, M., Li, Z., McGee, J., McManus, O., Riss, T., Saradjian, P., Sittampalam, G. S., Tarselli, M., Trask, O. J. Jr., Wang, Y., Weidner, J. R., Wildey, M. J., Wilson, K., Xia, M. and Xu, X. (Eds). *Assay Guidance Manual*. Eli Lilly and Company/National Center for Advancing Translational Sciences, Bethesda, MD.

Bang, J. S., Choi, N. Y., Lee, M., Ko, K., Lee, H. J., Park, Y. S., Jeong, D., Chung, H. M. and Ko, K. 2018. Optimization of episomal reprogramming for generation of human-induced pluripotent stem cells from fibroblasts. *Anim. Cells Syst. (Seoul)* 22(2): 132–139. doi: 10.1080/19768354.2018.1451367

Baquero-Perez, B., Kuchipudi, S. V., Nelli, R. K. and Chang, K. C. 2012. A simplified but robust method for the isolation of avian and mammalian muscle satellite cells. *BMC Cell Biol.* 13: 16. doi: 10.1186/1471-2121-13-16.

Ben-David, U. and Benvenisty, N. 2011. The tumorigenicity of human embryonic and induced pluripotent stem cells. *Nat. Rev. Cancer* 11(4): 268–277. doi: 10.1038/nrc3034.

Bhat, S., Viswanathan, P., Chandanala, S., Prasanna, S. J. and Seetharam, R. N. 2021. Expansion and characterization of bone marrow derived human mesenchymal stromal cells in culture-free conditions. *Sci. Rep.* 11(1): 3403. doi: 10.1038/s41598-021-83088-1.

Bindroos, B., Boucher, S., Chase, L., Kuokkanen, H., Huhtala, H., Haataja, R., Vemuri, M., Suuronen, R. and Miettinen, S. 2009. Serum-free, xeno-free culture media maintain the proliferation rate and multi-potentiality of adipose stem cells in vitro. *Cytotherapy* 11(7): 958–972. doi: 10.3109/14653240903233081.

Blanton, R. A., Perez-Reyes, N., Merrick, D. T. and McDougall, J. K. 1991. Epithelial cells immortalized by human papillomaviruses have pre-malignant characteristics in organotypic culture. *Am. J. Pathol.* 138(3): 673–685.

Bryan, T. M. and Reddel, R. R. 1994. SV40-induced immortalization of human cells. *Crit. Rev. Oncog.* 5(4): 331–357.

Cao, H., Chu, Y., Zhu, H., Sun, J., Pu, Y., Gao, Z., Yang, C., Peng, S., Dou, Z. and Hua, J. 2011. Characterization of immortalized mesenchymal stem cells derived from foetal porcine pancreas. *Cell Prolif.* 44(1): 19–32.

Chen, X., Qian, W., Song, Z., Li, Q. X. and Guo, S. 2020. Authentication, characterization and contamination detection of cell lines, xenografts and organoids by barcode deep NGS sequencing. *NAR Genom. Bioinform.* 2(3, September): lqaa060. doi: 10.1093/nargab/lqaa060.

Chen, Q., Luo, W., Veach, R. A., Hickman, A. B., Wilson, M. H. and Dyda, F. 2020. Structural basis of seamless excision and specific targeting by PiggyBac transposase. *Nat. Commun.* 11(1): 3446. doi: 10.1038/s41467-020-17128-1.

Chen, I. P., Fukuda, K., Fusaki, N., Iida, A., Hasegawa, M., Lichtler, A. and Reichenberger, E. J. 2013. Induced pluripotent stem cell reprogramming by integration-free Sendai virus vectors from peripheral blood of patients with craniometaphyseal dysplasia. *Cell Reprogram* 15(6): 503–513. doi: 10.1089/cell.2013.0037.

Choi, K.-H., Yoon, J. W., Kim, M., Lee, H. J., Jeong, J., Ryu, M., Jo, C. and Lee, C. K. 2020. Muscle stem isolation and *in vitro* culture for meat production: A methodological review. *Compr. Rev. Food Sci. Food Saf.* 20(1): 429–457. doi: 10.1111/1541-4337.12661.

Davis, R. L., Weintraub, H. and Lassar, A. B. 1987. Expression of a single transfected cDNA converts fibroblasts to myoblasts. *Cell* 51(6): 987–1000. doi: 10.1016/0092-8674(87)90585-X.

de Figuiredo Pessoa, L. V. F., Bressan, F. F. and Freude, K. K. 2019. Induced pluripotent stem cells throughout the animal kingdom: Availability and applications. *World J. Stem Cells* 11(8): 491–505.

Delia, D., Fontanella, E., Ferrario, C., Chessa, L. and Mizutani, S. 2003. DNA damage-induced cell-cycle phase regulation of p53 and p21WAF1 in normal and ATM-defective cells. *Oncogene* 22(49): 7866–7869.

ECJ 2018. Court of Justice of the European Union: Organisms obtained by mutagenesis are GMOs and are, in principle, subject to the obligations laid down by the GMO directive. Press Release No 111/18. Luxembourg. Jul 25, 2018. Available at: https://curia.europa.eu/jcms/upload/docs/application/pdf/2018-07/cp180111en.pdf.

Evans, M. J. and Kaufman, M. H. 1981. Establishment in culture of pluripotential cells from mouse embryos. *Nature* 292(5819): 154–156.

Ezashi, T., Telugu, B. P., Alexenko, A. P., Sachdev, S., Sinha, S. and Roberts, R. M. 2009. Derivation of induced pluripotent stem cells from pig somatic cells. *Proc. Natl Acad. Sci. U. S. A.* 106(27): 10993–10998. doi: 10.1073/pnas.0905284106.

Federation, A., Bradner, J. E. and Meissner, A. 2014. The use of small molecules in somatic-cell reprogramming. *Trends Cell Biol.* 24(3): 179–187. doi: 10.1016/j.tcb.2013.09.011.

Ferguson-Smith, M. A. and Richt, J. A. 2009. Rare BSE mutation raises concerns over risks to public health. *Nature* 457(7233): 1079. doi: 10.1038/4571079b.

Fridman, A. L. and Tainsky, M. A. 2008. Critical pathways in cellular senescence and immortalization revealed by gene expression profiling. *Oncogene* 27(46): 5975–5987.

Gao, X., Nowak-Imialek, M., Chen, X., Chen, D., Herrmann, D., Ruan, D., Chen, A. C. H., Eckersley-Maslin, M. A., Ahmad, S., Lee, Y. L., Kobayashi, T., Ryan, D., Zhong, J., Zhu, J., Wu, J., Lan, G., Petkov, S., Yang, J., Antunes, L., Campos, L. S., Fu, B., Wang, S., Yong, Y., Wang, X., Xue, S. G., Ge, L., Liu, Z., Huang, Y., Nie, T., Li, P., Wu, D., Pei, D., Zhang, Y., Lu, L., Yang, F., Kimber, S. J., Reik, W., Zou, X., Shang, Z., Lai, L., Surani, A., Tam, P. P. L., Ahmed, A., Yeung, W. S. B., Teichmann, S. A., Niemann, H. and Liu, P. 2019. Establishment of porcine and human-expanded potential stem cells. *Nat. Cell Biol.* 21(6): 687–699. doi: 10.1038/s41556-019-0333-2.

Gautier-Bouchardon, A. V. 2018. ASM journals, microbiology spectrum. *Antimicrob. Resist. Mycoplasma Spp* 6, No. 4. doi: 10.1128/microbiolspec.arba-0030-2018.

Genet, M. and Torres-Padilla, M. E. 2020. The molecular and cellular features of 2-cell-like cells: A reference guide. *Development* 147(16): dev189688.

Godfray, H. C. J., Aveyard, P., Garnett, T., Hall, J. W., Key, T. J., Lorimer, J., Pierrehumbert, R. T., Scarborough, P., Springmann, M. and Jebb, S. A. 2018. Meat consumption, health, and the environment. *Science* 361(6399). doi: 10.1126/science.aam5324.

Guan, J., Wang, G., Wang, J., Zhang, Z., Fu, Y., Cheng, L., Meng, G., Lyu, Y., Zhu, J., Li, Y., Wang, Y., Liuyang, S., Liu, B., Yang, Z., He, H., Zhong, X., Chen, Q., Zhang, X., Sun, S., Lai, W., Shi, Y., Liu, L., Wang, L., Li, C., Lu, S. and Deng, H. 2022. Chemical reprogramming of human somatic cells to pluripotent stem cells. *Nature* 605(7909): 325–331. doi: 10.1038/s41586-022-04593-5.

Hafner, A. L.. and Dani, C. 2014. Human-induced pluripotent stem cells: A new source for brown and white adipocytes. *World J. Stem Cells* 6(4): 467–472. doi: 10.4252/wjsc.v6.i4.467.

Holmes, D., Humbird, D., Dutkiewicz, J., Tejeda-Saldana, Y., Duffy, B. and Datar, I. 2022. Cultured meat needs a race to mission not a race to market. *Nat. Food* 3(10): 785–787. doi: 10.1038/s43016-022-00586-9.

Hong, T. K., Shin, D. M., Choi, J., Do, J. T. and Han, S. G. 2021. Current issues and technical advances in cultured meat production: a review. *Food Sci. Anim. Resour.* 41(3): 355–372. doi: 10.5851/kosfa.2021.e14.

Hou, P., Li, Y., Zhang, X., Liu, C., Guan, J., Li, H., Zhao, T., Ye, J., Yang, W., Liu, K., Ge, J., Xu, J., Zhang, Q., Zhao, Y. and Deng, H. 2013. Pluripotent stem cells induced from mouse somatic cells by small-molecule compounds. *Science* 341(6146): 651–654. doi: 10.1126/science.1239278.

Iberite, F., Gruppioni, E. and Ricotti, L. 2022. Skeletal muscle differentiation of human iPSCs meets bioengineering strategies: Perspectives and challenges. *NPJ Regen. Med.* 7(1): 23. doi: 10.1038/s41536-022-00216-9.

Kern, S., Eichler, H., Stoeve, J., Klüter, H. and Bieback, K. 2006. Comparative analysis of mesenchymal stem cells from bone marrow, umbilical cord blood or adipose tissue. *Stem Cells* 24(5): 1294–1301.

Kim, K., Doi, A., Wen, B., Ng, K., Zhao, R., Cahan, P., Kim, J., Aryee, M. J., Ji, H., Ehrlich, L. I., Yabuuchi, A., Takeuchi, A., Cunniff, K. C., Hongguang, H., McKinney-Freeman, S., Naveiras, O., Yoon, T. J., Irizarry, R. A., Jung, N., Seita, J., Hanna, J., Murakami, P., Jaenisch, R., Weissleder, R., Orkin, S. H., Weissman, I. L., Feinberg, A. P. and Daley, G. Q. 2010. Epigenetic memory in induced pluripotent stem cells. *Nature* 467(7313): 285–290. doi: 10.1038/nature09342.

Kim, K. H., Qiu, J. and Kuang, S. 2020. Isolation, culture and differentiation of primary myoblasts derived from muscle satellite cells. *Bio. Protoc* 10(14): e3686. doi: 10.21769/BioProtoc.3686.

Kolkmann, A. M., Van Essen, A., Post, M. J. and Moutsatsou, P. 2022. Development of a chemically defined medium for *in vitro* expansion of primary bovine satellite cells. *Front. Bioeng. Biotechnol.* 10: 895289. doi: 10.3389/fbioe.2022.895289.

Kuiper, H. A., Noordam, M. Y., van Dooren-Flipsen, M. M., Schilt, R. and Roos, A. H. 1998 Jan. Illegal use of beta-adrenergic agonists: European Community. *J. Anim. Sci.* 76(1): 195–207. doi: 10.2527/1998.761195x. PMID: 9464899.

Lee, S. Y., Kang, H. J., Lee, D. Y., Kang, J. H., Ramani, S., Park, S. and Hur, S. J. 2021. Principle protocols for the processing of cultured meat. *J. Anim. Sci. Technol.* 63(4): 673–680. doi: 10.5187/jast.2021.e40.

Lee, Y., Lee, H. J., Ham, S., Jeong, D., Lee, M., Lee, U., Lee, M., Kwon, T. H. and Ko, K. 2022 Jan. Plant-derived human recombinant growth factors and serum albumin maintain stemness of human-induced pluripotent stem cells. *Cell Biol. Int.* 46(1): 139–147. doi: 10.1002/cbin.11715.

Leslie, J. F., Moretti, A., Mesterházy, Á., Ameye, M., Audenaert, K., Singh, P. K., Richard-Forget, F., Chulze, S. N., Ponte, E. M. D., Chala, A., Battilani, P. and Logrieco, A. F. 2021. Key global actions for mycotoxin management in wheat and other small grains. *Toxins* 13(10): 725. doi: 10.3390/toxins13100725.

Li, B. J., Li, P. H., Huang, R. H., Sun, W. X., Wang, H., Li, Q. F., Chen, J., Wu, W. J. and Liu, H. L. 2015. Isolation, culture and identification of porcine skeletal muscle satellite cells. *Asian-Australas J. Anim. Sci.* 28(8): 1171–1177.

Lin, T., Ambasudhan, R., Yuan, X., Li, W., Hilcove, S., Abujarour, R., Lin, X., Hahm, H. S., Hao, E., Hayek, A. and Ding, S. 2009. A chemical platform for improved induction of human iPSCs. *Nat. Methods* 6(11): 805–808. doi: 10.1038/nmeth.1393.

Listrat, A., Lebret, B., Louveau, I., Astruc, T., Bonnet, M., Lefaucheur, L., Picard, B. and Bugeon, J. 2016. How muscle structure and composition influence meat and flesh quality. *Sci. World J.*: 3182746. (*Sci. World J.* 2016: 1–14. doi: 10.1155/2016/31827469.

Liu, T. M., Martina, M., Hutmacher, D. W., Hui, J. H., Lee, E. H. and Lim, B. 2007. Identification of common pathways mediating differentiation of bone marrow and adipose tissue-derived human mesenchymal stem cells into three mesenchymal lineages. *Stem Cells* 25(3): 750–760.

Liu, W., Hao, Z., Florkowski, W. J., Wu, L. and Yang, Z. 2022. A review of the challenges facing global commercialization of the artificial meat industry. *Foods* 11(22): 3609. doi: 10.3390/foods11223609.

MacDonald, M. L., Polson, S. W. and Lee, K. H. 2021. k-mer-based metagenomics tools provide a fast and sensitive approach for the detection of viral contaminants in biopharmaceutical and vaccine manufacturing applications using next-generation sequencing. *mSphere* 6(2): e01336-20. doi: 10.1128/mSphere.01336-20.

Malerich, M. and Bryant, C. 2022. Nomenclature of cell-cultivated meat & seafood products. *NPJ Sci. Food* 6(1): 56. doi: 10.1038/s41538-022-00172-0.

Marczynska, B., Khoobyarian, N., Chao, T. S. and Tao, M. 1991. Phorbol ester promotes growth and transformation of carcinogen-exposed nonhuman primate cells in vitro. *Anticancer Res.* 11(5): 1711–1717.

Markovic, O. and Markovic, N. 1998. Cell cross-contamination in cell cultures: the silent and neglected danger. *In Vitro Cell. Dev. Biol. Anim.* 34(1): 1–8. doi: 10.1007/s11626-998-0040-y.

Matsumura, T., Takesue, M., Westerman, K. A., Okitsu, T., Sakaguchi, M., Fukazawa, T., Totsugawa, T., Noguchi, H., Yamamoto, S., Stolz, D. B., Tanaka, N., Leboulch, P. and Kobayashi, N. 2004. Establishment of an immortalized human-liver endothelial cell line with SV40T and hTERT. *Transplantation* 77(9): 1357–1365.

Mazid, M. A., Ward, C., Luo, Z., Liu, C., Li, Y., Lai, Y., Wu, L., Li, J., Jia, W., Jiang, Y., Liu, H., Fu, L., Yang, Y., Ibañez, D. P., Lai, J., Wei, X., An, J., Guo, P., Yuan, Y., Deng, Q., Wang, Y., Liu, Y., Gao, F., Wang, J., Zaman, S., Qin, B., Wu, G., Maxwell, P. H., Xu, X., Liu, L., Li, W. and Esteban, M. A. 2022. Rolling back human pluripotent stem cells to an eight-cell embryo-like stage. *Nature* 605(7909): 315–324.

Miller, G. 1982. Immortalization of human lymphocytes by Epstein-Barr virus. *Yale J. Biol. Med.* 55(3–4): 305–310.

Mohamed-Ahmed, S., Fristad, I., Lie, S. A., Suliman, S., Mustafa, K., Vindenes, H. and Idris, S. B. 2018. Adipose-derived and bone marrow mesenchymal stem cells: A donor-matched comparison. *Stem Cell Res. Ther.* 9(1): 168.

Mohsen-Kanson, T., Hafner, A. L., Wdziekonski, B., Takashima, Y., Villageois, P., Carrière, A., Svensson, M., Bagnis, C., Chignon-Sicard, B., Svensson, P. A., Casteilla, L., Smith, A. and Dani, C. 2014. Differentiation of human-induced pluripotent stem cells into brown and white adipocytes: Role of Pax3. *Stem Cells* 32(6): 1459–1467. doi: 10.1002/stem.1607.

Morgan, J. E. and Partridge, T. A. 2003. Muscle satellite cells. *Int. J. Biochem. Cell Biol.* 35(8): 1151–1156. doi: 10.1016/S1357-2725(03)00042-6.

Musina, R. A., Bekchanova, E. S. and Sukhikh, G. T. 2005. Comparison of mesenchymal stem cells obtained from different human tissues. *Bull. Exp. Biol. Med.* 139(4): 504–509.

Nachtigal, M., Nagpal, M. L., Greenspan, P., Nachtigal, S. A. and Legrand, A. 1989. Characterization of a continuous smooth muscle cell line derived from rabbit aorta. *In Vitro Cell. Dev. Biol.* 25(10): 892–898.

Namba, M., Nishitani, K., Fukushima, F. and Kimoto, T. 1988. Multistep carcinogenesis of normal human fibroblasts. Human fibroblasts immortalized by repeated treatment with Co–60 gamma rays were transformed into tumorigenic cells with Ha-ras oncogenes. *Anticancer Res.* 8(5A): 947–958.

Nichols, J. and Smith, A. 2009. Naïve and primed plutipotent states. *Cell Stem Cell* 4(6): 487–492.

Noble, J. R., Zhong, Z. H., Neumann, A. A., Melki, J. R., Clark, S. J. and Reddel, R. R. 2004. Alterations in the p16(INK4a) and p53 tumor suppressor genes of hTERT-immortalized human fibroblasts. *Oncogene* 23(17): 3116–3121.

Nowak-Imialek, M. and Niemann, H. 2012. Pluripotent cells in farm animals: State of the art and future perspectives. *Reprod. Fertil. Dev.* 25(1): 103–128.

O'Neill, E. N., Ansel, J. C., Kwong, G. A., Plastino, M. E., Nelson, J., Baar, K. and Block, D. E. 2022. Spent media analysis suggests cultivated meat media will require species and cell type optimization. *NPJ Sci. Food* 6(1): 46. doi: 10.1038/s41538-022-00157-z.

Obinata, M. 2007. Immortalized cell lines with differentiation potentials: Their stablishmnt and possible application. *Cancer Sci.* 98(3): 275–283.

Ogorevc, J., Orehek, S. and Dovč, P. 2016. Cellular reprogramming in farm animals: An overview of iPSC generation J Animal Sci and Bio. *J. Anim. Sci. Biotechnol.* 7: 10. doi: 10.1186/s40104-016-0070-3.

Ong, K. J., Johnston, J., Datar, I., Sewalt, V., Holmes, D. and Shatkin, J. A. 2021. Food safety considerations and research priorities for the cultured meat and seafood industry. *Compr. Rev. Food Sci. Food Saf.* 20(6): 5421–5448. doi: 10.1111/1541-4337.12853.

Pamies, D., Leist, M., Coecke, S., Bowe, G., Allen, D. G., Gstraunthaler, G., Bal-Price, A., Pistollato, F., de Vries, R. B. M., Hogberg, H. T., Hartung, T. and Stacey, G. 2022. Guidance document on Good Cell and Tissue Culture Practice 2.0 (GCCP 2.0). *ALTEX Altern. Anim. Exp.* 39(1): 30–70. doi: 10.14573/altex.2111011.

Pasitka, L., Cohen, M., Ehrlich, A., Gildor, B., Reuveni, E., Ayyash, M., Wissotsky, G., Herscovici, A., Kaminker, R., Niv, A., Bitcover, R., Dadia, O., Rudik, A., Voloschin, A., Shimoni, M., Cinnamon, Y. and Nahmias, Y. 2023. Spontaneous immortalization of chicken fibroblasts generates stable, high-yield cell lines for serum-free production of cultured meat. *Nat. Food* 4(1): 35–50.

Pera, M. F. and Rossant, J. 2021. The exploration of pluripotency space: Charting cell state transitions in peri-implantation development. *Cell Stem Cell* 28(11): 1896–1906.

Pessôa, L. V. F., Bressan, F. F. and Freude, K. K. 2019. Induced pluripotent stem cells throughout the animal kingdom: availability and applications. *World J. Stem Cells* 11(8): 491–505. doi: 10.4252/wjsc.v11.i8.491.

Post, M. J., Levenberg, S., Kaplan, D. L., Genovese, N., Fu, J., Bryant, C. J., Negowetti, N., Verzijden, K. and Moutsatsou, P. 2020. Scientific, sustainability and regulatory challenges of cultured meat. *Nat. Food* 1(7): 403–415.

Rambthala, L., Chiu, C. P., Glickman, R. D. and Rowe-Rendleman, C. 2002. In vitro differentiation capacity of telomerase immortalized human RPE cells. *Invest. Ophthalmol. Vis. Sci.* 43(5): 1622–1630.

Risner, D., Li, F., Fell, J. S., Pace, S. A., Siegel, J. B., Tagkopoulos, I. and Spang, E. S. 2020. BioRxiv doi: 10.1101/2020.09.10.292144.

Risner, D., Kim, Y., Nguyen, C., Siegel, J. B. and Spang, E. S. 2023. Environmental impacts of cultured meat: A cradle-to-gate lifecycle assessment. Available at: https://www.biorxiv.org/content/10.1101/2023.04.21.537778v1.full.pdf.

Rubio, N. R., Xiang, N. and Kaplan, D. L. 2020. Plant-based and cell-based approaches to meat production. *Nat. Commun.* 11(1): 6276. doi: 10.1038/s41467-020-20061-y.

Saad, M. K., Yuen, J. S. K., Joyce, C. M., Li, X., Lim, T., Wolfson, T. L., Wu, J., Laird, J., Vissapragada, S., Calkins, O. P., Ali, A. and Kaplan, D. L. 2023. Continuous fish muscle cell line with capacity for myogenic and adipogenic-like phenotypes. *Sci. Rep.* 13(1): 5098.

Scarfone, R. A., Pena, S. M., Russell, K. A., Betts, D. H. and Koch, T. G. 2020. The use of induced pluripotent stem cells in domestic animals: A narrative review. *BMC Vet. Res.* 16(1): 477. doi: 10.1186/s12917-020-02696-7.

Schillberg, S. and ¨Finnern, R. 2021. Plant molecular farming for the production of valuable proteins – critical evaluation of achievements and future challenges. *J. Plant Physiol.* 258–259: 153359.

Schillberg, S. and Finnern, R. 258–259 2021. Plant molecular farming for the production of valuable proteins – Critical evaluation of achievements and future challenges. *J. Plant Physiol.* 0176-1617. doi: 10.1016/j.jplph.2020.153359.

Seo, B. J., Hong, Y. J. and Do, J. T. 2017. Cellular reprogramming using protein and cell-penetrating peptides. *Int. J. Mol. Sci.* 18(3): 553. doi: 10.3390/ijms18030552.

Sharun, K., Pawde, A. M., Kumar, R., Kalaiselvan, E., Kinjavdekar, P., Dhama, K. and Pal, A. 2021. Standardization and characterization of adipose-derived stromal vascular fraction from New Zealand white rabbits for bone tissue engineering. *Vet. World* 14(2): 508–514.

Shi, H., Qin, R., Wu, R., Rong, J., Jia, C. and Liu, R. 2021. Effect of cryoprotectants on the formation of advanced glycation end products and acrylamide in fried fish cakes. *Food Bioscience* 44(B): 101433. doi: 10.1016/j.fbio.2021.101433. ISSN: 2212-4292.

Soice, E. and Johnston, J. 2021. Immortalizing cells for human consumption. *Int. J. Mol. Sci.* 22(21): 1166.

Stout, A. J., Mirliani, A. B., Rittenberg, M. L., Shub, M., White, E. C., Yuen, J. S. K. and Kaplan, D. L. 2022. Simple and effective serum-free medium for sustained expansion of bovine satellite cells for cell cultured meat. *Commun. Biol.* 5(1): 466. doi: 10.1038/s42003-022-03423-8.

Stout, A. J., Arnett, M. J., Chai, K., Guo, T., Liao, L., Mirliani, A. B., Rittenberg, M. L., Shub, M., White, E. C., Yuen, J. S. K., Zhang, X. and Kaplan, D. L. 2023. Immortalized bovine satellite cells for cultured meat applications. *ACS Synth. Biol.* 12(5): 1567–1573.

Sugii, S., Wong, C. Y. Q., Lwin, A. K. O. and Chew, L. J. M. 2022. Reassessment of adipocyte technology for cellular agriculture of alternative fat. *Compr. Rev. Food Sci. Food Saf.* 21(5): 4146–4163. doi: 10.1111/1541-4337.13021.

Suleiman, K., Aljulidan, M. M., Hussein, G. E. M. and Alkhalaf, H. N. 2022. *Establishment of a Novel Fetal Ovine Cell Line by Spontaneous Cell Fusion BioRxiv Pre-print.*

Takahasi, K. and Yamanaka, S. 2006. Induction of pluripotent stem cells from mouse embryonic and adult fibroblast cultures by defined factors. *Cell* 126(4): 663–676. doi: 10.1016/j.cell.2006.07.024.

Thompson, J. A., Itskovitz-Eldor, J., Shapiro, S. S., Waknitz, M. A., Swiergiel, J. J., Marshall, V. S. and Jones, J. M. 1998. Embryonic stem cells derived from human blastocysts. *Science* 282(5391): 1145–1147. doi: 10.1126/science.282.5391.1145.

van de Lavoir, M. C. and Mather-Love, C. 2006. Avian embryonic stem cells. *Methods Enzymol.* 418: 38–64.

Vlahos, K., Sourris, K., Mayberry, R., McDonald, P., Bruveris, F. F., Schiesser, J. V., Bozaoglu, K., Lockhart, P. J., Stanley, E. G. and Elefanty, A. G. 2019. Generation of iPSC lines from peripheral blood mononuclear cells from 5 healthy adults. *Stem Cell Res.* 34: 101380. doi: 10.1016/j.scr.2018.101380.

Verma, R., Holland, M. K., Temple-Smith, P. and Verma, P. J. 2012. Inducing pluripotency in somatic cells from the snow leopard (Panthera uncia), an endangered felid. *Theriogenology* 77(1): 220–228. doi: 10.1016/j.theriogenology.2011.09.022.

Verma, R., Liu, J., Holland, M. K., Temple-Smith, P., Williamson, M. and Verma, P. J. 2013. Nanog is an essential factor for induction of pluripotency in somatic cells from endangered felids. *Biores Open Access* 2(1): 72–76. doi: 10.1089/biores.2012.0297.

Wan, S., Pan, Q., Wei, J., Zhu, C., Jing, J., Qin, S., Hu, R., Lou, M., Li, S. and Ling, Y. 2022. Isolation, culture and identification of sheep skeletal muscle satellite cells. *Czech J. Anim. Sci.* 67(10): 416–423.

Wang, A. Y. L. 2021. Application of modified mRNA in somatic reprogramming to pluripotency and directed conversion of cell fate. *Int. J. Mol. Sci.* 22(15): 8148. doi: 10.3390/ijms22158148.

Williams, R. L., Morrow, A., Allred, J., Tolar, J., Haining, W. N., Yosef, N. and Blazar, B. R. 2022. Retention of epigenetic and transcriptional memory among reprogrammed T cells (T-iPSCS) imparts selective advantage for T-cell redifferentiation. *J. Immunol.* 208(1_Supplement): 107.12. doi: 10.4049/jimmunol.208.Supp.107.12.

Winder, M. L.. and Trokovic, R. 2021. Chapter 3. Induced pluripotent stem cell derivation from myoblasts. In: Birbrair, A. (Ed.). *Advances in Stem Cell Biology, Cell Sources for iPSCs, Academic Press* (vol. 7), Pages 37–55, ISBN 9780128221358.

Woltjen, K., Michael, I. P., Mohseni, P., Desai, R., Mileikovsky, M., Hämäläinen, R., Cowling, R., Wang, W., Liu, P., Gertsenstein, M., Kaji, K., Sung, H. K. and Nagy, A. 2009. piggyBac transposition reprograms fibroblasts to induced pluripotent stem cells. *Nature* 458(7239): 766–770. doi: 10.1038/nature07863.

Xiong, Y., Zhuang, R., Zhao, G., Liu, Y., Su, Y., Wang, W., Xi, X., Yang, Y., Han, X., Xie, S. and Wang, H. 2022. Identification of the CKM gene as a potential muscle-specific safe habor locus in the pig genome. *Genes* 13(5): 92.

Yang, J., Ryan, D. J., Wang, W., Tsang, J. C., Lan, G., Masaki, H., Gao, X., Antunes, L., Yu, Y., Zhu, Z., Wang, J., Kolodziejczyk, A. A., Campos, L. S., Wang, C., Yang, F., Zhong, Z., Fu, B., Eckersley-Maslin, M. A., Woods, M., Tanaka, Y., Chen, X., Wilkinson, A. C., Bussell, J., White, J., Ramirez-Solis, R., Reik, W., Göttgens, B., Teichmann, S. A., Tam, P. P. L., Nakauchi, H., Zou, X., Lu, L. and Liu, P. 2017. Establishment of mouse expanded potential stem cells. *Nature* 550(7676): 393–397.

Yoshioka, N., Gros, E., Li, H. R., Kumar, S., Deacon, D. C., Maron, C., Muotri, A. R., Chi, N. C., Fu, X. D., Yu, B. D. and Dowdy, S. F. 2013. Efficient generation of human iPSCs by a synthetic self-replicative RNA. *Cell Stem Cell* 13(2): 246–254. doi: 10.1016/j.stem.2013.06.001.

Yu, J., Vodyanik, M. A., Smuga-Otto, K., Antosiewicz-Bourget, J., Frane, J. L., Tian, S., Nie, J., Jonsdottir, G. A., Ruotti, V., Stewart, R., Slukvin, I. I. and Thomson, J. A. 2007. Induced pluripotent stem cell lines derived from human somatic cells. *Science* 318(5858): 1917–1920. doi: 10.1126/science.1151526. Epub 2007 Nov 20. PMID: 18029452.

Zagury, Y., Ianovici, I., Landau, S., Lavon, N. and Levenberg, S. 2022. Engineered marble-like bovine fat tissue for cultured meat. *Commun. Biol.* 5(1): 927 (10.1038/s42003-022-03852-5).

Zanicotti, D. G. and Coates, D. E. 2017. Growing adipose-derived stem cells under serum-free conditions. *Methods Mol. Biol.* 1537: 439–446. doi: 10.1007/978-1-4939-6685-1_26.

Zhao, C., Meng, L., Hu, H., Wang, X., Shi, F., Wang, Y., Li, Q. and Lin, A. 2010. Spontaneously immortalized bovine mammary epithelial cells exhibit a distinct gene expression pattern from breast cancer cells. *BMC Mol. Cell Biol.* 11: 82. doi: 10.1186/1471-2121-11-82.

Zhu, H., Wu, Z., Ding, X., Post, M. J., Guo, R., Wang, J., Wu, J., Tang, W., Ding, S. and Zhou, G. 2022. Production of cultured meat from pig muscle stem cells. *Biomaterials* 287: 121650. doi: 10.1016/j.biomaterials.2022.121650.

Chapter 6

Developments in cell culture media for cultured meat applications

Andrew J. Stout, Tufts University, USA

1 Introduction

Culture media - or the solutions of nutrients and signaling molecules that are fed to cells *in vitro* - are the primary inputs of the cultured meat production process. These media promote cell survival, growth, and maturation and provide the chemical building blocks that cells use to proliferate and produce the muscle and fat biomass that comprise cultured meat. Culture media are fundamental to the generation of cell-based meat, and their development and optimization are essential for the field's advancement. In this chapter, we will discuss the requirements for media used in cultured meat production, current options and considerations for media selection, and trends and directions for further development.

Animal cell culture media provide three main functions *in vitro*. First, they maintain physiological pH, osmolarity, dissolved gas, and salt concentrations to facilitate cell survival. Second, they supply nutrients such as sugars, amino acids, vitamins, and minerals which fuel cell metabolism and function. Finally, they provide any necessary signaling molecules which instruct cell behavior

http://dx.doi.org/10.19103/AS.2023.0130.08

and guide cells down proliferative or differentiative pathways. Along with these three core functions, media can contain auxiliary components to improve culture processes, such as antibiotics to prevent contamination, surfactants to prevent clumping of cell suspensions, or anti-foaming agents to prevent foam formation in bioreactors.

The history of culture media for animal cells began in 1882, when Sydney Ringer demonstrated that a solution sodium chloride, potassium chloride, calcium chloride, and sodium bicarbonate was able to maintain the contractile function of amphibious cardiac ventricles *ex vivo* (Ringer, 1882, 1883). This success set off a flurry of research on the topic, which reached a key milestone in 1907 when Ross G. Harrison developed a medium that went beyond mere maintenance and supported the actual growth and proliferation of cells from frog embryos (Harrison et al., 1907). Notably, the key ingredient in this first true animal cell proliferation medium was a 'black-box' animal-derived component - frog lymph - which was added to Ringers' balanced salt solution. Thus was *ex vivo* cell proliferation achieved, and thus began a long and pervasive trend of relying on undefined animal-derived supplements as key components of animal cell growth media.

The term 'pervasive' is appropriate here, as it was and is evident that undefined and variable components such as lymph, plasma, or other inputs derived from animals were not ideally suited for highly controlled scientific inquiry (Boone et al., 1971). Because of this, the search soon turned toward the production of synthetic and/or fully defined cell culture media, which was first presented in something akin to the modern perception of culture medium in 1911 when Margaret and Warren Lewis cultured cells from chick embryos in media comprising a balanced salt solution supplemented with amino acids and sugars, all of which improved cell growth *in vitro* (Lewis and Lewis, 1911). However, they noted that with these defined media, 'while some of the growths are better than those in simple saline solutions, they are by no means equal to those in plasma or lymph either in extent or duration' (Lewis and Lewis, 1912). They were able to achieve some cell growth with a defined, animal component-free medium, but it was not an effective replacement for animal components. This fact has held for most animal cell cultures in the hundred years since, and indeed the majority of cell culture systems today still rely on animal-derived components such as fetal bovine serum (FBS) or others.

The use of these animal-derived components is limiting for several reasons. First, these components are undefined, meaning that the full details of their compositions are unknown, and so it is impossible to fully understand or control the culture environment which is experienced by the cells. Second, these components are highly variable animal-to-animal, changing with factors such as genetic diversity and diet (McCoy et al., 1970; Zheng et al., 2006). This results in further uncertainty about precise culture conditions and adds potentially confounding variability to cell culture experiments. Finally, for cultured meat specifically, the fact that these are animal-derived components in and of itself renders them unfit for purpose. Specifically, a meat production system that aims to eliminate our reliance on animals for their muscle and fat tissue must not itself rely on animals for their plasma or blood serum if it is to address the numerous challenges that derive from large-scale animal agriculture.

Because of this, the present chapter will focus on the developments of defined, serum and animal component-free media for cultured meat applications and closely related cell culture systems - for a more general overview of the history of cell culture media, the reader is pointed toward an excellent review by Tatsuma Yao and Yuta Asayama (Yao and Asayama, 2017). We will discuss the developments of serum-free media for various cell types and the specific medium requirements of muscle and fat cells. Additionally, because of the unique features of cell culture for cultured meat - namely the requirements for an affordable and edible final product - we will also discuss sourcing of components for the scalable production of low-cost cell culture media, and some considerations around nutrition and food safety. Finally, we will discuss future directions for research and development of media for cultured meat.

1.1 Moving toward serum-free media

While some early media - such as those developed by Sydney Ringer or Margaret and Warren Lewis - were free from animal-derived components, they were also ineffective over long-term culture (Lewis and Lewis, 1912; Ringer, 1883, 1882), and so cell culture convention turned to using animal sera such as horse serum or FBS. This convention did not preclude valuable insight and progress toward serum-free media, as researchers still endeavored to understand the specific nutrient requirements of cells. For instance, in a set of clever experiments, Albert Fischer and then Harry Eagle dialyzed horse serum to remove the small molecule compounds (e.g. vitamins and amino acids) and then supplemented those specific compounds back into the media in a controlled manner. In this way, they could elucidate the importance of specific small molecules (H. Eagle, 1955; Fischer, 1948). From this work, Harry Eagle was then able to develop a Minimal Essential Medium, which laid the foundation for many of the most used basal media formulations in animal cell culture (Eagle, 1959). Similarly motivated work by others studied fractionated serum with or without supplementation by known growth-promoting factors (e.g. insulin and fetuin) to find fractions or components which promoted cell growth (Lieberman and Ove, 1959; Lipton et al., 1972; Marr et al., 1962; Pierson Jr. and Temin, 1972), though the specific mechanisms involved with these factors often remained unknown.

With these and other advances, slow progress was made toward the development of chemically defined serum-free media throughout the middle of the twentieth century, with the first instances of true serum-free media for sustained cell expansion arising in the late 1950s with Charity Waymouth's MB752/1, Virginia Evans's NCTC109, and George Healy's CMRL1066 (Evans et al., 1956; Healy et al., 1955; Waymouth, 1959). However, these media were only effective for culturing Earle's strain L cells, an immortalized cell line produced through carcinogenic treatment of mouse fibroblastic cells, which presented an abnormal (i.e. mutated) cellular phenotype (Sanford et al., 1948). Serum-free media were next expanded to human cervical cancer (HeLa) cells and Chinese hamster ovary (CHO) cells in media consisting of basal nutrients supplemented with albumin and fetuin (α-globulin) extracted from animal sera (Fisher et al., 1959; Ham, 1965, 1962; Hamilton and Ham, 1977).

From here, serum-free media development expanded rapidly during the 1970s and 1980s to encompass many cell types, including hematopoietic cells, hybridomas, epithelial cells, fibroblasts, neural cells, and kidney cells, among others (Barnes and Sato, 1980; Bjare, 1992; Yao and Asayama, 2017). During this time, key discoveries were made to better elucidate specific components which were generally necessary for serum-free growth or which were particularly growth-promoting for specific cell types. Specifically, insulin, selenium, and transferrin were identified as key components for many cells – and indeed effective serum-substitutes on their own for some cell lines (Barnes and Sato, 1980; Breitman et al., 1980; Guilbert and Iscove, 1976; McKeehan et al., 1976). On the other hand, the roles of albumin and fetuin – which had previously been key components of many serum-free formulations – were better elucidated, and it was discovered that their roles were likely more due to their binding of other components (e.g. binding and presenting growth-promoting hormones, fatty acids, or adhesion factors, or binding and sequestering growth-inhibiting byproducts) rather than through direct interaction with cells either as nutrients or signaling molecules (Francis, 2010; Guilbert and Iscove, 1976; Hayman et al., 1985; Nilausen, 1978; Rizzino and Sato, 1978; Rockwell et al., 1980). Finally, while the direct growth-promoting effects of fetuin and albumin were called into question, the growth-promoting effects of other hormones and growth factors were studied and described (Yao and Asayama, 2017). Of particular relevance to cultured meat, several muscle cell growth and/or differentiation-promoting factors were described and studied during this time, including fibroblast growth factors (FGFs) (Allen et al., 1984; Allen and Boxhorn, 1989; Gospodarowicz, 1975, 1974; Gospodarowicz et al., 1976), insulin or insulin-like growth factors (IGFs) (Allen and Boxhorn, 1989; Dodson et al., 1985; Florini et al., 1977; Hill and Milner, 1985; Rinderknecht and Humbel, 1976), transforming growth factors (TGFs) (Allen and Boxhorn, 1989; de Larco and Todaro, 1978; Florini et al., 1986; Massagué et al., 1986; Olson et al., 1986; Roberts et al., 1980), and various hormones and steroids (e.g. growth hormone, testosterone, and dexamethasone) (Allen et al., 1983; Florini et al., 1977; Guerriero and Florini, 1978; Powers and Florini, 1975). Similarly, key factors were studied in the context of preadipocytes, most notably insulin, dexamethasone, and 3-Isobutyl-1-methylxanthine (IBMX) (Lanotte et al., 1982; Martin and Pohl, 1980; Morikawa et al., 1982; Reed et al., 1977; Reed and Lane, 1980; Rosen et al., 1979; Russell and Ho, 1976).

2 Serum-free media for muscle cells

Skeletal muscle is arguably the most important cell type for cultured meat, as it is both the main cell type of meat tissues by volume and mass and contributes essential functionality to meat, including fibrous texture, water- and fat-binding, key nutrients, and the shared 'meaty' flavor that is common across products (Hopkins, 2017; Mottram, 1998; Warner, 2017). As discussed previously in this book, skeletal muscle tissue undergoes two distinct phases in its development, proliferation of mononucleated muscle precursor cells and differentiation as these cells fuse into multinucleated myotubes. It is likely that cultured meat production processes will follow this paradigm, so serum-free media formulations must be developed for both the proliferation and differentiation phases of production. In this section, we will present several serum-free proliferation and differentiation media that have been developed for muscle stem cells, with a focus on those components that are specifically important for replacing animal sera (i.e. recombinant proteins and signaling molecules) and with less focus on the basal nutrient requirements of cells.

2.1 Muscle proliferation media

As with other cell types, muscle cell culture historically faced a long reliance on serum. While researchers were able to find certain supplements - such as insulin or various hormones - which could improve myogenesis, they faced the same challenges as Margaret and Warren Lewis did at the start of serum-free media exploration: nothing worked as well as serum (de La Haba et al., 1966). Indeed, it was not until 1979 that James Florini and Susan Roberts first expanded muscle progenitors (immortalized rat L6 cells) in serum-free media containing Ham's F-12 media supplemented with fetuin, insulin, and dexamethasone (Table 1). However, while this medium did achieve cell proliferation, it was only over 3 days and the authors did not verify cell myogenicity and differentiation after growth (Florini and Roberts, 1979). Two years later, Peter Dollenmeier et al. (1981) achieved growth of embryo-derived chick muscle cells over 10 days in a serum-free medium comprised of basal MCDB 201 supplemented with various proteins and hormones (Table 1). Here, the authors did verify the myogenicity

Table 1 Select muscle cell proliferation serum-free media compositions (ordered chronologically)

Basal	Supplements used in final media development	Key study details
Ham's F-12	Insulin (10^{-6} M); fetuin (10^{-5} M); dexamethasone (10^{-7} M)	Rat L6 cell growth over 3 days. No validation of myogenicity. Cells seeded in serum-containing medium (Florini and Roberts, 1979).
MCDB-201	Fibronectin (50 μg/mL); transferrin (10 μg/ mL); insulin (5 μg/mL); FGF (300 ng/ mL); glutamine (5 × 10^{-3} M); urea (10^{-1} M); cortisol-21-phosphate (10^{-3} M); β-hydroxybutyrate (5 × 10^{-2} M); $CoCl_2$ (10^{-5} M); glycerol (4 M); biotin (0.1 mg/mL); oleic acid (1 mg/mL)	Chick embryo-derived muscle cell growth over 10 days with subsequent validation of myotube formation. Fibronectin is necessary for cell attachment (Dollenmeier et al., 1981).
DMEM	Fetuin (0.5 mg/mL); insulin (10^{-6} M); dexamethasone (10^{-7} M)	Human satellite cell growth over 5 days with subsequent validation of myotube formation. Seeding in serum-containing medium is necessary (Yasin and van Beers, 1983).
DMEM; MCDB-104	Linoleic acid (1 μg/mL); fetuin (0.5 mg/mL); dexamethasone (10^{-7} M); FGF (100 ng/mL); insulin (10^{-6} M)	Rat primary satellite cell growth over 4 days with subsequent validation of myotube formation. Fibronectin-coated & seeding in serum-containing medium (Allen et al., 1985).
MCDB-120	Bovine serum albumin (BSA; 0.5 mg/mL); dexamethasone (10^{-6} M); EGF (10 ng/mL); fetuin (0.5 mg/mL); insulin (3 × 10^{-5} M)	Human satellite cell growth over 14 days with subsequent validation of myotube formation (Ham et al., 1988).
McCoys 5A[a]	Insulin (10^{-9} M); FGF (25 ng/mL); dexamethasone (10^{-7} M); BSA (10 ng/ mL); fetuin (10 μM); fibronectin (25 ng/ mL); transferrin (5 μg/mL); biotin (2 μg/mL); selenium (3.8 ng/mL); vitamin C (5 × 10^{-5} M); vitamin E (10 ng/mL); myoinositol (1 μg/mL); L-glutamine (500 ng/mL); linoleic acid (1 μg/ mL); sodium pyruvate (110 μg/mL); calcium chloride (1 mM)	Growth of primary rat and ovine satellite cells over 7 days with subsequent validation of myotube formation. Showed variations in growth kinetics and medium responses between rat and ovine cells (Dodson and Mathison, 1988).

(*Continued*)

Table 1 *(Continued)*

Basal	Supplements used in final media development	Key study details
McCoys 5A[a]	Insulin (0.5 mg/mL); FGF (25 ng/mL); fetuin (0.45 mg/mL); BSA (2 mg/mL); dexamethasone (10^{-6} M); linoleic acid (0.75 µg/mL); thymidine (726.6 µg/mL); adenine (405.3 µg/mL); d-α-tocopherol (0.129 µg/mL); various trace minerals (various concentrations)	Turkey satellite cell growth over 8 days. No subsequent validation of myotube formation (McFarland et al., 1991).
McCoys 5A[a]; MCDB-104	Fetuin (0.5 mg/mL); linoleic acid (1 µg/mL); dexamethasone (10^{-7} M); selenium (3×10^{-8} M); biotin (0.5 µg/mL); fibronectin (50 ng/mL); insulin (10^{-11} M); heparin (100 µg/mL); vitamin E (0.1 µg/mL); IGF-1 (various); FGF-2 (various); TGF-β (various)	Bovine satellite cell growth over 4 days followed by subsequent validation of myogenicity. Explored interactions between various growth factors (Greene and Allen, 1991).
MEM; MCDB-110	BSA (0.5 mg/mL); insulin (10^{-10} M); transferrin (100 µg/mL); dexamethasone (10^{-7} M); fetuin (125 µg/mL); FGF-2 (0–10 ng/mL); platelet-derived growth-factor BB (PDGF-BB; 0–10 ng/mL); EGF (0–10 ng/mL); IGF-1 (0–50 ng/mL)	Porcine satellite cell growth over 3 days with various combinations of growth factors. FGF-2 and IGF-1 alone shown to be sufficient to induce proliferation. No myogenic verification (Doumit et al., 1993).
DMEM	TCM (4%) - proprietary media supplement.	Murine C2C12 cell growth over 8 days with myotube validation (Yoshiko et al., 1996).
L15; M199	Vitronectin (100 ng/mL); B27 supplement (2%); FGF-2 (10 ng/mL); cardiotrophin 1 (CT1; 20 ng/mL); glial cell-line-derived neurotrophic factor (GDNF; 20 ng/mL); brain-derived neurotrophic factor (BDNF; 20 ng/mL); neurotrophin-3 (NT3; 20 ng/mL); Neurotrophin-4 (NT4; 20 ng/mL); sodium bicarbonate (1.4 mg/mL)	Proliferation of rat satellite cells with subsequent validation of myogenesis in both monocultures and in co-culture with nerve cells. Used growth-promoting substrates: with N-1[3-(trimethoxysilyl) propyl] diethylenetriamine (DETA) (Das et al., 2009).
DMEM/ F12	L-ascorbic acid 2-phosphate (200 µg/mL); insulin (20 µg/mL), transferrin (20 µg/mL), selenium (20 ng/mL), FGF-2 or heat-stable FGF2-G3 (Kuo et al., 2020) (40 ng/mL); neuregulin (NRG1; 0.1 ng/mL); TGFβ-3 (0.1 ng/mL); recombinant albumin (0.8-6.4 mg/mL)	Bovine satellite cell growth over 28 days & 7 passages with subsequent validation of myotube formation. Seeding with recombinant laminin or vitronectin and in the absence of recombinant albumin is necessary (Stout et al., 2022).

(Continued)

Table 1 (*Continued*)

Basal	Supplements used in final media development	Key study details
DMEM/ F12	L-ascorbic acid 2-phosphate (50 μg/mL); insulin (10 μg/mL), transferrin (5.5 μg/mL), selenium (6.7 ng/mL); ethanolamine (2 μg/mL); recombinant albumin (5 mg/mL); fibronectin (10 μg/mL); hydrocortisone (36 ng/mL); interleukin-6 (20 ng/mL); α-linolenic acid (1 μg/mL); FGF-2 (10 ng/mL); vascular endothelial growth factor (VEGF; 10 ng/mL); hepatocyte growth factor (HGF; 5 ng/mL); IGF-1 (100 ng/mL); platelet-derived growth factor BB (10 ng/mL)	Bovine myoblast growth over six passages with subsequent validation of myotube formation. Pre-coating with bovine collagen was utilized for cell seeding (Kolkmann et al., 2022).
DMEM/ F12	Fetuin (600 μg/mL); BSA (75 μg/mL); FGF-2 (2 ng/mL); insulin (10 μg/mL); transferrin (5.5 μg/mL); selenium (6.7 ng/mL); HGF (20 ng/mL)	Porcine, bovine, and murine satellite cell growth over multiple passages with subsequent validation of myotube formation (Skrivergaard et al., 2023).

[a]McCoys 5A medium contains Bacto-Peptone, an animal-origin component, so while these media can be serum-free, they cannot be animal-component-free. All other media can be made animal-component-free at least at the primary component level by using recombinant proteins.

of cells after proliferation in the serum-free medium and also discovered the outsized importance of FGF and insulin in muscle cell proliferation (Dollenmeier et al., 1981).

Following these initial studies, other serum-free media for muscle cells were tested for cells from various species (Table 1). These studies offered further insight into the additional benefits that can be conferred with the addition of numerous components, including dexamethasone, linoleic acid, and carrier proteins such as fetuin or albumin (Allen et al., 1985). However, none of these early media were demonstrated in a multi-passage culture system; rather, the standard method employed was to first plate cells in a serum-containing medium, and then change to a serum-free medium after a day or two to study the serum-free system through to terminal differentiation. Indeed, the only multi-passage serum-free study of muscle-derived cells used BC3H1 cells, a non-fusing cell line of muscle origin that does not retain function as a skeletal muscle precursor line (Gospodarowicz and Cheng, 1987). For cultured meat, though, it is essential to be able to culture cells in a serum-free system for long periods while retaining their myogenic capacity, so studying cell growth over multiple passages is essential for assessing media (Cosenza et al., 2021). Recently, a serum-free medium suitable for multi-passage culture has been demonstrated for bovine satellite cells (BSCs) (Stout et al., 2022). A key finding in developing this medium ('Beefy-9') was the necessary exclusion of one component - recombinant albumin - during cell seeding, which allowed for fully animal-component free multi-passage BSC culture. However, it is possible that alternative protocols (such as pre-coating plates or using different culture surfaces) would negate this requirement by adjusting the propensity of albumin to 'block' cell attachment (Martins Lima et al., 2020). Additionally, since the development of Beefy-9, several other serum-free media have been developed specifically for cultured meat (Table 1).

While Table 1 shows the use of many different components in serum-free media for muscle proliferation, there are a few consistent supplements involved in promoting proliferation, which also warrants special consideration when thinking about producing commodity-scale food products through cell culture. These include FGF-2, insulin and/or IGF-1, transferrin, dexamethasone, and carrier proteins such as fetuin or albumin. While not all of these components are strictly necessary, and indeed there are media formulations which exclude them, they are consistently included as core components of many serum-free media for muscle cells. Cultured meat research can therefore look toward these components as either factor to optimize (e.g. by engineering more stable or active forms of FGF-2 (Dvorak et al., 2018; Onuma et al., 2015) or IGF-1 (Francis et al., 1992)) or as cost-reduction targets (e.g. by decreasing the costs of recombinant production (Venkatesan et al., 2022) or finding low-cost analogs to components such as recombinant albumin (Farges-Haddani et al., 2006; Stout et al., 2023)). This is particularly true given the relatively high cost of these components compared with basal medium components (e.g. glucose, amino acids, fatty acids, and vitamins) that are also required for cell growth. The key functions and some considerations around these components are given in the following sections.

2.1.1 Fibroblast growth factor-2

FGF-2, also known as basic FGF or bFGF is 1 of 22 mammalian FGFs which are widely involved in tissue development and regeneration *in vivo* (Itoh and Ornitz, 2004). FGF-2 was originally isolated from bovine brain or pituitary tissue following its discovery as a potent mitogen in 1978 (Gospodarowicz, 1975; Gospodarowicz et al., 1978; Westall et al., 1978) and was first produced recombinantly a decade later in *E. coli* host systems (Fox et al., 1988). While the FGF-2 used in many early serum-free media formulations was animal derived, it has since been transitioned primarily to recombinant, animal component-free sources. As a medium component, FGF-2 is involved in several cellular processes. Most prominently, after binding cell surface FGF receptors (mostly FGFR-1, in the case of FGF-2 (Xie et al., 2020; Yablonka-Reuveni et al., 2015)), FGF-2 activates key pathways including the Phosphoinositide 3-kinase/

protein kinase B (P13k/Akt) pathway, the rat sarcoma virus/rapidly accelerated fibrosarcoma/mitogen-activated protein kinase/extracellular-signal-regulated kinase (Ras/Raf/MEK/ERK) pathway, and Jun N-terminal kinase which induce cell cycle progression (Xie et al., 2020). In muscle cells, FGF-2 is a key promoter of a growth phenotype and inhibits cell fusion and myogenic differentiation in favor of proliferation (Clegg et al., 1987). In cell culture media for cultured meat, challenges around FGF-2 include the short half-life for the wild-type protein (<7 h *in vitro* at 37°C), the relatively high cost in serum-free media, and the relatively high concentration requirements of FGF-2 in many serum-free media formulations (Dvorak et al., 2018; Stout et al., 2022). As mentioned, methods to stabilize and reduce the cost of FGF-2 have been explored and are promising options for addressing these challenges (Chen et al., 2012; Kuo et al., 2020; Nawrocka et al., 2020; Poudel et al., 2019; Ren et al., 2019; Song et al., 2013; Venkatesan et al., 2022). Additionally, cell engineering approaches can be used to abrogate FGF-2 requirements in muscle cells, by inducing cells to trigger the same signaling pathways in the absence of exogenous FGF-2 (Stout et al., 2023).

2.1.2 Insulin and insulin-like growth factor-1

While insulin and IGF-1 are distinct proteins with distinct roles *in vivo*, there is substantial cross-talk between these ligands and their receptors, and so they often are used to serve the same roles in serum-free media (De Meyts, 2002; Duan et al., 2010; Kim and Accili, 2002; Slaaby et al., 2006; Werner et al., 2008). Broadly, the differences between these ligands are categorized by insulin's general role in regulating metabolism and IGF-1's general role in regulating growth processes; however, each of these ligands is capable of binding the other receptor, and therefore stimulating the same signaling pathways (albeit with lower affinity than to their own respective receptors) (Bayne et al., 1989; De Meyts, 2002; Kim and Accili, 2002; Lawrence et al., 2007; Werner et al., 2008). Specifically, pathways that are subject to insulin and/or IGF-1 activation and regulation include metabolic pathways such as glucose uptake and utilization (Kim and Accili, 2002; O'Neill et al., 2015), as well as cell-cycle-progressing cascades such as P13k/Akt and Ras/Raf/MEK/ERK pathways (Chakravarthy et al., 2000; Duan et al., 2010; Han et al., 2008; Machida and Booth, 2004; Siddle, 2011). Interestingly, IGF-1 has been shown to stimulate both myogenic proliferation and differentiation, despite the fact that these are two mutually exclusive phases of muscle development (Engert et al., 1996; Florini et al., 1996; Florini et al., 1986). Indeed, insulin and IGF-1 are common components of both proliferation and differentiation media for muscle stem cells. It has been proposed that oxygen may play an important role in dictating whether IGF-1 acts in a pro-proliferative or pro-differentiative capacity, with hypoxia leading to an interaction between IGF-1 and hypoxia-inducible factor 1 (HIF-1) to promote proliferation, and normoxia promoting myogenic differentiation in the absence of HIF-1 (Duan et al., 2010). Ultimately, more research on the unique roles of insulin and IGFs in culture media is warranted, as is research on optimization and cost-saving strategies in the context of large-scale cultured meat production. These include the production of more stable or bioactive forms (such as the hyperactive LR3-IGF-1 (Francis et al., 1992)), exploring plant-derived analogs of insulin (Paula et al., 2017; Venâncio et al., 2003), or engineering muscle cells to ectopically produce IGF-1 and its associated receptor, which has been shown to eliminate exogenous IGF requirements in CHO cells (Pak et al., 1996; Pietrzkowski et al., 1992). As with FGF-2, insulin and IGF-1 were originally animal-derived in nature but have since been produced recombinantly.

2.1.3 Transferrin

As mentioned, transferrin's role in cell culture is less one of direct growth promotion through interaction with cell signaling cascades and more a result of the ability of the protein to bind iron and either deliver it in an efficient and controlled manner to cells – where the iron then acts as an essential cofactor in cellular processes – or sequester free iron and mitigate oxidative damage (Angoro et al., 2022; Brandsma et al., 2011; Puig et al., 2017). Toward this effect, transferrin works in cell culture by binding iron (thereby becoming holo-transferrin) before binding transferrin receptors on the cell surface. These complexes are then endocytosed, and the iron is transferred from the transferrin (thereby converting it to the iron-free apo-transferrin) and out of the endosomes into the cytoplasm, where it can be used for a range of cellular processes. The apo-transferrin proteins are then exocytosed back into the extracellular environment. The first recombinant transferrin was produced in 1993 in *E. coli*. Since then, a number of hosts have been used for its production and use in serum-free media, including yeast and plants (Brandsma et al., 2010; Huang et al., 2008; Keenan et al., 2006; Sargent et al., 2006; Steinlein and Ikeda, 1993). In the context of cultured meat, the costs associated with recombinant protein production are substantially higher for a protein like transferrin than for growth factors like FGF-2 or IGF-1, due to the relatively higher concentrations required (on the order of μg/mL vs ng/mL). As such, cost-saving strategies for transferrin will be extremely important for this application. Here, genetic engineering strategies can be explored as with FGF-2 and IGF-1, where ectopic expression of transferrin in CHO cells has ameliorated exogenous requirements (Pak et al., 1996). To avoid genetic engineering, other strategies could be using enhanced strains for recombinant production (Finnis et al., 2010; Zhang et al., 2010), the use of transferrin or transferrin-like proteins from non-animal sources (e.g. algae or plants) (Bai et al., 2016; Fisher et al., 1998, 1997), or the use of other iron-containing compounds such as ferric citrate or iron chelating agents (Eto et al., 1991; Gallego-Murillo et al., 2021).

2.1.4 Dexamethasone

Dexamethasone is a synthetic glucocorticoid that is a more potent and stable analog of the natural steroid hormone cortisol (also known as hydrocortisone) (Adcock and Mumby, 2017). From the first serum-free media for muscle cells that was described, dexamethasone has been a common additive due to its ability to promote myoblast proliferation *in vitro* (Florini and Roberts, 1979; Guerriero and Florini, 1980, 1978; Syverud et al., 2016). Interestingly, however, glucocorticoids such as dexamethasone have also been shown to directly oppose P13k/Akt signaling induced by insulin and IGF-1 and to regulate insulin-dependent glucose uptake rates, which are themselves important pathways for promoting muscle stem cell proliferation (Bruno et al., 2021; Kuo et al., 2012). One possible explanation for this could lie in the fact that dexamethasone treatment has been shown to increase IGF receptor expression in muscle and other cells (Giorgino and Smith, 1995; Urban et al., 1994), and therefore inhibition of IGF-1 or insulin-modulated P13k/Akt signaling by dexamethasone could induce IGF-1 receptor overexpression in response, thereby in actuality inducing enhanced signaling overall. This theory is supported by the fact that in the serum-free muscle media described above where dexamethasone proved particularly effective, concentrations of insulin were relatively low compared to media where dexamethasone was not as impactful (e.g. <10 μg/mL vs >20 μg/mL). Therefore, dexamethasone could be simply enhancing insulin IGF-1 receptor signaling in media that do not offer as many ligands for those receptors. In this sense, dexamethasone and insulin/IGF-1 could be considered together in designing serum-free media for cultured meat, weighing their specific advantages and disadvantages. One such disadvantage lies in the fact that dexamethasone is a synthetic steroid that is often used as an oral therapeutic, and as such might offer concerns about consumer safety and perception, and regulatory approval in a cell-cultured food product (Barbera et al., 2018; Medicine, 2022; Reig and Toldrá, 2008). One possible option to overcome this concern is to use natural glucocorticoids such as hydrocortisone to achieve the same effect. Indeed, in some instances, hydrocortisone has been shown to increase myoblast proliferation when added to cell culture media at the same levels of physiological normal cortisol in the blood (Carrasco-García et al., 2020; Kolkmann et al., 2022; Vierck et al., 2003). It might therefore be reasonably assumed that this level of addition to culture media would see regulatory approval, but studies of hormone residues in the final product may still be required for regulators and consumers. Ultimately, the value of glucocorticoids in serum-free media for cultured meat should be considered in the context of regulators and consumers, and the potential for achieving the same cellular effects through different additives or approaches.

2.1.5 Albumin and fetuin

Albumin and fetuin are abundant serum proteins with diverse roles in cell culture media. *In vivo,* fetuin is more prevalent in fetal serum, while albumin is more prevalent in adult serum. *In vitro*, fetuin was historically used as an additive to many early serum-free media formulations for muscle cells as both a growth-promoting and a cell attachment factor (Allen et al., 1985; Florini and Roberts, 1979). It was later suggested that for many cells – including muscle cells – the actual efficacy of fetuin in cell culture media most likely came from contaminants present after extraction of the protein, rather than the protein itself (Brown et al., 1992; Nie, 1992; Nie et al., 1991; Spiro, 1960), though this has been a somewhat contested claim and fetuin is still considered to be a potential cell adhesive factor for *in vitro* serum-free culture systems (Devireddy et al., 2019; Jirak et al., 2019; Sakwe et al., 2010; Sheikhpour, 2017; Watson et al., 2012). The ultimate utility of fetuin for cultured meat remains to be seen, as non-animal-derived fetuin (e.g. recombinant fetuin) must be explored in order to elucidate the actual function of the fetuin protein without associated contaminants, as well as to explore fetuin in a context that is suitable for cultured meat production (Arnaud and Kalabay, 2002).

Compared with fetuin, the biological functions of albumin in cell culture media are relatively well understood, though these too are somewhat obfuscated by the large diversity of functions that albumin can play *in vitro*. Specifically, albumin acts to bind, carry, stabilize, and present numerous compounds in cell culture, including fatty acids, metal ions, signaling molecules such as growth factors, amino acids, and other factors (Francis, 2010; Quinlan et al., 2005). In this, way, albumin can both stabilize and deliver beneficial components to cells (such as linoleic acid or growth factors) (Baur et al., 2005; Ishima et al., 2008; Nilausen, 1978; Quinlan et al., 2005; Raoufinia et al., 2016), and sequester chemical species from redox or other degradation reactions that are toxic to cells, thereby protecting cells *in vitro* and improving growth (Merlot et al., 2014; Roche et al., 2008).

Interestingly, despite the importance of albumin in many serum-free media for muscle cell expansion, it was recently shown that albumin can act counter to initial cell attachment onto cell culture substrates, even in the presence of adhesive factors such as laminin or vitronectin (Stout et al., 2022). As such, it was determined that recombinant albumin should be withheld from the cell culture media during plating and/or passaging and should instead only be added once the cells have attached to a substrate. It remains to be seen how this impacts large-scale culture processes. If single-cell suspension cultures can be achieved for muscle stem cells, it may not be much of a concern; however, if adhesion-culture remains a requirement for these cells, alternative solutions may need to be developed.

Additionally, along with this consideration, recombinant albumin is likely cost prohibitive to large-scale cultured meat production, as it is one of the primary cost drivers for those media in which it is an essential component (Stout et al., 2022). Options for overcoming this challenge could include the use of low-cost alternatives, such as plant-derived protein hydrolysates which have been shown to replace albumin in CHO cell culture (Chabanon et al., 2008; Farges-Haddani et al., 2006; Francis, 2010; Gorfien et al., 1998), or reducing the cost of recombinant production. Recently, bulk protein isolates from various oilseeds including soy and rapeseed were shown to effectively replace recombinant albumin in Beefy-9 serum-free medium for cultured meat production, thereby drastically lowering the cost of this medium (Stout et al., 2023). Similarly, methylcellulose has shown promise as a potential albumin alternative in Beefy-9 (Schenzle et al., 2022). Ultimately, for many media formulations, albumin is one of the key factors that must be optimized - or preferably replaced - in order to reach the economic viability of media formulations and cultured meat products (Hubalek et al., 2022).

2.1.6 Other factors

Despite the described formulations, serum-free media for satellite cells are not yet comparable to serum-containing media both in terms of growth rate and in terms of maintaining cellular myogenicity (Gospodarowicz and Cheng, 1987; Kolkmann et al., 2020; Stout et al., 2022). Significant optimization is still required to improve cell proliferation. This could include the exploration of additional growth factors beyond those discussed earlier which have shown promise as muscle cell culture supplements in some cultures, though not ubiquitous efficacy across culture systems. These supplements include proteins such as hepatocyte growth factor (Sheehan and Allen, 1999; Schenzle et al., 2022), leukemia inhibitory factor (Spangenburg and Booth, 2002), Interleukin-6 (Serrano et al., 2008; Stout et al., 2022), epidermal growth factor (EGF) (Doumit et al., 1993; Kelvin et al., 1989), platelet-derived growth factor-BB (PDGF-BB) (Doumit et al., 1993), pigment epithelial-derived factor (Ho et al., 2015), hormones such as prostaglandins (Mo et al., 2015), plant-derived compounds such as flavonoids or vitamins (Duran et al., 2019; Guo et al., 2022), small-molecule regulators of muscle stem cell stemness and myogenic capacity (Billin et al., 2016; Ding et al., 2018; Tierney et al., 2014), and a host of other potentially pro-mitotic compounds.

2.2 Muscle differentiation media

As mentioned, muscle stem cells undergo two distinct processes during their growth and maturation: proliferation and differentiation. These are in many ways antagonistic processes, and pro-mitotic components such as FGF-2 often act directly counter to myotube fusion and differentiation (Clegg et al., 1987). If cultured meat production processes expect to recapitulate the fibrous tissue architecture of meat through cellular structures, separately optimized media will be required for the proliferation and differentiation phases.

Fortunately, as with muscle stem cell proliferation, substantial research has gone into replacing serum in muscle differentiation medium. This is, in many ways, an easier task than proliferation, as a key feature of most muscle differentiation media is a *reduced* serum concentration, often achieved through serum starvation, in order to remove mitotic factors such as FGF-2 from the media. Indeed, simple cessation of feeding is often enough to induce robust differentiation, as mitotic factors such as FGF-2 have such short half-lives in media that they will degrade over time and muscle stem cells will subsequently be allowed to progress down their differentiation pathways (Stout et al., 2020). Even more simply, at sufficient confluency, muscle stem cells can be so overwhelmingly inclined to differentiate that they will form myotubes even in the presence of mitotic agents such as FGF-2. Indeed, many of the early serum-free media experiments with muscle stem cells did not change media formulations between proliferation and differentiation phases and were still able to see robust myotube formation (Allen et al., 1985; Allen and Boxhorn, 1989; Dodson and Mathison, 1988; Dollenmeier et al., 1981; Ham et al., 1988; Yasin and van Beers, 1983).

These studies aimed mostly to explore serum-free media for muscle cell proliferation, and so were not as focused on achieving robust differentiation. For cultured meat, it is possible that rapid and robust differentiation will be essential both for product quality (to achieve desired textures and nutrient profiles) and for favorable production economics (to reduce bioprocess time and bring down energy, labor, and input costs) (Chriki and Hocquette, 2020; Fraeye et al., 2020; Risner et al., 2020). To inform the requisite optimization to achieve this, several dedicated serum-free differentiation media have been developed for muscle cells (Table 2).

Here, the most consistent additives are EGF and insulin/IGF-1.

2.2.1 Insulin and insulin-like growth factor -1

As discussed earlier in this chapter, insulin and IGF-1 show both pro-proliferative and pro-differentiative effects for muscle stem cells, and so it is reasonable to expect these growth factors to be valuable components of serum-free differentiation media (Engert et al., 1996; Florini et al., 1996; Florini et al.,

Table 2 Select muscle cell differentiation serum-free media compositions

Basal media	Supplements used in final media development	Key study details
DMEM/F12	Dexamethasone (10^{-6} M); linoleic acid (1 μg/mL); insulin (10^{-7} M)	Bovine satellite cell fusion and myotube formation after 72 h of differentiation (Will et al., 2015).
MCDB-120	EGF (10 ng/mL); insulin (10 μg/mL); BSA (0.5 mg/mL)	Human satellite cell fusion with robust striation and spontaneous contraction (St. Clair et al., 1992).
Neurobasal; L15	EGF (100 ng/mL); IGF-1 (10 ng/mL)	Functional myotube formation over 5 days of primary rat satellite cells capable of contraction *in vitro* (McAleer et al., 2015). Additionally effective for bovine satellite cells following long-term expansion in serum-free proliferation medium and differentiation over 4 days (Stout et al., 2022).
DMEM/F12	Sodium bicarbonate (543 μg/mL); L-ascorbic acid 2-phosphate (unspecified); EGF (unspecified); MEM amino acids (unspecified); BSA (unspecified); transferrin (135 × 10^{-9} M); insulin (1.8 × 10^{-6} M); lysophosphatidic acid (LPA, 10^{-6} M)	Inferred supplements from analysis of surface receptor upregulation during differentiation. Bovine satellite cell differentiation over 4 days, including following growth in serum-free proliferation medium (Messmer et al., 2022).

1986). Specifically, insulin and/or IGF-1 activation of the p13k/Akt pathway have been shown to be key factors in their ability to induce muscle hypertrophy both via activation of mammalian target of rapamycin (mTOR) signaling and via inhibition of glycogen synthase kinase-3 (GSK3) and forkhead box protein O1 (FOXO1) signaling (Jin et al., 2020; Litwiniuk et al., 2016; Rommel et al., 2001; Xu et al., 2017). As discussed, dexamethasone treatment of cells is known to increase IGF-1 receptor levels, and so the inclusion and efficacy of this component in one of the above formulations could be likewise related to insulin and/or IGF-1 signaling (Giorgino and Smith, 1995; Urban et al., 1994). The challenges and opportunities around recombinant insulin and IGF-1 for muscle cell differentiation are the same as those for proliferation and are discussed previously.

2.2.2 Epidermal growth factor

EGF is another factor that can act to enhance both proliferation and differentiation in myogenic cultures (Doumit et al., 1993; St. Clair et al., 1992; Wroblewski et al., 2021). In muscle differentiation, EGF acts to enhance myosin heavy chain expression, enhance myotube size and density, enhance the formation of robust sarcomeric structures, and enable more contractile function and strength in *in vitro* constructs (Boularaoui et al., 2018; Wroblewski et al., 2021). While the mechanisms of this are still not fully characterized, EGF has several known downstream targets that could be factors in muscle differentiation. For instance, EGF is a potent activator of src-type kinase, which has been shown to be necessary for myogenin expression and myotube fusion (Lu et al., 2002; Osherov and Levitzki, 1994). Additionally, like IGF-1, EGF activates the p13k/Akt pathway, and so may enhance myotube fusion through this signaling cascade as well (Matheny et al., 2012; Shah et al., 2006). Indeed, there has been demonstrated interaction and cross-talk between the EGF receptor and IGF-1 receptor, so these factors may be acting synergistically or redundantly, and further analysis of their importance in serum-free differentiation media is warranted. Lastly, while EGF is also an activator of the Ras/Raf/MEK/ERK pathway, it is thought that this is more relevant to EGF's pro-proliferative effects, and does not induce a significant impact on myotube differentiation (Jones et al., 2001; Roberts and Der, 2007). The challenges and research opportunities of EGF in muscle cell differentiation media are similar to those present for all recombinant proteins discussed in this chapter and focus mainly on reducing the cost of production or finding low-cost alternatives (i.e. small molecules or cheap protein analogs).

2.2.3 Other additives

As with proliferation, muscle cell differentiation can be enhanced by the inclusion of other compounds, an exhaustive survey of which is beyond the scope of this chapter. To give a brief overview, though, these include hormones such as growth hormone (Ge et al., 2012; Sotiropoulos et al., 2006), steroids such as testosterone (Hughes et al., 2016), lipids such as LPA or cholesterol (Cornell et al., 1980; Messmer et al., 2022), or various small-molecule GSK3 or FOXO1 inhibitors (Kurgan et al., 2019; Ragozzino et al., 2019; Xu et al., 2017). Additionally, ERK inhibition has recently been shown to dramatically enhance and quicken robust differentiation of primary mouse myoblasts, which could shorten the time required for cultured meat differentiation and thereby dramatically reduce production costs of differentiated muscle (Eigler et al., 2021; Risner et al., 2020).

3 Serum-free media for fat cells

While skeletal muscle is the largest component of meat by volume and mass, adipose tissue is essential for generating the proper flavor and juiciness of meat products. Indeed, just as muscle contributes to the shared 'meaty' flavor of all meat products, fats are responsible for imparting species-specific flavors to products (Mottram, 1998). Because of this, the possibility of incorporating cell-based fats with plant-based proteins has been proposed as a promising initial way of generating hybrid plant-based / cultured meat products with reduced costs. As with muscle, media formulations for adipogenic cells must also be designed for the two steps of adipogenesis: proliferation and lipid accumulation. In the first of these phases, preadipocytes are expanded similarly to primary muscle stem cells. In this phase, the cells are typically grown to confluency *in vitro*. In the second phase, growth halts and cellular energy is turned toward the production and storage of lipids in lipid droplets within cells, which ultimately fuse to form one large lipid droplet in uniocular cells. As with muscle cells, this can happen spontaneously without the need to change media formulation, though such formulation changes can accelerate and enhance differentiation and lipid accumulation. Often, this second stage is broken down into two sub-stages: one in which the adipogenic process is initiated in an 'induction' medium and a following one in which lipid accumulation is carried out in a paired down 'maintenance' medium (J. S. K. Yuen et al., 2022). As with muscle, media formulations should be designed specifically to maximize these two distinct processes in a cultured meat production system.

The development of serum-free media for preadipocytes follows a similar trajectory as that for most mammalian cell types. In 1974, the 3T3-L1 cell line was established: a preadipogenic mouse fibroblast subline capable of robust lipid accumulation (Green and Kehinde, 1974). This set off a flurry of activity in the following years around trying to understand media compositions for these and other preadipocyte cell lines. Much useful insight was gained during this time, such as the roles of insulin, IBMX, dexamethasone, prostaglandins (e.g. E1 or F2α), growth hormone, and various other hormones in promoting growth and lipid accumulation in 3T3-L1s (Green and Kehinde, 1975; Hiragun et al., 1980; Mackall et al., 1976; Morikawa et al., 1982; Murray, 1980; Reed et al., 1977; Russell and Ho, 1976), and the converse capacity of growth factors such as FGF and PDGF to suppress adipogenic induction (Hayashi et al., 1981). This work was also plagued by reliance on serum and sometimes confounding evidence of the importance of unknown serum factors in preadipocyte growth and differentiation (Barnes et al., 1980; Kuri-Harcuch and Green, 1978). A useful review and interesting contemporary perspective on this early work was published by Gary Hausman, Dennis Campion, and Roy Martin in 1980 (Hausman et al., 1980).

In 1982, out of this progress, Ginette Serrero and John Khoo first demonstrated serum-free growth and lipid accumulation of a triglyceride-producing teratoma-derived cell line (1246 cells) (Darmon et al., 1981). Specifically, they demonstrated growth in a serum-free medium containing insulin, transferrin, FGF, selenium, and fibronectin and achieved subsequent differentiation by adding dexamethasone (10^{-6} M) and IBMX (2×10^{-4} M) to this medium for 48 h (the 'induction phase') followed by a return to the original serum-free medium (the 'maintenance phase') to allow for lipid accumulation over time (Serrero and Khoo, 1982). From here, several further serum-free media were developed for various cell lines from various species (Table 3).

Table 3 Select preadipocyte serum-free media compositions (ordered chronologically)

Proliferation medium	Induction medium	Maintenance medium	Key study details
DMEM/F12 + fibronectin (10 μg/mL); insulin (10 μg/mL); transferrin (10 μg/mL); FGF (100 ng/mL); selenium (3×10^{-10} M)	Proliferation medium + dexamethasone (10^{-6} M); IBMX (2×10^{-4} M)	Proliferation medium	1246 cells treated with serum-free induction components accumulated ~4× more triglyceride than no treatment but ~3× less than induction plus fetal calf serum (Serrero and Khoo, 1982).
MCDB 202 + insulin (1 μg/mL); dexamethasone (10^{-6} M); FGF (100 ng/mL); lipid mixture (20 μg/mL; [comprising soybean lecithin, cholesterol, sphingomyelin, vitamin E, and vitamin E acetate at a weight ratio of 6:3:1:0.06:0.2]) - one media used for all three stages of adipogenesis			Sheep preadipocyte proliferation achieved at 50% the rate of 10% FBS but with substantially higher lipid accumfulation (Broad and Ham, 1983).
DMEM/F12 + fibronectin (2.5 μg/mL); insulin (10 μg/mL); transferrin (10 μg/mL); FGF (25 μg/mL) - one media used for all three stages of adipogenesis			Rat preadipocytes isolated from inguinal fat pads and cultured for 7 days show improved adipogenesis compared with medium containing 10% FBS (Serrero and Mills, 1987).
DMEM/F12 + HEPES buffer (pH 7.4); biotin (33×10^{-6} M); pantothenate (17×10^{-6} M); insulin (800×10^{-9} M); transferrin (10 μg/mL); triiodothyronine (200×10^{-12} M)	Proliferation medium + dexamethasone (10^{-7} M)	Induction medium (it was shown that dexamethasone was only required for 2-3, albeit with reduced lipid accumulation; therefore, it was kept in the final formulation)	Rabbit preadipocytes were differentiated in induction medium after 14 days of proliferation. The necessity of dexamethasone or cortisol was shown for lipid accumulation in these cells (Reyne et al., 1989).

(*Continued*)

Table 3 (*Continued*)

Proliferation medium	Induction medium	Maintenance medium	Key study details
ASF301 [containing EGF (0.01 μg/mL); insulin (1 μg/mL); transferrin (10 μg/mL); triiodothyronine (0.2 ng/mL) as additives] (Yamane et al., 1981) + PDGF (20 ng/mL)	DMEM +10% fetal calf serum (FCS)[a]; dexamethasone (0.25 × 10^{-6} M); IBMX (0.5 × 10^{-3} M); insulin (10 μg/mL)	DMEM + 10% FCS[b]; insulin (5 μg/mL)	Serum-free proliferation of 3T3-L1 cells was achieved with the maintenance of an adipogenic phenotype. Serum-free differentiation was not explored (Kawada et al., 1990).
DMEM + transferrin (2 μg/mL); insulin (10^{-6} M); arachidonic acid (10^{-6} M); PDGF (0.25 × 10^{-9} M)	DMEM + insulin (10^{-9} M); corticosterone (0.5 × 10^{-6} M); IBMX (200 × 10^{-6} M); PDGF (10^{-9} M)	DMEM + insulin (10^{-9} M)	Fully serum-free growth and differentiation of 3T3-L1 cells (Staiger and Loffler, 1998).
DMEM + 10% FCS[c]; FGF-2 (2.5 ng/mL)	DMEM/F12 + ascorbic acid (10^{-7} M); insulin (0.85 × 10^{-6} M); sodium selenite (20 × 10^{-9} M); triiodothyronine (0.2 × 10^{-9} M); dexamethasone (10^{-6} M); IBMX (10^{-4} M); rosiglitazone (10^{-6} M)	Induction medium without dexamethasone or IBMX	Human adipose-derived stem cells showed robust lipid accumulation with these serum-free induction and maintenance media, though the proliferation medium relied on serum. The high impact of rosiglitazone was shown (Rodriguez et al., 2004).
Knockout DMEM + albumin (10 mg/mL); insulin (10 μg/mL); transferrin (8 μg/mL); FGF-2 (8 ng/mL); activin A (5 ng/mL) + supplementary amino acids, vitamins, fatty acids, and trace elements (various) (Rajala et al., 2010)	Proliferation medium + biotin (33 × 10^{-6} M); dexamethasone (10^{-6} M); insulin (0.1 × 10^{-6} M); pantothenate (17 × 10^{-6} M)		Robust lipid accumulation of differentiated human adipose-derived stem cells (ASCs). Proliferated cells were also capable to differentiation down osteogenic and chondrogenic lineages (Rajala et al., 2010).

(*Continued*)

Table 3 (*Continued*)

Proliferation medium	Induction medium	Maintenance medium	Key study details
XF/SF-MSCGM (proprietary medium from PELOBiotech)	Adipocyte basal medium (PELOBiotech) + dexamethasone (10^{-6} M); insulin (1 μg/mL); IBMX (0.5×10^{-3} M); indomethacin (10^{-4} M) + proprietary supplement (PELOBiotech)	Adipocyte Basal Medium (PELOBiotech) + dexamethasone (10^{-6} M); insulin (1 μg/mL); biotin (8 μg/mL); pantothenate (5 μg/mL) + proprietary supplement (PELOBiotech)	Human adipose-derived stem cells (ASCs) grown and differentiated in a fully defined system, albeit with a reliance on proprietary supplements which limits our ability to understand the specific components contributing to adipogenesis (Volz and Kluger, 2018).
DMEM/F12 + supplements previously developed for satellite cell proliferation, and provided in Table 1 (Kolkmann et al., 2022)	DMEM/F12 + HEPES (4.9×10^{-3} M); lipid concentrate (0.1%); putrescine (57×10^{-6} M); progesterone (17.8×10^{-9} M); hydrocortisone (25×10^{-9} M); calcium chloride (10^{-3} M); L-ascorbic acid 2-phosphate (227×10^{-6} M); Glucose (17×10^{-3} M); FGF-2 (2 ng/mL); EGF1 (2 ng/mL); bone morphogenic protein 4 (BMP4; 10 ng/mL); insulin (10 μg/mL); rosiglitazone (5×10^{-6} M)		Bovine, porcine, and ovine precursors grown and differentiated in defined systems. IBMX and dexamethasone were determined to be unnecessary for differentiation, while insulin and rosiglitazone were necessary. Progesterone and hydrocortisone were found to be helpful, though not strictly necessary (Mitić et al., 2023).

Additionally, a number of proprietary serum-free media for preadipocytes or adipose-derived stem cells have been developed (Lee et al., 2017; Panella et al., 2021), though these are excluded from the Table 3 due to their inability to elucidate the importance of specific factors or formulations.
[a]Not serum-free for induction.
[b]Not serum-free for maintenance.
[c]Not serum-free for proliferation.

As with muscle cells, it is clear that the above media utilize a range of ingredients, with several factors particularly effective for inducing adipogenesis. Namely, insulin, glucocorticoids such as dexamethasone, and IBMX are particularly important (Gregoire et al., 1998). Interestingly, some of these - such as insulin and dexamethasone - appear to be important for both muscle and fat tissue growth and differentiation. Here, we will discuss these common factors as they relate to adipogenic biology *in vitro*.

3.1 Insulin

The importance of insulin in adipogenic differentiation was discovered quickly after the development of the 3T3-L1 cell line when it was discovered that insulin receptors were greatly upregulated during adipogenesis (Reed et al., 1977). This is in contrast to IGF-1 receptors, which have been shown to decrease in differentiating 3T3-L1 cells (Zizola et al., 2002). As with muscle, there is some question as to the degree to which IGF/insulin cross-reactivity is obfuscating the relative importance of one or the other factor (Fernyhough et al., 2007). Either way, insulin and/or IGF-1 act in preadipocytes primarily through activation of ras signaling, which subsequently activates proliferator-activated receptor gamma (PPARγ), the master regulator of adipogenesis (Fernyhough et al., 2007; Fish et al., 2020). PPARγ fulfills this role by promoting gene expression of a host of downstream adipogenic genes which induce growth arrest and lipogenesis, and reduce lipolysis, the key final steps for lipid accumulation and adipogenesis (Fernyhough et al., 2007; Gregoire et al., 1998; Morrison and Farmer, 2000; Wabitsch et al., 1995). Considerations around the use of insulin or IGFs at scale are discussed previously. Fortunately, solving these challenges for either muscle or fat will offer parallel benefits to the other system, and so improving the scalability of insulin or insulin analog production will have a high impact on lowering the costs of both cultured muscle and fat production systems.

3.2 Dexamethasone

As mentioned previously, dexamethasone interacts closely with IGF signaling pathways and induces an increase in IGF-1 receptor expression. In adipogenesis, glucocorticoids such as dexamethasone have also been shown to directly promote expression of CCAAT/enhancer binding proteins (C/EBPs), which are key upstream adipogenic factors that induce subsequent expression of PPARγ (Hausman, 2000). In this way, C/EBPs act upstream of key adipocyte genes, such as aP2, SCD1, GLUT-4, leptin, and insulin receptors, and are therefore thought to be both necessary and sufficient for inducing adipogenic differentiation (Gregoire et al., 1998). As upstream factors, activation of C/EBPs is particularly essential for the 'induction' phase of adipogenesis but may not be as essential for the maintenance stage. Indeed, it can be seen in Table 3 that many induction media require dexamethasone, while it is often absent from maintenance media. The challenges around using synthetic glucocorticoids like dexamethasone for cultured meat production are discussed previously in the context of muscle cells and would likely be present for fat cells as well, as would the possible methods for overcoming those challenges. As with insulin, there is a parallel benefit for both muscle and fat in addressing these challenges, which could point toward a particularly effective area of research and development. In the context of fat production, the fact that dexamethasone might only be required for a brief time during the induction phase could lower the consumer and regulatory barriers to its use, though this remains to be seen.

3.3 3-Isobutyl-1-methylxanthine

IBMX is routinely used for promoting differentiation in preadipocytes. As with dexamethasone, IBMX acts by promoting the expression of C/EBP-β and subsequently PPARγ (Hausman, 2000). Indeed, it has been shown that ectopic expression of C/EBP-β in 3T3 cells mitigates the requirements for IBMX for adipogenic differentiation, which supports the premise that this is the primary target of this media factor (Gregoire et al., 1998; Wu et al., 1996). Interestingly, this same ectopic expression could not alleviate the requirements for dexamethasone. As with dexamethasone, the primary role of IBMX in promoting C/EBP expression correlates with this component's frequent use in induction media, but not maintenance media. The main challenges associated with IBMX lie in its acute oral toxicity (National Center for Biotechnology Information, 2022). While the use of IBMX solely for induction (and not for the maintenance phase) could result in negligible residual concentrations in the final cultured fat product, this would need to be thoroughly validated to ensure consumer safety. To bypass this need, alternative compounds can be explored. Also, as mentioned, it has previously been reported that the ectopic expression of C/EBPs can mitigate the requirements for IBMX in adipogenic induction, and so inducible genetic systems could offer promising low-cost opportunities for replacing this media component, though this would inevitably raise other concerns or questions around the use of genetic engineering.

3.4 Other factors

Along with these factors that are commonly used in the above serum-free media, some other factors have garnered attention in the fields of tissue engineering and cultured meat for their ability to promote adipogenic differentiation. Specifically, PPARγ agonists have been used to enhance adipogenic differentiation *in vitro*. A common agonist used in cell culture is rosiglitazone; however, this is a pharmaceutical drug (anti-diabetic) which may pose challenges in food-safe cell cultures. Fortunately, other food-safe PPARγ agonists have been described, including retinoic acid and free fatty acids (Fernyhough et al., 2007; García-Rojas et al., 2010; Li et al., 2014; Madsen et al.,

2005; Mehta et al., 2019; J. S. K. Yuen et al., 2022). These components may offer promising additives to improve or induce adipogenesis for food-safe cultured meat applications. Additionally, other factors such as bone morphogenic protein 4 (BMP4), FGF-1, ascorbic acid, and others have been shown to enhance adipogenesis and could be explored for cultured meat applications (He et al., 2016; Liu et al., 2017; Modica and Wolfrum, 2017; Scott et al., 2011).

Ultimately, as with muscle cells, there is a substantial body of evidence which can be used to inform the serum-free proliferation and differentiation of adipogenic cells for use in cultured meat production, and also a substantial amount of research and development that should be pursued in order to bring down costs while improving efficacy and food-safety of relevant media compositions. A key component that will be required of this development will be not merely inducing lipid accumulation, but inducing lipid accumulation that can provide aroma and flavor profiles which are desired and expected by consumers – either by directly tuning fatty acid profiles through media composition (J. S. Yuen et al., 2022), or by understanding and leveraging the different adipose biologies inherent to different species (Bergen and Mersmann, 2005).

4 Serum-free media for other potential cell types: induced pluripotent stem cells

While the earlier sections focus on muscle- and adipose-specific precursors, pluripotent stem cells such as embryonic stem cells (ESCs) or induced pluripotent stem cells (iPSCs) are also promising options for cultured meat. Fortunately, for these cell types, there are more generally accepted or 'canonical' serum-free media formulations that have been used to great effect over the last decade. These include TeSR, essential 8 (E8) an optimized and low-cost version of E8, termed B8, and others (Chen et al., 2011; Hua et al., 2022; Kuo et al., 2020). E8, in particular, has been widely used for ESC and iPSC culture, in part due to its simplicity (comprising only eight components: DMEM/F12 supplemented with insulin, selenium, transferrin, ascorbic acid, FGF-2, TGFβ, and a sodium bicarbonate buffer system (Chen et al., 2011).

Because iPSCs are pluripotent, they can be used for both muscle and fat generation. To induce differentiation, different genetic strategies or media formulations can be used. For instance, the robust generation of skeletal myocytes from iPSCs was recently achieved through the induction of transgenic MYOD1 with media supplementation by retinoic acid (Pawlowski et al., 2017). Additionally, iPSC myogenesis can be achieved via media supplementation, as reviewed in a thorough recent publication (Iberite et al., 2022). Similarly, iPSCs can be differentiated into adipose tissue through ectopic expression of key factors such as PPARγ and C/EBP-β in serum-free media (Ahfeldt et al.,

2012). Together, these possibilities point toward iPSCs as a promising option for cultured meat – though currently one that may face regulatory challenges if genetic reprograming of cells is used. Fortunately, barring those challenges, iPSCs offer a potentially more straightforward or well-established system for serum-free culture and so could offer an easier system from the perspective of the cell culture medium.

5 Basal media

As mentioned, the earlier sections focus on supplements that have specific roles in replacing animal sera for muscle cell growth; however, the importance of basal medium components should not be overlooked – both in promoting cell proliferation and in contributing to media cost. An in-depth look into the functions of the various components in basal media is beyond the scope of this chapter, but the key categories of components to consider include sugars (particularly glucose), amino acids (particularly L-glutamine), vitamins, inorganic salts, and auxiliary components (Table 4).

Glucose is the key carbon and energy source in cell culture media, though it can be combined with galactose and/or pyruvate in some instances. In proliferating muscle cells, glucose is primarily used to generate adenosine triphosphate through glycolysis, a less efficient but faster form of energy generation that is utilized in proliferating muscle through a phenomenon known as the Warburg effect (Fu et al., 2015; Relaix et al., 2021; Ryall, 2013). For cultured meat, this preference toward glycolysis instead of the more efficient tricarboxylic acid cycle poses some challenges in that it results in faster accumulation of the byproduct lactic acid, which can build up to toxic levels in cell culture. This makes glucose and glucose consumption (and subsequent lactate production) a key parameter to consider when developing cultured meat production systems. Indeed, recent techno-economic assessments of cultured meat production have suggested that glucose and glucose consumption rates will be key metrics for the economic realities of cultured meat production, and so optimizing concentrations for efficient consumption and limited metabolic waste build-up will be an important area of development for cultured meat.

Amino acids are key building blocks of cells, with 13 amino acids being considered essential for muscle cell growth *in vitro*. The most prevalent of these in culture media – and perhaps the most important – is glutamine, which is a major nitrogen source *in vitro,* and which is used as a major building block of nucleic acids and proteins, as well as a substrate to produce other amino acids in cells (DeBerardinis and Cheng, 2010) – particularly asparagine, aspartic acid, glutamic acid, and proline (Eagle, 1959). While glutamine is often considered an essential component of cell culture medium, it is not strictly necessary in all cell culture and, in some instances, can be replaced by high levels of glutamic

Table 4 Minimum and maximum components of demonstrative muscle serum-free media. Components in bold are those present at some level in every basal media formulation listed.

Amino acids								
Component	Minimum (mg/L)	Maximum (mg/L)	Component	Minimum (mg/L)	Maximum (mg/L)	Component	Minimum (mg/L)	Maximum (mg/L)
L-alanine	0	175	**L-glutamine**	**146**	**1461.5**	**L-methionine**	**4.479**	**60**
L-arginine	0	375	**Glycine**	**2.25**	**162.5**	**L-phenylalanine**	**4.959**	**100**
L-arginine HCl	**17.5**	**211**	L-histidine	0	187.5	L-proline	0	34.5
L-asparagine	0	187.5	**L-histidine HCl**	**5.47**	**42**	**L-serine**	**10.5**	**156.25**
L-asparagine-H2O	0	150	L-hydroxyproline	0	2.5	**L-threonine**	**11.9**	**232.5**
L-aspartic acid	0	19.97	**L-isoleucine**	**4**	**197.5**	**L-tryptophan**	**2.04**	**17.5**
L-cysteine HCl	0	90.025	**L-leucine**	**13.1**	**131.17**	L-tyrosine	0	225
L-cystine-2HCl	0	63	L-lysine	0	56.25	L-tyrosine disodium salt dihydrate	0	104
L-glutamic acid	0	22.1	**L-lysine HCl**	**17.5**	**181.65**	**L-valine**	**11.7**	**117.15**
Vitamins								
Component	Minimum (mg/L)	Maximum (mg/L)	Component	Minimum (mg/L)	Maximum (mg/L)	Component	Minimum (mg/L)	Maximum (mg/L)
Alpha tocopherol	0	0.0025	**Myo-inositol**	**1.5125**	**36**	**Thiamine HCl**	**0.0025**	**4**
Ascorbic acid	0	0.5	**Niacinamide**	**0.036**	**6.11**	Thiamine monophosphate	0	0.75
Biotin	0	0.2	Nicotinic acid	0	0.5	Vitamin A	0	0.025
Choline chloride	**0.875**	**14**	Para-aminobenzoic acid	0	1	Vitamin B12	0	2
D-calcium pantothenate	**0.156**	**23.82**	Pyridoxal HCl	0	0.5	Vitamin D2	0	0.025

Folic acid	0	10	**Pyridoxine HCl**	**0.06**	**4**			
Menadione (vitamin K3)	0	0.0025	**Riboflavin**	**0.00376**	**0.4**			
			Inorganic salts					
Component	Minimum (mg/L)	Maximum (mg/L)	Component	Minimum (mg/L)	Maximum (mg/L)	Component	Minimum (mg/L)	Maximum (mg/L)
Ammonium metavanadate	0	0.00059	Magnesium sulfate heptahydrate	0	246.38	Sodium metasillicate nonahydrate	0	2.842
Ammonium molybdate tetrahydrate	0	0.00371	Manganese sulphate	0	0.00024	Sodium phosphate dibasic (anhyd.)	0	142.5
Calcium chloride (anhyd.)	0	200	Nickel chloride hexahydrate	0	0.00007	Sodium phosphate dibasic	0	201.075
Calcium chloride dihydrate	**33.2**	**294**	**Potassium chloride**	**223.6**	**400**	Sodium phosphate monobasic	0	580
Cupric sulfate pentahydrate	0	0.0025	Potassium phosphate monobasic	0	45	Sodium selenite	0	0.00087
Ferric nitrate nonahydrate	0	0.175	Selenic acid	0	0.00387	Stannous chloride dihydrate	0	0.00003
Ferric sulfate heptahydrate	0	1.668	Sodium bicarbonate	0	3700	Zinc sulfate heptahydrate	0	0.864
Magnesium chloride (anhyd.)	0	70.275	**Sodium chloride**	**6306.25**	**7700**			
Magnesium sulfate (anhyd.)	0	200	Sodium hydroxide	0	260			

(*Continued*)

Table 4 (*Continued*)

Other components								
Component	Minimum (mg/L)	Maximum (mg/L)	Component	Minimum (mg/L)	Maximum (mg/L)	Component	Minimum (mg/L)	Maximum (mg/L)
2-deoxy-D-ribose	0	0.125	Glutathione (reduced)	0	0.5	Sodium acetate	0	12.5
Adenine HCl	0	1.72	Guanine HCl	0	0.075	Sodium pyruvate	0	412.5
Adenine sulfate	0	2.5	HEPES	0	8340.5	Thymidine	0	0.7
Adenosine 5'-phosphate	0	0.05	Hypoxanthine Na	0	4.77	Thymine	0	0.075
Adenosine 5'-triphosphate	0	0.25	Linoleic acid	0	0.084	Tween 80	0	5
Bacto-Peptone (animal origin)	0	600	Lipoic acid	0	0.21	Uracil	0	0.075
Cholesterol	0	0.05	**Phenol red**	**1.2**	**15**	Xanthine-Na	0	0.085
D-Galactose	0	135	Putrescine 2HCl	0	0.161			
D-Glucose (dextrose)	**250**	**4500**	Ribose	0	0.125			

acid in conjunction with overexpression of glutamine synthetase - an enzyme responsible for converting glutamic acid and ammonia into glutamine - which can be overexpressed through genetic engineering or as a natural cellular response to excess glutamic acid (DeBerardinis et al., 2007; Eagle, 1959; Genovese et al., 2021; Noh et al., 2017). This approach could be valuable for cultured meat production due both to the oversized requirements for glutamine in many cell culture media and due to glutamine metabolism and the role of decomposition in generating another toxic byproduct in culture media: ammonia (Eagle, 1959; Harry Eagle, 1955; Yao and Asayama, 2017). Along with the aforementioned approach to limiting the toxic buildup of ammonia, media formulations often utilize glutamine dipeptides such as L-alanyl-L-glutamine ('GlutaMAX'), which can improve peptide stability (Yao and Asayama, 2017). In the context of cultured meat, amino acids in the cell culture media may play the additional important role of impacting amino acid ratios in the final product, thus impacting product nutrition (O'Neill et al., 2021).

Vitamins have broad functions in cell culture media, acting as cofactors, antioxidants, and cell signaling factors. Many are typically included in basal media (Table 4) or else added as supplements for serum-free media (Table 1). As with amino acids, an interesting consideration around vitamins for cell-cultured meat media lies in the nutritional requirements of the potential future consumers of cellular agriculture products. Indeed, some vitamins should likely be included in cell culture media both to encourage cell growth and as a means of ensuring the nutritional completeness of cultured meat products. For instance, vitamin B12 is a flagship nutrient of animal products but is not produced by animal cells themselves - rather by ruminal bacteria in the case of ruminants (Watanabe and Bito, 2018). It will therefore be essential to ensure that cultured meat products have sufficient vitamin B12 to satisfy consumer expectations. This could be achieved by post-harvest fortification or by cellular uptake of B12 from the media during cell growth. The best option here will depend on the vitamin levels, vitamin bioavailability, and other process impacts (e.g. cell growth rates) of each specific method, and warrants dedicated exploration.

The importance of inorganic salts has been evident in cell culture ever since Sydney Ringer first demonstrated the necessity of sodium chloride, potassium chloride, and calcium chloride for maintaining tissues *ex vivo* (Ringer, 1883, 1882). While these three salts maintain primacy as the only salts that are strictly included in each of the aforementioned serum-free media formulations (Tables 1 and 4), others can offer value as buffering systems (e.g. sodium bicarbonate) or as sources of trace minerals, which play vital roles in cellular metabolism and function (Arigony et al., 2013; Hamilton and Ham, 1977). At times, these minerals can be presented in compounded form, such as iron-bound transferrin or vitamin B12 containing cobalt; however, presentation in inorganic salts could be valuable for cultured meat media formulation due to the relatively low cost of these salts. One trace element that receives particular attention in the context of serum-free media is selenium, which is typically included in media as either sodium selenite or selenic acid, and which is a cofactor in various enzymes that has become a key component of many serum-free media, either in the basal medium or as an additional supplement (Guilbert and Iscove, 1976; McKeehan et al., 1976).

Lastly, basal media often contain other miscellaneous components which play specific roles in cell culture, such as phenol red for pH monitoring of cultures, HEPES buffer for maintaining pH, nucleic acid precursors to serve as cellular building blocks, or lipids and fatty acids to do the same. One component of note in the context of cellular agriculture is linoleic acid, which has been shown to have a substantial dose-dependent impact on satellite cell proliferation *in vitro* in serum-free media (Allen et al., 1985; Stout et al., 2022). Therefore, linoleic acid is liable to be an important component for developing serum-free media for cultured meat.

Ultimately, the optimization of basal media for cultured meat presents a daunting scientific challenge due to the huge number of components that could possibly be tested and optimized, and the likelihood of variation in optimal formulation between species, cell type, and growth stage. Here, the computational and statistical design of experiments will be valuable tools for reducing the number of experiments required to optimize basal media, as will be discussed later in this chapter (Cosenza et al., 2022, 2021; O'Neill et al., 2021).

6 Reducing the cost and impact of serum-free media at scale

When looking at the cost of cultured media currently, it is clear that cost is dominated by signaling molecules and other recombinant proteins, which comprise over 95% of the cost of serum-free media like E8 or Beefy-9 (Stout et al., 2022). However, as these components are made less expensive through economies of scale, process optimization, or the discovery of low-cost alternatives, the other components of the media are likely to become significant price pressures on the economic outlook of cultured meat products. Therefore, it is prudent to take an early look at how all media components can be sourced at scale and for a low cost while not compromising efficacy. These include signaling molecules and recombinant proteins, carbon sources (e.g. glucose), amino acids, and trace vitamins and minerals.

6.1 Alternatives to recombinant proteins

As mentioned, the current primary driver of media cost lies in recombinant proteins, such as FGF-2 and albumin in the case of muscle cell expansion media (Stout et al., 2022). Finding ways to reduce the cost and impact of these proteins is essential for the economic viability of cultured meat. One option includes lowering the cost of recombinant production through innovation or economies of scale, as discussed previously, though the limits of how inexpensive these factors can get through process optimization are not yet known. This is particularly true for high-concentration proteins like albumin and transferrin, and so alternatives to recombinant proteins may be necessary in these instances (Hubalek et al., 2022). For instance, small peptide mimetics of FGF-2 have been shown to interact with FGFR-1 and could offer lower-cost options for inducing proliferation (Casey-Sawicki et al., 2014). Alternately, plant protein isolates or plant protein hydrolysates could provide specific growth-promoting functions *in vitro*, either as general growth-promoting agents in protein-free culture (Christel Andreassen et al., 2020; Franěk et al., 2000) or as specific replacements of recombinant proteins such as albumin (Chabanon et al., 2008; Farges-Haddani et al., 2006; George et al., 2009; Stout et al., 2023) or insulin (Barnes et al., 2015; Venâncio et al., 2003). Lastly, as discussed, genetic strategies could be exploited to overcome cellular requirements for recombinant proteins, though these bring associated complications in terms of regulatory approval and consumer acceptance (Pak et al., 1996, p.).

6.2 Carbon and amino acid sources

After the cost of growth factors, one of the next key culture media cost factors is the use – for most cell culture media – of highly pure sources of sugars (i.e. glucose) and amino acids (Humbird, 2021). The use of low-cost sources is necessary. Some ground may be gained here through the simple use of food-grade sugars and amino acids compared with the commonly used pharma-grade components. This could yield a nearly 100-fold cost reduction in some cases – such as with sugar (Hubalek et al., 2022). Additionally, complex sources of these components are potentially promising. For instance, molasses, corn steep liquor, yeast extract, or plant hydrolysates can offer free amino acids and sugars without requiring intensive purification processes, as covered in a recent review by O'Neill et al. (O'Neill et al., 2021). As one potential proof-of-concept, the potential efficacy of microalgal extracts for supplying the necessary sugars and amino acids to primary BSCs was demonstrated for both proliferation and differentiation stages (Okamoto et al., 2022, 2020). These strategies could offer significant cost-savings for cultured meat production, though their efficacy for animal cell culture media has yet to be proven as a robust and scalable solution (Humbird, 2021).

6.3 Media recycling

A promising option for lowering the cost of cell culture media is to develop media recycling systems that can enable the removal of toxic cellular byproducts (e.g. lactate and ammonia) from the culture media while retaining high-cost components (e.g. growth factors). An example of this was recently demonstrated for human iPSCs, in which a simple dialysis method allowed for the removal of lactic acid while retaining >80% of FGF and TGF in the media (Nath et al., 2017). Additionally, co-culture systems with lactate- or ammonia-consuming cells may offer beneficial *in situ* media recycling opportunities. A preliminary demonstration of this has been shown with murine muscle cells co-cultured with microalgae, which reduced anaerobic glycolysis-associated lactate and carbon dioxide production and consumed ammonia *in vitro* (Haraguchi and Shimizu, 2021a,b). Another possible strategy has been proposed which involves co-culture of hepatocytes with muscle cells, where the hepatocytes act to consume lactic acid and convert it back to glucose through gluconeogenesis (Datar and Betti, 2010; Hanyu and Kawashima, 2019). Finally, the removal of lactic acid or ammonia from cell culture media could be achieved through membrane filtration strategies, after which valuable compounds (e.g. glucose and growth factors) could be added to the medium as required while the collected waste (lactate and ammonia) could be valorized into products such as fertilizers or useful polymers (Kameda et al., 2021; Sikdar and Sawant, 1994). While these potential media recycling systems range in their technical readiness - with membrane-based filtration potentially being the closest to implementation - it is likely that some form of media recycling will be required to achieve the economic viability of cultured meat products.

6.4 Environmental impact of media

So far, the environmental impact analysis of cultured meat has relied on preliminary life-cycle assessments which must often rely on informed assumptions - rather than empirical data - to predict the impact of various cultured meat production systems (Lynch and Pierrehumbert, 2019; Carolyn S. Mattick et al., 2015; Carolyn S Mattick et al., 2015; Sinke and Odegard, 2021; Tuomisto and Teixeira de Mattos, 2011). While these results are therefore somewhat speculative, it has been suggested that culture medium will represent the second largest environmental impact of cultured meat production, after the energy required for various bioprocesses (Sinke and Odegard, 2021). Opportunities for reducing this environmental impact should be explored. These include many of the aforementioned optimization strategies, including media recycling or the use of agricultural byproducts (such as rapeseed meal, a byproduct of canola oil production) to generate media components (Farges-Haddani et al., 2006; Stout et al., 2023). As more data become available, the environmental impact should be assessed for all of the discussed media optimization strategies in order to ensure that cultured meat delivers on its promise of generating products in a more environmentally sustainable manner.

6.5 Animal component-free media

The distinction should be made between serum-free media, which do not include FBS or other animal sera, and animal component-free media, which do not contain any animal components at all. For instance, the early serum-free media for muscle cells did still rely on animal-derived fetuin and fibronectin (Allen et al., 1985; Dollenmeier et al., 1981; Florini and Roberts, 1979). For cultured meat applications, it will ultimately be important for media to be completely animal component-free, either by using recombinant proteins or by using alternative functional ingredients as discussed earlier. Here, it is necessary to look upstream through the media production process and ensure that no animal-derived components are used at any step of that process. For instance, recombinant proteins produced by animal cells (such as CHO cells) could rely on FBS in that cell culture process and so could still be reliant on animal-derived components even if the end media itself does not contain them. Recombinant proteins produced in animal component-free settings (e.g. through microbial fermentation or animal-component-free eukaryotic bioprocesses) offer one possible option for ensuring this, as are alternatives to those proteins as mentioned previously. Ultimately, it will be up to producers and regulators to look upstream within production processes in order to ensure that the ethical and environmental premise of cultured meat is being sustained throughout the process (i.e. by using wholly animal component-free inputs).

7 Media impact on cellular composition: safety and nutrition

Even in some of the earliest experiments exploring serum-free media, it was clear that media composition could drastically affect cellular composition. For instance, in 1987, it was shown that fatty acid composition varied significantly when mammalian cells were cultured in protein and lipid-free chemically defined media compared with serum-containing medium (Oshima et al., 1987). The impact of media on cellular composition will be an essential component in determining tissue composition, which will have implications on safety and nutrition.

7.1 Food safety

It is often said by food and nutrition experts that 'if it isn't safe, it isn't food.' As such, food safety is perhaps the starkest hurdle for cultured meat technology to clear in order to achieve market viability. While regulatory and safety considerations have been discussed in depth in prior chapters, and are thoroughly detailed in a recent review (Ong et al., 2021), it is likely that culture media composition will seriously impact the food safety features of products and so some discussion of these issues will be given here.

First, any component of cell culture medium has the potential to become part of the final product, and so both known components (i.e. those intentionally added to the medium as ingredients) and unexpected components (i.e. unintentional components added to the medium as part of, say, complex ingredients or cell secretions) should be considered from a food-safety perspective. These could be present as internalized chemicals within cells, components bound to cell surface receptors, or residual components left in the interstitial fluid after harvesting. When thinking about known components discussed previously in this chapter, the most obvious potential hazards are additives which have demonstrated oral toxicity, such as IBMX. As mentioned, it is possible that exclusion of this component from the final 'maintenance' phase of lipid accumulation would result in negligible quantities in the final product, though the low toxic dose (LD50 = 44 mg/kg in mice) of IBMX suggests that elimination of this component from the production process entirely is likely to be a more attractive option for regulators and consumers (National Center for Biotechnology Information, 2022).

Following the removal of acutely toxic compounds, therapeutic agents such as dexamethasone or rosiglitazone should be addressed. As mentioned, alternative PPARγ agonists such as free fatty acids could potentially replace rosiglitazone, and natural glucocorticoids such as cortisol could replace dexamethasone. Here, dietary cortisol could still impact consumer physiology and so should be assessed and compared with cortisol found in conventional meat products (Yoshioka et al., 2005). Finally, considering other constitutive signaling molecules discussed (i.e. growth factors), it has been shown that oral IGF-1 administration can impact serum IGF-1 and physiology in mice (Kim et al., 2006). Whether or not a relevant level of growth factors such as IGF-1 remains in a final cultured meat product must be determined, both due to the short half-life of growth factors (<7 h for FGF-2 and <20 h for IGF-1) and their heat labile nature, which could result in degradation during aging and cooking of cultured meat products (Dvorak et al., 2018; Mizuno et al., 2001; Yun et al., 2007). Ultimately, the levels and stability of these components should be assessed and compared with conventional meat products, particularly if stabilized forms are used. Finally, when using complex ingredients such as plant hydrolysates, care should be taken to understand and avoid (or at least label) potential allergens such as those present in soy or wheat (Radošević et al., 2016).

When thinking about potential unexpected components, one could imagine that contaminants such as pesticide residues may reside on the plant protein meals or sugars that are used to make hydrolysates or serve as complex carbon sources in cell culture media (Jiang et al., 2012). These components should be assessed for accumulation and presence in cultured meat products. Alternatively, unexpected components could arise from the cellular secretion of growth factors or other byproducts. Particularly if media recycling systems are used, these components could accumulate in the cell culture medium, and therefore be present in the final product (Ong et al., 2021). As such, this possibility should also be considered when assessing the food safety of cultured meat.

Next, microbial contamination and the prevention thereof are important considerations for cell-cultured meats. In small-scale laboratory experiments, it is common practice to use antibiotics (i.e. penicillin, streptomycin, and fungizone) to prevent contamination; however, this is likely to be untenable for large-scale cultivated meat production. The main safety consideration here is the concern that antibiotics could end up in the final product and disrupt consumer microbiomes with negative impacts on health (Jeong et al., 2010). Cultured meat production systems must either: ensure adequate sterility to mitigate microbial contamination, validate that any antibiotics used at any point during the production process do not find their way into the final product in any non-negligible quantity, or use alternative microbial management strategies. These alternative strategies could include components such as antimicrobial proteins or peptides that could confer antibiotic activities *in vitro* but then be degraded during cooking and digestion (Huan et al., 2020). As one potentially interesting example, lactoferrin has been shown to have broad antimicrobial properties, while also acting as a transferrin analog *in vitro* (Huang et al., 2008; Jenssen and Hancock, 2009). The substitution of transferrin with recombinant lactoferrin could offer commensurate benefits in preventing microbial contamination in culture, though efficacy and recombinant protein costs would need to be considered further in this system. Ultimately, careful monitoring of microbial and viral contamination will be required for all cultured meat production systems, though media formulation may offer opportunities for helping to mitigate this risk.

Finally, it might be reasonable to assume that the primary burden of food safety will fall on the final culture media that is used (i.e. on differentiation media for muscle or maintenance media for adipose), and so the media used prior to that may have fewer food safety requirements. Similarly, there could be fewer restrictions for media used earlier in the production process, such as in the 'seed train' of bioreactors, than for media used in the final production batch. For instance, if it is shown that dexamethasone used in the initial seed train of reactors does not leave any residues in the final product, so long as it is excluded from the final production run and/or differentiation step, then it is possible that regulators would approve its use in those early stages. Ultimately, though, thorough safety analyses will be required for all new cultured meat production processes in order to ensure food safety, regulatory approval, and consumer trust.

7.2 Nutrition

After safety, nutrition is likely to be a key feature that dictates the regulation, consumer acceptance, labeling, and market impact of cultured meat products. Here, media composition is likely to play a key role, as the nutrients fed to cells have a substantial impact on the molecular composition of the resulting tissue. For instance, along with the aforementioned example from 1987, recent studies have shown that media composition significantly impacts the lipidomic profile of cultured adipose cells. Specifically, the inclusion of intralipid - a lipid emulsion - in the maintenance media of lipid-accumulating porcine preadipocytes resulted in a significant increase in polyunsaturated fats and a significant decrease in monounsaturated fats in the resulting tissue, which could have beneficial impacts on cultured fat nutrition (Schwingshackl et al., 2021; J. S. Yuen et al., 2022). It is unclear whether similar tuning could be achieved for the amino acid composition of cultured meat products, though preliminary analyses suggest that without such tuning, the amino acid compositions of cultured meat may differ from conventional meat products (Joo et al., 2022). At the same time, though, it is possible that bioprocesses will be better served if the amino acid composition is tuned to enhance cellular growth rate (i.e. by matching cellular amino acid requirements and metabolism) rather than targeting nutritional outcomes (Xing et al., 2011).

In terms of micronutrients such as vitamins and minerals, there are opportunities for matching and improving upon the nutrition of cultured meat via media formulation. For instance, as mentioned, vitamin B12 is a flagship nutrient in meat that must be delivered to cells in the cell culture media, as cells lack the machinery for endogenous synthesis (Watanabe and Bito, 2018). However, the dynamics of B12 uptake and accumulation in muscle or fat cells remains unclear. Specifically, the uptake of B12 relies on specialized carrier proteins - transcobalamin's - which are produced and secreted at varying degrees by different cell types (Finkler and Hall, 1967; Hall, 1975; Rabinowitz et al., 1982), and the secretion of which could itself be dependent on culture conditions and media formulation (Florin et al., 2020). While it might be possible to enhance or tune the concentrations of vitamins in cell-cultured meat, these properties should be studied for each specific production system. The same is true for minerals such as iron, which are significantly more bioavailable when bound to heme proteins such as hemoglobin or myoglobin (López and Martos, 2004). It has been shown that adding heme proteins to cell culture media can improve cell growth and food quality features of cultured meat, and so this is one possible option for tuning media to enhance nutritional properties (i.e. iron bioavailability), though the economic considerations of adding additional recombinant proteins to the medium must be considered (Simsa et al., 2019). Ultimately, there is substantial room to explore how media can impact cultured meat nutrition; however, as covered in a recent review, there is more that is currently unknown than known (Fraeye et al., 2020).

8 Other considerations and future trends

It should be noted that the media formulations and considerations given in this chapter have a distinct focus on terrestrial species - particularly mammalian. This has been done for the sake of scope and practicality; however, cultured meat technologies can be used to generate products from other species as well, including fish, birds, crustaceans, and insects. Even between different species of mammals, media requirements can be significantly different (O'Neill et al., 2022). Thus, it is likely that cells from different taxonomic classes will need highly diverse media formualtions. Considerations around media development for these non-mammalian species have been discussed previously in the context of cultured meat, and the reader is directed to these publications for more insight (Potter et al., 2020; N. Rubio et al., 2019; N. R. Rubio et al., 2019).

It is likely that computational tools and automated high-throughput research strategies will be essential for realizing this goal. Specifically, statistical tools such as design-of-experiments can be used to efficiently test over a large design space of multiple factors, such as is present when optimizing culture media containing >30 components which are prone to interaction and cross-talk (Cosenza et al., 2022, 2021; O'Neill et al., 2021). Another valuable computational tool for media optimization could exist in the form of genome-scale metabolic models of relevant cell types, which can be used to discover growth-limiting metabolites *in vitro*, although challenges still exist around parameterizing these models to be predictive of cellular activity (Becker and Palsson, 2008; Carvalho et al., 2019; Pérez-Fernández et al., 2021). A third promising computational tool in serum-free media development could exist in computer-driven protein design and prediction, such as has recently been developed through the AlphaFold system (Jumper et al., 2021). These systems could allow for an improved understanding of protein structure and function and so could help in the development of hyperactive or hyperstable growth factors, or novel agonists to cell surface receptors. Last, data-driven -omics strategies, such as transcriptomics and proteomics, can be leveraged to better understand cellular activity and requirements *in vitro* and so could help generate media that are well suited to each stage of cell growth and differentiation. Indeed, a recent serum-free differentiation medium for bovine muscle stem cells used this strategy to identify cell surface receptors that were upregulated in differentiating cells and subsequently deliver increased concentrations of those receptors' ligands in the culture medium to improve differentiation (Messmer et al., 2022). Ultimately, these and other computational tools will likely accelerate media development in order to drive down cost and increase efficacy.

9 Conclusion and designing new media

Ultimately, the development of a robust, low-cost animal-component-free medium is required across species, cell types, cell growth phases, and bioprocess stages. Indeed, without effective and low-cost media, cultured meat is inherently nonviable or limited to specialty or novelty products. This is a key area of research and development for the field. Indeed, serum-free media for food-relevant cell types still, for the most part, do not perform as well as serum-containing media and still are not producible at prices that will enable price-parity with commodity meats (Hubalek et al., 2022). Fortunately, to improve on current formulations, cultured meat can leverage a long history of muscle and fat cell culture media development that begins with Sydney Ringer and Margaret and Warren Lewis and continues with the contributions of Richard Ham, James Florini, Ronald Allen, Denis Gospodarowicz, and many others. Fortunately, too, cultured meat can leverage opportunities for creative new media formulations that were impossible or impractical for biomedical contexts. Cultured meat can explore the use of food-grade inputs, complex inputs, unexplored media ingredients or cell lines, or novel cell line engineering strategies in order to achieve cell culture at an unprecedented scale.

Putting this into practice, though, poses its own challenges. Culture media are complex mixtures of components which interact both molecularly in solution and intracellularly in the metabolic and signaling pathways which they affect. As such, changing one component can impact the effect of others. More, culture media could theoretically be composed of an essentially infinite number of compounds, both natural and synthetic. Indeed, FBS alone is thought to have over 1000 components which exist in various forms (i.e. bound, unbound, vesicular, etc.) and which may or may not have a measurable impact on cell growth *in vitro* (Aswad et al., 2016). It is therefore an understandably daunting task to try and replace FBS, let alone move beyond FBS to truly optimize and maximize cell growth and development. This is a problem that is made even more difficult due to the deficiency of short-term growth experiments in predicting long-term growth outcomes, and the labor-intensive nature of long-term growth experiments (Cosenza et al., 2022; Stout et al., 2022). That said, there are three primary ways to try and rise to these challenges, which are described here and in Fig. 1.

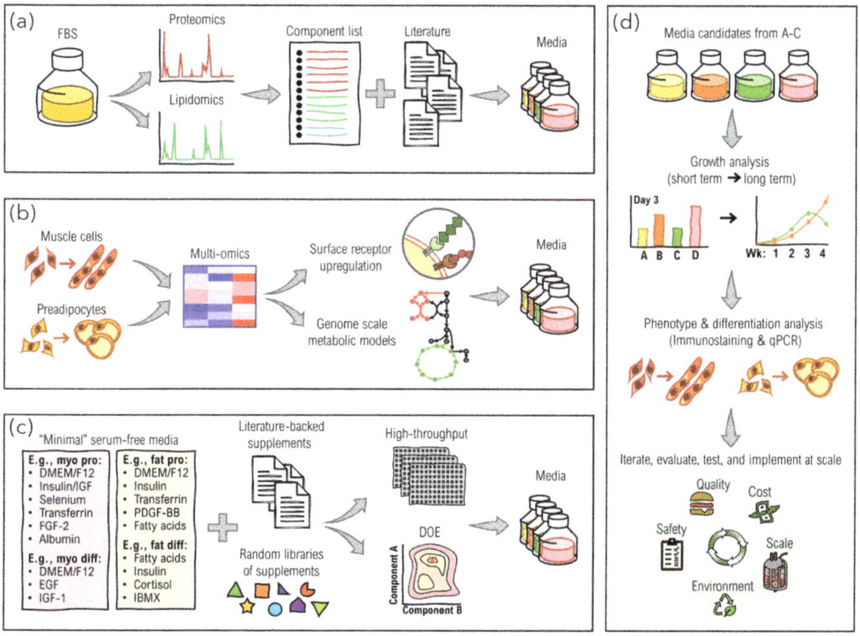

Figure 1 Summary of strategies for media development. (a) 'Serum-down' approach, in which characterization of animal-derived components (e.g. fetal bovine serum) and literature evidence informs formulation of 'synthetic FBS' that can be used to generate media formulations. (b) 'Cell-down' approach, in which cell requirements are characterized through metabolomics, transcriptomics, and proteomics of cells grown in serum-contain or serum-free media. These datasets are then used to determine which cell surfaces receptors, signaling pathways, and metabolic pathways are being over- or under-expressed in cells during proliferation and/or differentiation. With this information, media formulations can be generated to meet the requirements of cells and/or release signaling or metabolic bottlenecks that are present. (c) 'Media-up' approach in which minimally suitable serum-free media formulations, such as those described in this chapter, are used as the foundation for high-throughput screens of randomly generated or literature-suggested supplements. Here, automation and design-of-experiments can be used to reduce the labor involved and generate candidate media formulations. Some possible minimal serum-free components are provided for muscle ('myo') and fat proliferation and differentiation, based on literature evidence presented in this chapter. (d) Flow diagram for testing candidate media formulations generated by approaches (a)–(c). First, short-term growth experiments such as metabolic assays (e.g. MTT or PrestoBlue) or dsDNA quantification assays (e.g. Cyquant or FluoReporter) can be used to narrow down candidates for long-term growth experiments, which are required and should explore cell growth at least over several weeks and passages (e.g. through cell counting). After and during this process, cell phenotype and differentiation capacity should be continually assessed via immunostaining and/or quantitative PCR to look for phenotype-specific markers. Finally, developed media should be iteratively improved (i.e. through the processes in (b) or (c)). Throughout this process, considerations around cost, safety, nutrition and quality, scalability, and environmental impact should be incorporated into the design and optimization process.

The first is the 'Serum-down' approach, which involves looking at the components of blood and FBS, combining this with the study of or understanding of the mechanisms of certain factors, and trying to reconstruct a relevant and effective mimic. Arguably, this is not the best approach for cultured meat development, as it is time- and resource-intensive, and may lead to the incorporation of redundant components, or might miss lower-cost alternative to serum components (e.g. plant-derived or small molecule analogs).

The second approach is the 'Cell-down' approach, which involves looking at cell behavior in serum-containing media or *in vivo* through tools like genome-scale metabolic models or -omics technologies, and then designing a media to match the metabolic requirements and signaling receptors that are expressed – in essence delivering those components that the cell is 'asking for.' As mentioned, these techniques have been used to some effect in previous serum-free media development studies, and so offer a compelling approach that is less likely to result in the addition of redundant components (Messmer et al., 2022; Pérez-Fernández et al., 2021). Still, though, this approach might miss unexpected beneficial additives, as the media will only ever be as good as the models are predictive or the -omics data is understood (i.e. robust and annotated).

The final approach is the 'Media-up' approach, which would start with a minimum serum-free media that has been developed to date and is able to encourage at least a baseline level of cell viability or growth and would then screen additives to build a new media component-by-component. This method can screen components that are supported by literature as potentially effective additives, as well as unexpected components such as libraries of small molecules, randomly assembled peptides or proteins, or complex plant extracts and fractions. Here, computational and automation tools will be required to enable efficient high-throughput screening of components, but this approach offers value in its ability to discover novel or unexpected additives with low cost and high effectiveness *in vitro* by looking outside normal cell biology or cell-media interactions – one of the main potential advantages that cultured meat has over medical tissue engineering.

A graphical summary of these three distinct approaches is given in Fig. 1, though ultimately a combination of all of them will likely prove most fruitful and effective for cultured meat.

Ultimately, too, a coordinated effort across numerous sectors and research models will likely benefit media development for the field. These include academia, industry, and alternative models which have been proposed for addressing large, targeted research efforts to broadly advance a field (Marblestone et al., 2022). It is worth considering the advantages and disadvantages of these specific models (discussed elsewhere in this book) when looking at challenges as complex as media development for cultured meat. For instance, while culture media formulations are obviously valuable pieces of companies' intellectual property portfolios, and therefore attract substantial research investment in the private sector, they are also the foundation on which so much broad research rests and so could potentially be well served as publicly available resources for researchers across sectors (Holmes et al., 2022; Stout et al., 2023). Indeed, it is interesting to consider that many of the first serum-free media formulations on which so much animal cell culture rests were developed openly in non-profit research institutes (such as CMRL1066, developed at Connaught Medical Research Laboratories). As with scientific approaches to media development, it is likely that a combination of R&D infrastructure approaches will best serve the rapid and simultaneously broad advancement of the field, an outcome which is to be desired if effective and equitable development of cultured meat technologies is to be realized.

10 Where to look for further information

Recently, regulatory approval in the USA has been granted for two cultured meat companies: UPSIDE Foods and GOOD Meat. This includes both FDA approval through a 'letter of no further questions', and USDA approval through a grant of inspection. The documents supporting the FDA's decisions can be highly informative of approved scientific processes that are likely to reach consumers (U.S. Food and Drug Administration, 2023). The final submission from GOOD Meat is particularly informative in the context of this chapter, as it includes the firm's own data on some media components which are included in their production process (Good Meat, Inc., Polsinelli PC, AR Toxicology, Burdock Group Consultants, 2022). As more companies are granted approval in the USA and other jurisdictions, it will be valuable to keep an eye on the associated scientific memos in order to gain further insight into how both companies and regulators address the questions of cell culture media for cultured meat.

Additionally, the reader is directed to the following review articles for deep insight into media development, both historically and in the context of cultured meat:

- Animal-cell culture media: History, characteristics, and current issues (Yao and Asayama, 2017).
- Animal cell culture media (Vijayasankaran et al., 2010).
- Understanding adipocyte differentiation (Gregoire et al., 1998).
- Perspectives on scaling production of adipose tissue for food applications (J. S. K. Yuen et al., 2022).
- Considerations for the development of cost-effective cell culture media for cultivated meat production (O'Neill et al., 2021).

11 References

Adcock, I. M. and Mumby, S. 2017. Glucocorticoids. In: *Handb. Exp. Pharmacol.* Page, C. P. and Barnes, P. J. (Eds). Pharmacology and Therapeutics of Asthma and COPD. Springer International Publishing, Cham, 171–196. https://doi.org/10.1007/164_2016_98.

Ahfeldt, T., Schinzel, R. T., Lee, Y.-K., Hendrickson, D., Kaplan, A., Lum, D. H., Camahort, R., Xia, F., Shay, J., Rhee, E. P., Clish, C. B., Deo, R. C., Shen, T., Lau, F. H., Cowley, A., Mowrer, G., Al-Siddiqi, H., Nahrendorf, M., Musunuru, K., Gerszten, R. E., Rinn, J. L. and Cowan, C. A. 2012. Programming human pluripotent stem cells into white and brown adipocytes. *Nat. Cell Biol.* 14(2), 209–219. https://doi.org/10.1038/ncb2411.

Allen, R. E. and Boxhorn, L. K. 1989. Regulation of skeletal muscle satellite cell proliferation and differentiation by transforming growth factor-beta, insulin-like growth factor I, and fibroblast growth factor. *J. Cell. Physiol.* 138(2), 311–315. https://doi.org/10.1002/jcp.1041380213.

Allen, R. E., Dodson, M. V. and Luiten, L. S. 1984. Regulation of skeletal muscle satellite cell proliferation by bovine pituitary fibroblast growth factor. *Exp. Cell Res.* 152(1), 154–160. https://doi.org/10.1016/0014-4827(84)90239-8.

Allen, R. E., Dodson, M. V., Luiten, L. S. and Boxhorn, L. K. 1985. A serum-free medium that supports the growth of cultured skeletal muscle satellite cells. *In Vitro Cell. Dev. Biol.* 21(11), 636–640.

Allen, R. E., Masak, K. C., McAllister, P. K. and Merkel, R. A. 1983. Effect of growth hormone, testosterone and serum concentration on actin synthesis in cultured satellite cells. *J. Anim. Sci.* 56(4), 833–837. https://doi.org/10.2527/jas1983.564833x.

Andreassen, R. C., Pedersen, M. E., Kristoffersen, K. A. and Beate Rønning, S. 2020. Screening of by-products from the food industry as growth promoting agents in serum-free media for skeletal muscle cell culture. *Food Funct.* 11(3), 2477–2488. https://doi.org/10.1039/C9FO02690H.

Angoro, B., Motshakeri, M., Hemmaway, C., Svirskis, D. and Sharma, M. 2022. Non-transferrin bound iron. *Clin. Chim. Acta* 531, 157–167. https://doi.org/10.1016/j.cca.2022.04.004.

Arigony, A. L. V., de Oliveira, I. M., Machado, M., Bordin, D. L., Bergter, L., Prá, D. and Pêgas Henriques, J. A. 2013. The influence of micronutrients in cell culture: A reflection on viability and genomic stability. *BioMed Res. Int.* 2013, 597282. https://doi.org/10.1155/2013/597282.

Arnaud, P. and Kalabay, L. 2002. Alpha2-HS glycoprotein: a protein in search of a function. *Diabetes Metab. Res. Rev.* 18(4), 311–314. https://doi.org/10.1002/dmrr.315.

Aswad, H., Jalabert, A. and Rome, S. 2016. Depleting extracellular vesicles from fetal bovine serum alters proliferation and differentiation of skeletal muscle cells in vitro. *BMC Biotechnol.* 16, 32. https://doi.org/10.1186/s12896-016-0262-0.

Bai, L., Qiao, M., Zheng, R., Deng, C., Mei, S. and Chen, W. 2016. Phylogenomic analysis of transferrin family from animals and plants. *Comp. Biochem. Physiol. Part D Genomics Proteomics* 17, 1–8. https://doi.org/10.1016/j.cbd.2015.11.002.

Barbera, S., Tarantola, M., Sala, G. and Nebbia, C. 2018. Canonical discriminant analysis and meat quality analysis as complementary tools to detect the illicit use of dexamethasone as a growth promoter in Friesian bulls. *Vet. J.* 235, 54–59. https://doi.org/10.1016/j.tvjl.2018.03.006.

Barnes, D. and Sato, G. 1980. Methods for growth of cultured cells in serum-free medium. *Anal. Biochem.* 102(2), 255–270. https://doi.org/10.1016/0003-2697(80)90151-7.

Barnes, D., Wolfe, R., Serrero, G., McClure, D. and Sato, G. 1980. Effects of a serum spreading factor on growth and morphology of cells in serum-free medium. *J. Supramol. Struct.* 14(1), 47–63. https://doi.org/10.1002/jss.400140106.

Barnes, M. J., Uruakpa, F. O. and Udenigwe, C. C. 2015. Influence of cowpea (Vigna unguiculata) peptides on insulin resistance. *J. Nutr. Health Food Sci.* 3. https://doi.org/10.15226/jnhfs.2015.00144.

Baur, A., Reski, R. and Gorr, G. 2005. Enhanced recovery of a secreted recombinant human growth factor using stabilizing additives and by co-expression of human serum albumin in the moss Physcomitrella patens. *Plant Biotechnol. J.* 3(3), 331–340. https://doi.org/10.1111/j.1467-7652.2005.00127.x.

Bayne, M. L., Applebaum, J., Underwood, D., Chicchi, G. G., Green, B. G., Hayes, N. S. and Cascieri, M. A. 1989. The C region of human insulin-like growth factor (IGF) I is required for high affinity binding to the Type 1 IGF receptor. *J. Biol. Chem.* 264(19), 11004–11008. https://doi.org/10.1016/S0021-9258(18)60418-1.

Becker, S. A. and Palsson, B. O. 2008. Context-specific metabolic networks are consistent with experiments. *PLoS Comput. Biol.* 4(5), e1000082. https://doi.org/10.1371/journal.pcbi.1000082.

Bergen, W. G. and Mersmann, H. J. 2005. Comparative aspects of lipid metabolism: impact on Contemporary Research and use of animal models. *J. Nutr.* 135(11), 2499–2502. https://doi.org/10.1093/jn/135.11.2499.

Billin, A. N., Bantscheff, M., Drewes, G., Ghidelli-Disse, S., Holt, J. A., Kramer, H. F., McDougal, A. J., Smalley, T. L., Wells, C. I., Zuercher, W. J. and Henke, B. R. 2016. Discovery of novel small molecules that activate satellite cell proliferation and enhance repair of damaged muscle. *ACS Chem. Biol.* 11(2), 518–529. https://doi.org/10.1021/acschembio.5b00772.

Bjare, U. 1992. Serum-free cell culture. *Pharmacol. Ther.* 53(3), 355–374. https://doi.org/10.1016/0163-7258(92)90056-6.

Boone, C. W., Mantel, N., Caruso, T. D., Kazam, E. and Stevenson, R. E. 1971. Quality control studies on fetal bovine serum used in tissue culture. *In Vitro* 7(3), 174–189. https://doi.org/10.1007/BF02617963.

Boularaoui, S. M., Abdel-Raouf, K. M. A., Alwahab, N. S. A., Kondash, M. E., Truskey, G. A., Teo, J. C. M. and Christoforou, N. 2018. Efficient transdifferentiation of human dermal fibroblasts into skeletal muscle. *J. Tissue Eng. Regen. Med.* 12(2), e918–e936. https://doi.org/10.1002/term.2415.

Brandsma, M. E., Diao, H., Wang, X., Kohalmi, S. E., Jevnikar, A. M. and Ma, S. 2010. Plant-derived recombinant human serum transferrin demonstrates multiple functions. *Plant Biotechnol. J.* 8(4), 489–505. https://doi.org/10.1111/j.1467-7652.2010.00499.x.

Brandsma, M. E., Jevnikar, A. M. and Ma, S. 2011. Recombinant human transferrin: beyond iron binding and transport. *Biotechnol. Adv.* 29(2), 230–238. https://doi.org/10.1016/j.biotechadv.2010.11.007.

Breitman, T. R., Collins, S. J. and Keene, B. R. 1980. Replacement of serum by insulin and transferrin supports growth and differentiation of the human promyelocytic cell line, HL-60. *Exp. Cell Res.* 126(2), 494–498. https://doi.org/10.1016/0014-4827(80)90296-7.

Broad, T. E. and Ham, R. G. 1983. Growth and adipose differentiation of sheep preadipocyte fibroblasts in serum-free medium. *Eur. J. Biochem.* 135(1), 33–39. https://doi.org/10.1111/j.1432-1033.1983.tb07614.x.

Brown, W. M., Saunders, N. R., Møllgård, K. and Dziegielewska, K. M. 1992. Fetuin - an old friend revisited. *BioEssays* 14(11), 749–755. https://doi.org/10.1002/bies.950141105.

Bruno, N. E., Nwachukwu, J. C., Srinivasan, S., Nettles, C. C., Izard, T., Jin, Z., Nowak, J., Cameron, M. D., Boregowda, S. V., Phinney, D. G., Elemento, O., Liu, X., Ortlund, E. A., Houtman, R., Stavreva, D. A., Hager, G. L., Kamenecka, T. M., Kojetin, D. J. and Nettles, K. W. 2021. Chemical systems biology reveals mechanisms of glucocorticoid receptor signaling. *Nat. Chem. Biol.* 17(3), 307–316. https://doi.org/10.1038/s41589-020-00719-w.

Carrasco-García, A. A., Pardío-Sedas, V. T., León-Banda, G. G., Ahuja-Aguirre, C., Paredes-Ramos, P., Hernández-Cruz, B. C. and Murillo, V. V. 2020. Effect of stress during slaughter on carcass characteristics and meat quality in tropical beef cattle. *Asian-Australas. J. Anim. Sci.* 33(10), 1656–1665. https://doi.org/10.5713/ajas.19.0804.

Carvalho, M., Nikdel, A., Riesberg, J., Lyons, D. and Budman, H. 2019. Identification of a dynamic metabolic flux model for a mammalian cell culture. IFAC-pap., 12th IFAC Symposium on Dynamics and Control of Process Systems, including Biosystems DYCOPS 52, 88–93. https://doi.org/10.1016/j.ifacol.2019.06.042.

Casey-Sawicki, K., Zhang, M., Kim, S., Zhang, A., Zhang, S. B., Zhang, Z., Singh, R., Yang, S., Swarts, S., Vidyasagar, S., Zhang, L., Zhang, A. and Okunieff, P. 2014. A basic fibroblast growth factor analog for protection and mitigation against acute radiation syndromes. *Health Phys.* 106(6), 704–712. https://doi.org/10.1097/HP.0000000000000095.

Chabanon, G., Alves da Costa, L., Farges, B., Harscoat, C., Chenu, S., Goergen, J. L., Marc, A., Marc, I. and Chevalot, I. 2008. Influence of the rapeseed protein hydrolysis process on CHO cell growth. *Bioresour. Technol.* 99(15), 7143–7151. https://doi.org/10.1016/j.biortech.2007.12.070.

Chakravarthy, M. V., Abraha, T. W., Schwartz, R. J., Fiorotto, M. L. and Booth, F. W. 2000. Insulin-like growth factor-I extends in VitroReplicative life span of skeletal muscle satellite cells by enhancing G1/S cell cycle progression via the activation of phosphatidylinositol 3′-kinase/Akt signaling pathway. *J. Biol. Chem.* 275(46), 35942–35952. https://doi.org/10.1074/jbc.M005832200.

Chen, G., Gulbranson, D. R., Hou, Z., Bolin, J. M., Ruotti, V., Probasco, M. D., Smuga-Otto, K., Howden, S. E., Diol, N. R., Propson, N. E., Wagner, R., Lee, G. O., Antosiewicz-Bourget, J., Teng, J. M. C. and Thomson, J. A. 2011. Chemically defined conditions for human iPSC derivation and culture. *Nat. Methods* 8(5), 424–429. https://doi.org/10.1038/nmeth.1593.

Chen, G., Gulbranson, D. R., Yu, P., Hou, Z. and Thomson, J. A. 2012. Thermal stability of fibroblast growth factor protein is a determinant factor in regulating self-renewal, differentiation, and reprogramming in human pluripotent stem cells. *Stem Cells* 30(4), 623–630. https://doi.org/10.1002/stem.1021.

Chriki, S. and Hocquette, J.-F. 2020. The myth of cultured meat: a review. *Front. Nutr.* 7 (7), 1-9.

Clegg, C. H., Linkhart, T. A., Olwin, B. B. and Hauschka, S. D. 1987. Growth factor control of skeletal muscle differentiation: commitment to terminal differentiation occurs in

G1 phase and is repressed by fibroblast growth factor. *J. Cell Biol.* 105(2), 949–956. https://doi.org/10.1083/jcb.105.2.949.

Cornell, R. B., Nissley, S. M. and Horwitz, A. F. 1980. Cholesterol availability modulates myoblast fusion. *J. Cell Biol.* 86(3), 820–824. https://doi.org/10.1083/jcb.86.3.820.

Cosenza, Z., Astudillo, R., Frazier, P., I., Baar, K. and Block, D. E. 2022. Multi-information source Bayesian optimization of culture media for cellular agriculture. *Biotechnol. Bioeng.*, 1–12. https://doi.org/10.1002/bit.28132.

Cosenza, Z., Block, D. E. and Baar, K. 2021. Optimization of muscle cell culture media using nonlinear design of experiments. *Biotechnol. J.* 16(11), 2100228. https://doi.org/10.1002/biot.202100228.

Darmon, M., Serrero, G., Rizzino, A. and Sato, G. 1981. Isolation of myoblastic, fibro-adipogenic, and fibroblastic clonal cell lines from a common precursor and study of their requirements for growth and differentiation. *Exp. Cell Res.* 132(2), 313–327. https://doi.org/10.1016/0014-4827(81)90107-5.

Das, M., Rumsey, J. W., Bhargava, N., Gregory, C., Reidel, L., Kang, J. F. and Hickman, J. J. 2009. Developing a novel serum-free cell culture model of skeletal muscle differentiation by systematically studying the role of different growth factors in myotube formation. *In Vitro Cell. Dev. Biol. Anim.* 45(7), 378–387. https://doi.org/10.1007/s11626-009-9192-7.

Datar, I. and Betti, M. 2010. Possibilities for an in vitro meat production system. *Innov. Food Sci. Emerg. Technol.* 11, 13–22. https://doi.org/10.1016/j.ifset.2009.10.007.

de La Haba, G. D. L., Cooper, G. W. and Elting, V. 1966. Hormonal requirements for myogenesis of striated muscle in vitro: insulin and somatotropin. *Proc. Natl. Acad. Sci. USA* 56(6), 1719–1723. https://doi.org/10.1073/pnas.56.6.1719.

de Larco, J. E. and Todaro, G. J. 1978. Growth factors from murine sarcoma virus-transformed cells. *Proc. Natl. Acad. Sci. U. S. A.* 75(8), 4001–4005. https://doi.org/10.1073/pnas.75.8.4001.

De Meyts, P. 2002. Insulin and insulin-like growth factors: the paradox of signaling specificity. *Growth Horm. IGF Res.* 12(2), 81–83. https://doi.org/10.1054/ghir.2002.0264.

DeBerardinis, R. J. and Cheng, T. 2010. Q's next: the diverse functions of glutamine in metabolism, cell biology and cancer. *Oncogene* 29(3), 313–324. https://doi.org/10.1038/onc.2009.358.

DeBerardinis, R. J., Mancuso, A., Daikhin, E., Nissim, I., Yudkoff, M., Wehrli, S. and Thompson, C. B. 2007. Beyond aerobic glycolysis: transformed cells can engage in glutamine metabolism that exceeds the requirement for protein and nucleotide synthesis. *Proc. Natl. Acad. Sci. U. S. A.* 104(49), 19345–19350. https://doi.org/10.1073/pnas.0709747104.

Devireddy, L. R., Myers, M., Screven, R., Liu, Z. and Boxer, L. 2019. A serum-free medium formulation efficiently supports isolation and propagation of canine adipose-derived mesenchymal stem/stromal cells. *PLoS ONE* 14(2), e0210250. https://doi.org/10.1371/journal.pone.0210250.

Ding, S., Swennen, G. N. M., Messmer, T., Gagliardi, M., Molin, D. G. M., Li, C., Zhou, G. and Post, M. J. 2018. Maintaining bovine satellite cells stemness through p38 pathway. *Sci. Rep.* 8(1), 10808. https://doi.org/10.1038/s41598-018-28746-7.

Dodson, M. V., Allen, R. E. and Hossner, K. L. 1985. Ovine somatomedin, multiplication-stimulating activity, and insulin promote skeletal muscle satellite cell proliferation in vitro. *Endocrinology* 117(6), 2357–2363. https://doi.org/10.1210/endo-117-6-2357.

Dodson, M. V. and Mathison, B. A. 1988. Comparison of ovine and rat muscle-derived satellite cells: response to insulin. *Tissue Cell* 20(6), 909–918. https://doi.org/10.1016/0040-8166(88)90032-8.

Dollenmeier, P., Turner, D. C. and Eppenberger, H. M. 1981. Proliferation and differentiation of chick skeletal muscle cells cultured in a chemically defined medium. *Exp. Cell Res.* 135(1), 47–61. https://doi.org/10.1016/0014-4827(81)90298-6.

Doumit, M. E., Cook, D. R. and Merkel, R. A. 1993. Fibroblast growth factor, epidermal growth factor, insulin-like growth factors, and platelet-derived growth factor-BB stimulate proliferation of clonally derived porcine myogenic satellite cells. *J. Cell. Physiol.* 157(2), 326–332. https://doi.org/10.1002/jcp.1041570216.

Duan, C., Ren, H. and Gao, S. 2010. Insulin-like growth factors (IGFs), IGF receptors, and IGF-binding proteins: roles in skeletal muscle growth and differentiation. *Gen. Comp. Endocrinol.*, Symposium on Fish Growth and Reproduction 167(3), 344–351. https://doi.org/10.1016/j.ygcen.2010.04.009.

Duran, B. O. S., Góes, G. A., Zanella, B. T. T., Freire, P. P., Valente, J. S., Salomão, R. A. S., Fernandes, A., Mareco, E. A., Carvalho, R. F. and Dal-Pai-Silva, M. 2019. Ascorbic acid stimulates the in vitro myoblast proliferation and migration of pacu (Piaractus mesopotamicus). *Sci. Rep.* 9(1), 2229. https://doi.org/10.1038/s41598-019-38536-4.

Dvorak, P., Bednar, D., Vanacek, P., Balek, L., Eiselleova, L., Stepankova, V., Sebestova, E., Kunova Bosakova, M., Konecna, Z., Mazurenko, S., Kunka, A., Vanova, T., Zoufalova, K., Chaloupkova, R., Brezovsky, J., Krejci, P., Prokop, Z., Dvorak, P. and Damborsky, J. 2018. Computer-assisted engineering of hyperstable fibroblast growth factor 2. *Biotechnol. Bioeng.* 115(4), 850–862. https://doi.org/10.1002/bit.26531.

Eagle, H. 1959. Amino acid metabolism in mammalian cell cultures. *Science* 130(3373), 432–437. https://doi.org/10.1126/science.130.3373.432.

Eagle, H. 1955. The specific amino acid requirements of a mammalian cell (strain L) in tissue culture. *J. Biol. Chem.* 214(2), 839–852.

Eagle, H. 1955. Nutrition needs of mammalian cells in tissue culture. *Science* 122(3168), 501–514. https://doi.org/10.1126/science.122.3168.501.

Eigler, T., Zarfati, G., Amzallag, E., Sinha, S., Segev, N., Zabary, Y., Zaritsky, A., Shakked, A., Umansky, K. B., Schejter, E. D., Millay, D. P., Tzahor, E. and Avinoam, O. 2021. ERK1/2 inhibition promotes robust myotube growth via CaMKII activation resulting in myoblast-to-myotube fusion. *Dev. Cell* 56(24), 3349-3363.e6. https://doi.org/10.1016/j.devcel.2021.11.022.

Engert, J. C., Berglund, E. B. and Rosenthal, N. 1996. Proliferation precedes differentiation in IGF-I-stimulated myogenesis. *J. Cell Biol.* 135(2), 431–440. https://doi.org/10.1083/jcb.135.2.431.

Eto, N., Yamada, K., Shito, T., Shirahata, S. and Murakami, H. 1991. Development of a protein-free medium with ferric citrate substituting transferrin for the cultivation of mouse-mouse hybridomas. *Agric. Biol. Chem.* 55(3), 863–865. https://doi.org/10.1080/00021369.1991.10870670.

Evans, V. J., Bryant, J. C., McQuilkin, W. T., Fioramonti, M. C., Sanford, K. K., Westfall, B. B. and Earle, W. R. 1956. Studies of nutrient media for tissue cells in vitro II. An improved protein-free chemically defined medium for long-term cultivation of strain L-929 cells. *Cancer Res.* 16(1), 87–94.

Farges-Haddani, B., Tessier, B., Chenu, S., Chevalot, I., Harscoat, C., Marc, I., Goergen, J. L. and Marc, A. 2006. Peptide fractions of rapeseed hydrolysates as an alternative to animal proteins in CHO cell culture media. *Process Biochem.* 41(11), 2297-2304. https://doi.org/10.1016/j.procbio.2006.06.002.

Fernyhough, M. E., Okine, E., Hausman, G., Vierck, J. L. and Dodson, M. V. 2007. PPARγ and GLUT-4 expression as developmental regulators/markers for preadipocyte differentiation into an adipocyte. *Domest. Anim. Endocrinol.* 33(4), 367–378. https://doi.org/10.1016/j.domaniend.2007.05.001.

Finkler, A. E. and Hall, C. A. 1967. Nature of the relationship between vitamin B12 binding and cell uptake. *Arch. Biochem. Biophys.* 120(1), 79–85. https://doi.org/10.1016/0003-9861(67)90600-5.

Finnis, C. J., Payne, T., Hay, J., Dodsworth, N., Wilkinson, D., Morton, P., Saxton, M. J., Tooth, D. J., Evans, R. W., Goldenberg, H., Scheiber-Mojdehkar, B., Ternes, N. and Sleep, D. 2010. High-level production of animal-free recombinant transferrin from Saccharomyces cerevisiae. *Microb. Cell Factories* 9, 87. https://doi.org/10.1186/1475-2859-9-87.

Fischer, A. 1948. Amino-acid metabolism of tissue cells in vitro. *Nature* 161(4104), 1008–1008. https://doi.org/10.1038/1611008a0.

Fish, K. D., Rubio, N. R., Stout, A. J., Yuen, J. S. K. and Kaplan, D. L. 2020. Prospects and challenges for cell-cultured fat as a novel food ingredient. *Trends Food Sci. Technol.* 98, 53–67. https://doi.org/10.1016/j.tifs.2020.02.005.

Fisher, H. W., Puck, T. T. and Sato, G. 1959. Molecular growth requirements of single mammalian cells. III. Quantitative colonial growth of single S3 cells in a medium containing synthetic small molecular constituents and two purified protein fractions. *J. Exp. Med.* 109(6), 649–660.

Fisher, M., Gokhman, I., Pick, U. and Zamir, A. 1997. A structurally novel transferrin-like protein accumulates in the plasma membrane of the unicellular green alga Dunaliella salina Grown in high salinities. *J. Biol. Chem.* 272(3), 1565–1570. https://doi.org/10.1074/jbc.272.3.1565.

Fisher, M., Zamir, A. and Pick, U. 1998. Iron uptake by the halotolerant alga Dunaliellals mediated by a plasma membrane transferrin. *J. Biol. Chem.* 273(28), 17553–17558. https://doi.org/10.1074/jbc.273.28.17553.

Florin, A., Lambert, C., Sanchez, C., Zappia, J., Durieux, N., Tieppo, A. M., Mobasheri, A. and Henrotin, Y. 2020. The secretome of skeletal muscle cells: A systematic review. *Osteoarthr. Cartil. Open* 2(1), 100019. https://doi.org/10.1016/j.ocarto.2019.100019.

Florini, J. R., Ewton, D. Z. and Coolican, S. A. 1996. Growth hormone and the insulin-like growth factor system in myogenesis. *Endocr. Rev.* 17(5), 481–517. https://doi.org/10.1210/edrv-17-5-481.

Florini, J. R., Ewton, D. Z., Falen, S. L. and Van Wyk, J. J. 1986. Biphasic concentration dependency of stimulation of myoblast differentiation by somatomedins. *Am. J. Physiol.* 250(5 Pt 1), C771–C778. https://doi.org/10.1152/ajpcell.1986.250.5.C771.

Florini, J. R., Nicholson, M. L. and Dulak, N. C. 1977. Effects of peptide anabolic hormones on growth of myoblasts in culture. *Endocrinology* 101(1), 32–41. https://doi.org/10.1210/endo-101-1-32.

Florini, J. R., Roberts, A. B., Ewton, D. Z., Falen, S. L., Flanders, K. C. and Sporn, M. B. 1986. Transforming growth factor-beta. A very potent inhibitor of myoblast differentiation, identical to the differentiation inhibitor secreted by Buffalo rat liver cells. *J. Biol. Chem.* 261(35), 16509–16513. https://doi.org/10.1016/S0021-9258(18)66595-0.

Florini, J. R. and Roberts, S. B. 1979. A serum-free medium for the growth of muscle cells in culture. *In Vitro* 15(12), 983–992. https://doi.org/10.1007/BF02619157.

Fox, G. M., Schiffer, S. G., Rohde, M. F., Tsai, L. B., Banks, A. R. and Arakawa, T. 1988. Production, biological activity, and structure of recombinant basic fibroblast growth factor and an analog with cysteine replaced by serine. *J. Biol. Chem.* 263(34), 18452–18458.

Fraeye, I., Kratka, M., Vandenburgh, H. and Thorrez, L. 2020. Sensorial and nutritional aspects of cultured meat in comparison to traditional meat: much to be inferred. *Front. Nutr.* 7, 35.

Francis, G. L. 2010. Albumin and mammalian cell culture: implications for biotechnology applications. *Cytotechnology* 62(1), 1–16. https://doi.org/10.1007/s10616-010-9263-3.

Francis, G. L., Ross, M., Ballard, F. J., Milner, S. J., Senn, C., McNeil, K. A., Wallace, J. C., King, R. and Wells, J. R. E. 1992. Novel recombinant fusion protein analogues of insulin-like growth factor (IGF)-I indicate the relative importance of IGF-binding protein and receptor binding for enhanced biological potency. *J. Mol. Endocrinol.* 8(3), 213–223. https://doi.org/10.1677/jme.0.0080213.

Franěk, F., Hohenwarter, O. and Katinger, H. 2000. Plant protein hydrolysates: preparation of defined peptide fractions promoting growth and production in animal cells cultures. *Biotechnol. Prog.* 16(5), 688–692. https://doi.org/10.1021/bp0001011.

Fu, X., Zhu, M. J., Dodson, M. V. and Du, M. 2015. AMP-activated protein kinase stimulates Warburg-like glycolysis and activation of satellite cells during muscle regeneration. *J. Biol. Chem.* 290(44), 26445–26456. https://doi.org/10.1074/jbc.M115.665232.

Gallego-Murillo, J. S., Yağcı, N., Pinho, E. M., Abeijón-Valle, A., Wahl, A., van den Akker, E. and von Lindern, M. 2021. *Iron-Loaded Deferiprone Can Support Full Hemoglobinization of Cultured Red Blood Cells in the Absence of Transferrin*. https://doi.org/10.1101/2021.08.02.454758.

García-Rojas, P., Antaramian, A., González-Dávalos, L., Villarroya, F., Shimada, A., Varela-Echavarría, A. and Mora, O. 2010. Induction of peroxisomal proliferator-activated receptor gamma and peroxisomal proliferator-activated receptor gamma coactivator 1 by unsaturated fatty acids, retinoic acid, and carotenoids in preadipocytes obtained from bovine white adipose tissue1,2. *J. Anim. Sci.* 88(5), 1801–1808. https://doi.org/10.2527/jas.2009-2579.

Ge, X., Yu, J. and Jiang, H. 2012. Growth hormone stimulates protein synthesis in bovine skeletal muscle cells without altering insulin-like growth factor-I mRNA expression1. *J. Anim. Sci.* 90(4), 1126–1133. https://doi.org/10.2527/jas.2011-4358.

Genovese, N. J., Schulze, E. N. and DESMET, D. N. 2021. *Compositions and Methods for Increasing the Efficiency of Cell Cultures Used for Food Production*. US20210340570A1.

George, F., Kerschen, D., Nuffel, A. V., Rees, J. F., Donnay, I., George, F., Kerschen, D., Nuffel, A. V., Rees, J. F. and Donnay, I. 2009. Plant protein hydrolysates (plant peptones) as substitutes for animal proteins in embryo culture medium. *Reprod. Fertil. Dev.* 21(4), 587–598. https://doi.org/10.1071/RD08147.

Giorgino, F. and Smith, R. J. 1995. Dexamethasone enhances insulin-like growth factor-I effects on skeletal muscle cell proliferation. Role of specific intracellular signaling pathways. *J. Clin. Invest.* 96(3), 1473–1483. https://doi.org/10.1172/JCI118184.

Good Meat, Inc., Polsinelli PC, AR Toxicology, Burdock Group Consultants. 2022. Dossier in support of the safety of good meat cultured chicken as a human food ingredient. Available at: https://www.fda.gov/media/166346/download.

Gorfien, S. F., Dzimian, J. L., Tilkins, M. L., Godwin, G. P. and Fike, R. 1998. Recombinant protein production by CHO cells cultured in a chemically defined medium. In: Nagai, K. and Wachi, M. (Eds). *Animal Cell Technology: Basic & Applied Aspects*. Proceedings of the Ninth Annual Meeting of the Japanese Association for Animal Cell Technology, Yokohama, Japan, September 1-4, 1996, Animal Cell Technology: Basic & Applied Aspects. Springer Netherlands, Dordrecht, pp. 247-252. https://doi.org/10.1007/978-94-011-5161-0_42.

Gospodarowicz, D. 1975. Purification of a fibroblast growth factor from bovine pituitary. *J. Biol. Chem.* 250(7), 2515–2520. https://doi.org/10.1016/S0021-9258(19)41631-1.

Gospodarowicz, D. 1974. Localisation of a fibroblast growth factor and its effect alone and with hydrocortisone on 3T3 cell growth. *Nature* 249(453), 123-127. https://doi.org/10.1038/249123a0.

Gospodarowicz, D., Bialecki, H. and Greenburg, G. 1978. Purification of the fibroblast growth factor activity from bovine brain. *J. Biol. Chem.* 253(10), 3736-3743. https://doi.org/10.1016/S0021-9258(17)34863-9.

Gospodarowicz, D. and Cheng, J. 1987. Growth of myoblasts in lipoprotein-supplemented, serum-free medium: regulation of proliferation by acidic and basic fibroblast growth factor. *In Vitro Cell. Dev. Biol.* 23(7), 507-514. https://doi.org/10.1007/BF02628421.

Gospodarowicz, D., Weseman, J., Moran, J. S. and Lindstrom, J. 1976. Effect of fibroblast growth factor on the division and fusion of bovine myoblasts. *J. Cell Biol.* 70(2 pt 1), 395-405. https://doi.org/10.1083/jcb.70.2.395.

Green, H. and Kehinde, O. 1975. An established preadipose cell line and its differentiation in culture II. Factors affecting the adipose conversion. *Cell* 5, 19-27. https://doi.org/10.1016/0092-8674(75)90087-2.

Green, H. and Kehinde, O. 1974. Sublines of mouse 3T3 cells that accumulate lipid. *Cell* 1(3), 113-116.

Greene, E. A. and Allen, R. E. 1991. Growth factor regulation of bovine satellite cell growth in vitro1. *J. Anim. Sci.* 69(1), 146-152. https://doi.org/10.2527/1991.691146x.

Gregoire, F. M., Smas, C. M. and Sul, H. S. 1998. Understanding adipocyte differentiation. *Physiol. Rev.* 78(3), 783-809. https://doi.org/10.1152/physrev.1998.78.3.783.

Guerriero, V. and Florini, J. R. 1980. Dexamethasone effects on myoblast proliferation and differentiation. *Endocrinology* 106(4), 1198-1202. https://doi.org/10.1210/endo-106-4-1198.

Guerriero, V. and Florini, J. R. 1978. Stimulation by glucocorticoids of myoblast growth at low cell densities. *Cell Biol. Int. Rep.* 2(5), 441-446. https://doi.org/10.1016/0309-1651(78)90095-4.

Guilbert, L. J. and Iscove, N. N. 1976. Partial replacement of serum by selenite, transferrin, albumin and lecithin in haemopoitec cell cultures. *Nature* 263(5578), 594-595. https://doi.org/10.1038/263594a0.

Guo, Y., Ding, S. J., Ding, X., Liu, Z., Wang, J. L., Chen, Y., Liu, P. P., Li, H. X., Zhou, G. H. and Tang, C. B. 2022. Effects of selected flavonoids oncellproliferation and differentiation of porcine muscle stem cells for cultured meat production. *Food Res. Int.* 160, 111459. https://doi.org/10.1016/j.foodres.2022.111459.

Hall, C. A. 1975. Transcobalamins I and II as natural transport proteins of vitamin B12. *J. Clin. Invest.* 56(5), 1125-1131.

Ham, R. G. 1965. Clonal growth of mammalian cells in a chemically defined, synthetic medium. *Proc. Natl. Acad. Sci. U. S. A.* 53(2), 288-293.

Ham, R. G. 1962. Clonal growth of diploid Chinese hamster cells in a synthetic medium supplemented with purified protein fractions. *Exp. Cell Res.* 28, 489–500. https://doi.org/10.1016/0014-4827(62)90253-7.

Ham, R. G., St Clair, J. A., Webster, C. and Blau, H. M. 1988. Improved media for normal human muscle satellite cells: serum-free clonal growth and enhanced growth with low serum. *In Vitro Cell. Dev. Biol.* 24(8), 833–844. https://doi.org/10.1007/BF02623656.

Hamilton, W. G. and Ham, R. G. 1977. Clonal growth of Chinese hamster cell lines in protein-free media. *In Vitro* 13(9), 537–547. https://doi.org/10.1007/BF02627849.

Han, B., Tong, J., Zhu, M. J., Ma, C. and Du, M. 2008. Insulin-like growth factor-1 (IGF-1) and leucine activate pig myogenic satellite cells through mammalian target of rapamycin (mTOR) pathway. *Mol. Reprod. Dev.* 75(5), 810–817. https://doi.org/10.1002/mrd.20832.

Hanyu, Y. and Kawashima, I. 2019. *Growth Induction System, Growth Induction Control Apparatus, Growth Induction Control Method, and Growth Induction Control Program*. US20190160461A1.

Haraguchi, Y. and Shimizu, T. 2021a. Three-dimensional tissue fabrication system by co-culture of microalgae and animal cells for production of thicker and healthy cultured food. *Biotechnol. Lett.* 43(6), 1117–1129. https://doi.org/10.1007/s10529-021-03106-0.

Haraguchi, Y. and Shimizu, T. 2021b. Microalgal culture in animal cell waste medium for sustainable 'cultured food' production. *Arch. Microbiol.* 203(9), 5525–5532. https://doi.org/10.1007/s00203-021-02509-x.

Harrison, R. G., Greenman, M. J., Mall, F. P. and Jackson, C. M. 1907. Observations of the living developing nerve fiber. *Anat. Rec.* 1(5), 116–128. https://doi.org/10.1002/ar.1090010503.

Hausman, G. J. 2000. The influence of dexamethasone and insulin on expression of CCAAT/enhancer binding protein isoforms during preadipocyte differentiation in porcine stromal-vascular cell cultures: evidence for very early expression of C/EBPα1. *J. Anim. Sci.* 78(5), 1227–1235. https://doi.org/10.2527/2000.7851227x.

Hausman, G. J., Campion, D. R. and Martin, R. J. 1980. Search for the adipocyte precursor cell and factors that promote its differentiation. *J. Lipid Res.* 21(6), 657–670. https://doi.org/10.1016/S0022-2275(20)34794-5.

Hayashi, I., Nixon, T., Morikawa, M. and Green, H. 1981. Adipogenic and anti-adipogenic factors in the pituitary and other organs. *Proc. Natl. Acad. Sci. USA* 78(6), 3969–3972. https://doi.org/10.1073/pnas.78.6.3969.

Hayman, E. G., Pierschbacher, M. D., Suzuki, S. and Ruoslahti, E. 1985. Vitronectin--a major cell attachment-promoting protein in fetal bovine serum. *Exp. Cell Res.* 160(2), 245–258. https://doi.org/10.1016/0014-4827(85)90173-9.

He, J., Chen, D. L., Samocha-Bonet, D., Gillinder, K. R., Barclay, J. L., Magor, G. W., Perkins, A. C., Greenfield, J. R., Yang, G. and Whitehead, J. P. 2016. Fibroblast growth factor-1 (FGF-1) promotes adipogenesis by downregulation of carboxypeptidase A4 (CPA4) - a negative regulator of adipogenesis implicated in the modulation of local and systemic insulin sensitivity. *Growth Factors* 34(5–6), 210–216. https://doi.org/10.1080/08977194.2017.1285764.

Healy, G. M., Fisher, D. C. and Parker, R. C. 1955. Nutrition of animal cells in tissue culture. X. Synthetic Medium No. 858. *Proc. Soc. Exp. Biol. Med. Soc. Exp. Biol. Med.* N. Y. N 89, 71–77. https://doi.org/10.3181/00379727-89-21719.

Hill, D. J. and Milner, R. D. G. 1985. Insulin as a growth factor. *Pediatr. Res.* 19(9), 879–886. https://doi.org/10.1203/00006450-198509000-00001.

Hiragun, A., Sato, M. and Mitsui, H. 1980. Establishment of a clonal cell line that differentiates into adipose cells in vitro. *In Vitro* 16(8), 685–693. https://doi.org/10.1007/BF02619198.

Ho, T. C., Chiang, Y. P., Chuang, C. K., Chen, S. L., Hsieh, J. W., Lan, Y. W. and Tsao, Y. P. 2015. PEDF-derived peptide promotes skeletal muscle regeneration through its mitogenic effect on muscle progenitor cells. *Am. J. Physiol. Cell Physiol.* 309(3), C159–C168. https://doi.org/10.1152/ajpcell.00344.2014.

Holmes, D., Humbird, D., Dutkiewicz, J., Tejeda-Saldana, Y., Duffy, B. and Datar, I. 2022. Cultured meat needs a race to mission not a race to market. *Nat. Food* 3(10), 785. https://doi.org/10.1038/s43016-022-00586-9.

Hopkins, D. L. 2017. Chapter 12. The eating quality of meat. In: Toldra, F. (Ed.) *Lawrie's Meat Science, Woodhead Publishing Series in Food Science, Technology and Nutrition* (8th edn.) (vol. II)–Tenderness. Woodhead Publishing, pp. 357–381. https://doi.org/10.1016/B978-0-08-100694-8.00012-1.

Hua, Y., Yoshimochi, K., Li, J., Takekita, K., Shimotsuma, M., Li, L., Qu, X., Zhang, J., Sawa, Y., Liu, L. and Miyagawa, S. 2022. Development and evaluation of a novel xeno-free culture medium for human-induced pluripotent stem cells. *Stem Cell Res. Ther.* 13(1), 223. https://doi.org/10.1186/s13287-022-02879-z.

Huan, Y., Kong, Q., Mou, H. and Yi, H. 2020. Antimicrobial peptides: classification, design, application and research progress in multiple fields. *Front. Microbiol.* 11, 582779.

Huang, N., Bethell, D., Card, C., Cornish, J., Marchbank, T., Wyatt, D., Mabery, K. and Playford, R. 2008. Bioactive recombinant human lactoferrin, derived from rice, stimulates mammalian cell growth. *In Vitro Cell. Dev. Biol. Anim.* 44(10), 464–471. https://doi.org/10.1007/s11626-008-9136-7.

Hubalek, S., Post, M. J. and Moutsatsou, P. 2022. Towards resource- and cost- efficient cultured meat. *Curr. Opin. Food Sci.* 47, 100885. https://doi.org/10.1016/j.cofs.2022.100885.

Hughes, D. C., Stewart, C. E., Sculthorpe, N., Dugdale, H. F., Yousefian, F., Lewis, M. P. and Sharples, A. P. 2016. Testosterone enables growth and hypertrophy in fusion impaired myoblasts that display myotube atrophy: deciphering the role of androgen and IGF-I receptors. *Biogerontology* 17(3), 619–639. https://doi.org/10.1007/s10522-015-9621-9.

Humbird, D. 2021. Scale-up economics for cultured meat. *Biotechnol. Bioeng.* 118(8), 3239–3250. https://doi.org/10.1002/bit.27848.

Iberite, F., Gruppioni, E. and Ricotti, L. 2022. Skeletal muscle differentiation of human iPSCs meets bioengineering strategies: perspectives and challenges. *NPJ Regen. Med.* 7(1), 23. https://doi.org/10.1038/s41536-022-00216-9.

Ishima, Y., Akaike, T., Kragh-Hansen, U., Hiroyama, S., Sawa, T., Suenaga, A., Maruyama, T., Kai, T. and Otagiri, M. 2008. S-Nitrosylated human serum albumin-mediated cytoprotective activity is enhanced by fatty acid binding. *J. Biol. Chem.* 283(50), 34966–34975. https://doi.org/10.1074/jbc.M807009200.

Itoh, N. and Ornitz, D. M. 2004. Evolution of the Fgf and Fgfr gene families. *Trends Genet.* 20(11), 563–569. https://doi.org/10.1016/j.tig.2004.08.007.

Jenssen, H. and Hancock, R. E. W. 2009. Antimicrobial properties of lactoferrin. *Biochimie* 91(1), 19–29. https://doi.org/10.1016/j.biochi.2008.05.015.

Jeong, S. H., Kang, D., Lim, M. W., Kang, C. S. and Sung, H. J. 2010. Risk assessment of growth hormones and antimicrobial residues in meat. *Toxicol. Res.* 26(4), 301–313. https://doi.org/10.5487/TR.2010.26.4.301.

Jiang, Y., Li, Y., Jiang, Y., Li, J. and Pan, C. 2012. Determination of multiresidues in rapeseed, rapeseed oil, and rapeseed meal by acetonitrile extraction, low-temperature cleanup, and detection by liquid chromatography with tandem mass spectrometry. *J. Agric. Food Chem.* 60(20), 5089–5098. https://doi.org/10.1021/jf3004064.

Jin, C. L., Zhang, Z. M., Song, Z. W., Gao, C. Q., Yan, H. C. and Wang, X. Q. 2020. mTORC1-mediated satellite cell differentiation is required for lysine-induced skeletal muscle growth. *J. Agric. Food Chem.* 68(17), 4884–4892. https://doi.org/10.1021/acs.jafc.0c01275.

Jirak, P., Stechemesser, L., Moré, E., Franzen, M., Topf, A., Mirna, M., Paar, V., Pistulli, R., Kretzschmar, D., Wernly, B., Hoppe, U. C., Lichtenauer, M. and Salmhofer, H. 2019. Chapter Three. Clinical implications of fetuin-A. In: Makowski, G. S., et al. (Ed.) *Advances in Clinical Chemistry*. Elsevier, pp. 79–130. https://doi.org/10.1016/bs.acc.2018.12.003.

Jones, N. C., Fedorov, Y. V., Rosenthal, R. S. and Olwin, B. B. 2001. ERK1/2 is required for myoblast proliferation but is dispensable for muscle gene expression and cell fusion. *J. Cell. Physiol.* 186(1), 104–115. https://doi.org/10.1002/1097-4652(200101)186:1<104::AID-JCP1015>3.0.CO;2-0.

Joo, S. T., Choi, J. S., Hur, S. J., Kim, G. D., Kim, C. J., Lee, E. Y., Bakhsh, A. and Hwang, Y. H. 2022. A comparative study on the taste characteristics of satellite cell cultured meat derived from chicken and cattle muscles. *Food Sci. Anim. Resour.* 42(1), 175–185. https://doi.org/10.5851/kosfa.2021.e72.

Jumper, J., Evans, R., Pritzel, A., Green, T., Figurnov, M., Ronneberger, O., Tunyasuvunakool, K., Bates, R., Žídek, A., Potapenko, A., Bridgland, A., Meyer, C., Kohl, S. A. A., Ballard, A. J., Cowie, A., Romera-Paredes, B., Nikolov, S., Jain, R., Adler, J., Back, T., Petersen, S., Reiman, D., Clancy, E., Zielinski, M., Steinegger, M., Pacholska, M., Berghammer, T., Bodenstein, S., Silver, D., Vinyals, O., Senior, A. W., Kavukcuoglu, K., Kohli, P. and Hassabis, D. 2021. Highly accurate protein structure prediction with AlphaFold. *Nature* 596(7873), 583–589. https://doi.org/10.1038/s41586-021-03819-2.

Kameda, T., Kikuchi, H., Kitagawa, F., Kumagai, S., Saito, Y., Kondo, M., Jimbo, Y. and Yoshioka, T. 2021. Lactate adsorption by layered double hydroxides in aqueous solution and cell culture medium. *Colloids Surf. Physicochem. Eng. ASp* 612, 125975. https://doi.org/10.1016/j.colsurfa.2020.125975.

Kawada, T., Aoki, N., Kawai, R. and Sugimoto, E. 1990. Proliferation of 3T3-L1 preadipocytes in a completely defined serum-free medium. *Cell Biol. Int. Rep.* 14, 567–574. https://doi.org/10.1016/0309-1651(90)91184-6.

Keenan, J., Pearson, D., O'Driscoll, L., Gammell, P. and Clynes, M. 2006. Evaluation of recombinant human transferrin (DeltaFerrinTM) as an iron chelator in serum-free media for mammalian cell culture. *Cytotechnology* 51(1), 29–37. https://doi.org/10.1007/s10616-006-9011-x.

Kelvin, D. J., Simard, G. and Connolly, J. A. 1989. FGF and EGF act synergistically to induce proliferation in BC3H1 myoblasts. *J. Cell. Physiol.* 138(2), 267–272. https://doi.org/10.1002/jcp.1041380207.

Kim, J. J. and Accili, D. 2002. Signalling through IGF-I and insulin receptors: where is the specificity? *Growth Horm. IGF Res.* 12(2), 84–90. https://doi.org/10.1054/ghir.2002.0265.

Kim, W. K., Ryu, Y. H., Seo, D. S., Lee, C. Y. and Ko, Y. 2006. Effects of oral administration of insulin-like growth factor-I on circulating concentration of insulin-like growth factor-I and growth of internal organs in weanling mice. *Biol. Neonate.* 89(3), 199–204. https://doi.org/10.1159/000089796.

Kolkmann, A. M., Essen, A., Post, M. J. and Moutsatsou, P. 2022. Development of a chemically defined medium for in vitro expansion of primary bovine satellite cells. *Front. Bioeng. Biotechnol.* 10, 895289. https://doi.org/10.3389/fbioe.2022.895289.

Kolkmann, A. M., Post, M. J., Rutjens, M. A. M., van Essen, A. L. M. and Moutsatsou, P. 2020. Serum-free media for the growth of primary bovine myoblasts. *Cytotechnology* 72(1), 111–120. https://doi.org/10.1007/s10616-019-00361-y.

Kolkmann, A. M., Van Essen, A., Post, M. J. and Moutsatsou, P. 2022. Development of a chemically defined medium for in vitro expansion of primary bovine satellite cells. *Front. Bioeng. Biotechnol.* 10, 895289. https://doi.org/10.3389/fbioe.2022.895289.

Kuo, H. H., Gao, X., DeKeyser, J. M., Fetterman, K. A., Pinheiro, E. A., Weddle, C. J., Fonoudi, H., Orman, M. V., Romero-Tejeda, M., Jouni, M., Blancard, M., Magdy, T., Epting, C. L., George, A. L. and Burridge, P. W. 2020. Negligible-cost and weekend-free chemically defined human iPSC culture. *Stem Cell Rep.* 14(2), 256–270. https://doi.org/10.1016/j.stemcr.2019.12.007.

Kuo, T., Lew, M. J., Mayba, O., Harris, C. A., Speed, T. P. and Wang, J.-C. 2012. Genome-wide analysis of glucocorticoid receptor-binding sites in myotubes identifies gene networks modulating insulin signaling. *Proc. Natl. Acad. Sci. USA* 109(28), 11160–11165. https://doi.org/10.1073/pnas.1111334109.

Kurgan, N., Whitley, K. C., Maddalena, L. A., Moradi, F., Stoikos, J., Hamstra, S. I., Rubie, E. A., Kumar, M., Roy, B. D., Woodgett, J. R., Stuart, J. A. and Fajardo, V. A. 2019. A low-therapeutic dose of lithium inhibits GSK3 and enhances myoblast fusion in C2C12 cells. *Cells* 8(11), 1340. https://doi.org/10.3390/cells8111340.

Kuri-Harcuch, W. and Green, H. 1978. Adipose conversion of 3T3 cells depends on a serum factor. *Proc. Natl. Acad. Sci. U. S. A.* 75(12), 6107–6109. https://doi.org/10.1073/pnas.75.12.6107.

Lanotte, M., Scott, D., Dexter, T. M. and Allen, T. D. 1982. Clonal preadipocyte cell lines with different phenotypes derived from murine marrow stroma: factors influencing growth and Adipogenesis in vitro. *J. Cell. Physiol.* 111(2), 177–186. https://doi.org/10.1002/jcp.1041110209.

Lawrence, M. C., McKern, N. M. and Ward, C. W. 2007. Insulin receptor structure and its implications for the IGF-1 receptor. *Curr. Opin. Struct. Biol.* 17(6), 699–705. https://doi.org/10.1016/j.sbi.2007.07.007.

Lee, M. S., Youn, C., Kim, J. H., Park, B. J., Ahn, J., Hong, S., Kim, Y. D., Shin, Y. K. and Park, S. G. 2017. Enhanced cell growth of adipocyte-derived mesenchymal stem cells using chemically-defined serum-free media. *Int. J. Mol. Sci.* 18(8), 1779. https://doi.org/10.3390/ijms18081779.

Lewis, M. R. and Lewis, W. H. 1911. The cultivation of tissues in salt solutions. *J. Am. Med. Assoc.* LVI, 1795–1796. https://doi.org/10.1001/jama.1911.02560240025011.

Lewis, W. H. and Lewis, M. R. 1912. The cultivation of chick tissues in media of known chemical constitution. *Anat. Rec.* 6(5), 207–211. https://doi.org/10.1002/ar.1090060503.

Li, G., Yao, W. and Jiang, H. 2014. Short-chain fatty acids enhance adipocyte differentiation in the stromal vascular fraction of porcine adipose tissue. *J. Nutr.* 144(12), 1887–1895. https://doi.org/10.3945/jn.114.198531.

Lieberman, I. and Ove, P. 1959. Growth factors for mammalian cells in culture. *J. Biol. Chem.* 234(10), 2754–2758. https://doi.org/10.1016/S0021-9258(18)69775-3.

Lipton, A., Paul, D., Henahan, M., Klinger, I. and Holley, R. W. 1972. Serum requirements for survival of 3T3 and SV3T3 fibroblasts. *Exp. Cell Res.* 74(2), 466–470. https://doi.org/10.1016/0014-4827(72)90402-8.

Litwiniuk, A., Pijet, B., Pijet-Kucicka, M., Gajewska, M., Pająk, B. and Orzechowski, A. 2016. FOXO1 and GSK-3β are main targets of insulin-mediated myogenesis in C2C12 muscle cells. *PLoS One* 11(1), e0146726. https://doi.org/10.1371/journal.pone.0146726.

Liu, C., Huang, K., Li, G., Wang, P., Liu, C., Guo, C., Sun, Z. and Pan, J. 2017. Ascorbic acid promotes 3T3-L1 cells adipogenesis by attenuating ERK signaling to upregulate the collagen VI. *Nutr. Metab. (Lond)* 14, 79. https://doi.org/10.1186/s12986-017-0234-y.

López, M. A. A. and Martos, F. C. 2004. Iron availability: an updated review. *Int. J. Food Sci. Nutr.* 55(8), 597–606. https://doi.org/10.1080/09637480500085820.

Lu, H., Shah, P., Ennis, D., Shinder, G., Sap, J., Le-Tien, H. and Fantus, I. G. 2002. The differentiation of skeletal muscle cells involves a protein-tyrosine phosphatase-α-mediated C-Src signaling pathway *. *J. Biol. Chem.* 277(48), 46687–46695. https://doi.org/10.1074/jbc.M209643200.

Lynch, J. and Pierrehumbert, R. 2019. Climate impacts of cultured meat and beef cattle. *Front. Sustain. Food Syst.* 3.

Machida, S. and Booth, F. W. 2004. Insulin-like growth factor 1 and muscle growth: implication for satellite cell proliferation. *Proc. Nutr. Soc.* 63(2), 337–340. https://doi.org/10.1079/PNS2004354.

Mackall, J. C., Student, A. K., Polakis, S. E. and Lane, M. D. 1976. Induction of lipogenesis during differentiation in a "preadipocyte" cell line. *J. Biol. Chem.* 251(20), 6462–6464. https://doi.org/10.1016/S0021-9258(20)81883-3.

Madsen, L., Petersen, R. K. and Kristiansen, K. 2005. Regulation of adipocyte differentiation and function by polyunsaturated fatty acids. Acta BBA - Mol. Basis Dis. *Carotenoids Diet. Lipids* 1740(2), 266–286. https://doi.org/10.1016/j.bbadis.2005.03.001.

Marblestone, A., Gamick, A., Kalil, T., Martin, C., Cvitkovic, M. and Rodriques, S. G. 2022. Unblock research bottlenecks with non-profit start-ups. *Nature* 601(7892), 188–190. https://doi.org/10.1038/d41586-022-00018-5.

Marr, A. G. M., Owen, J. A. and Wilson, G. S. 1962. Studies on the growth-promoting glycoprotein fraction of foetal calf serum. *Biochim. Biophys. Acta* 63(2), 276–285. https://doi.org/10.1016/0006-3002(62)90681-9.

Martin, M. S. and Pohl, S. L. 1980. Effects of chronic exposure to insulin on lipid synthesis in 3T3-L1 fatty fibroblasts. *Mol. Cell. Biochem.* 33(3), 161–164. https://doi.org/10.1007/BF00225289.

Martins Lima, A., Saint Auguste, D. S., Cuenot, F., Martins Cavaco, A. C., Lachkar, T., Khawand, C. M. E., Fraga-Silva, R. A. and Stergiopulos, N. 2020. Standardization and validation of fluorescence-based quantitative assay to study human platelet adhesion to extracellular-matrix in a 384-well plate. *Int. J. Mol. Sci.* 21(18), 6539. https://doi.org/10.3390/ijms21186539.

Massagué, J., Cheifetz, S., Endo, T. and Nadal-Ginard, B. 1986. Type beta transforming growth factor is an inhibitor of myogenic differentiation. *Proc. Natl. Acad. Sci. U. S. A.* 83(21), 8206–8210. https://doi.org/10.1073/pnas.83.21.8206.

Matheny, R. W., Lynch, C. M. and Leandry, L. A. 2012. Enhanced Akt phosphorylation and myogenic differentiation in PI3K p110β-deficient myoblasts is mediated by

PI3K p110α and mTORC2. *Growth Factors* 30(6), 367–384. https://doi.org/10.3109/08977194.2012.734507.

Mattick, C. S., Landis, A. E. and Allenby, B. R. 2015. A case for systemic environmental analysis of cultured meat. *J. Integr. Agric.* 14(2), 249–254. https://doi.org/10.1016/S2095-3119(14)60885-6.

Mattick, C. S., Landis, A. E., Allenby, B. R. and Genovese, N. J. 2015. Anticipatory life cycle analysis of in vitro biomass cultivation for cultured meat production in the United States. *Environ. Sci. Technol.* 49(19), 11941–11949. https://doi.org/10.1021/acs.est.5b01614.

McAleer, C. W., Rumsey, J. W., Stancescu, M. and Hickman, J. J. 2015. Functional myotube formation from adult rat satellite cells in a defined serum-free system. *Biotechnol. Prog.* 31(4), 997–1003. https://doi.org/10.1002/btpr.2063.

McCoy, G. C., Reneau, J. K., Hunter, A. G. and Williams, J. B. 1970. Effects of diet and time on blood serum proteins in the newborn Calf1. *J. Dairy Sci.* 53, 358–362. https://doi.org/10.3168/jds.S0022-0302(70)86209-9.

McFarland, D. C., Pesall, J. E., Norberg, J. M. and Dvoracek, M. A. 1991. Proliferation of the turkey myogenic satellite cell in a serum-free medium. *Comp. Biochem. Physiol. A* 99(1–2), 163–167. https://doi.org/10.1016/0300-9629(91)90252-8.

McKeehan, W. L., Hamilton, W. G. and Ham, R. G. 1976. Selenium is an essential trace nutrient for growth of WI-38 diploid human fibroblasts. *Proc. Natl. Acad. Sci. U. S. A.* 73(6), 2023–2027.

Medicine, C. for V 2022. *Steroid Hormone Implants Used for Growth in Food-Producing Animals*. FDA.

Mehta, F., Theunissen, R. and Post, M. J. 2019. Adipogenesis from bovine precursors. In: *Myogenesis: Methods and Protocols, Methods in Molecular Biology. Springer Nature*, pp. 111–125.

Merlot, A. M., Kalinowski, D. S. and Richardson, D. R. 2014. Unraveling the mysteries of serum albumin–more than just a serum protein. *Front. Physiol.* 5, 299.

Messmer, T., Klevernic, I., Furquim, C., Ovchinnikova, E., Dogan, A., Cruz, H., Post, M. J. and Flack, J. E. 2022. A serum-free media formulation for cultured meat production supports bovine satellite cell differentiation in the absence of serum starvation. *Nat. Food* 3(1), 74–85. https://doi.org/10.1038/s43016-021-00419-1.

Mitić, R., Cantoni, F., Börlin, C. S., Post, M. J. and Jackisch, L. 2023. A simplified and defined serum-free medium for cultivating fat across species. *iScience* 26(1), 105822. https://doi.org/10.1016/j.isci.2022.105822.

Mizuno, N., Kato, Y., Iwamoto, M., Urae, A., Amamoto, T., Niwa, T. and Sugiyama, Y. 2001. Kinetic analysis of the disposition of insulin-like growth factor 1 in healthy volunteers. *Pharm. Res.* 18(8), 1203–1209. https://doi.org/10.1023/A:1010991313633.

Mo, C., Zhao, R., Vallejo, J., Igwe, O., Bonewald, L., Wetmore, L. and Brotto, M. 2015. Prostaglandin E2 promotes proliferation of skeletal muscle myoblasts via EP4 receptor activation. *Cell Cycle* 14(10), 1507–1516. https://doi.org/10.1080/15384101.2015.1026520.

Modica, S. and Wolfrum, C. 2017. The dual role of BMP4 in adipogenesis and metabolism. *Adipocyte* 6(2), 141–146. https://doi.org/10.1080/21623945.2017.1287637.

Morikawa, M., Nixon, T. and Green, H. 1982. Growth hormone and the adipose conversion of 3T3 cells. *Cell* 29(3), 783-789. https://doi.org/10.1016/0092-8674(82)90440-8.

Morrison, R. F. and Farmer, S. R. 2000. Hormonal signaling and transcriptional control of adipocyte differentiation. *J. Nutr.* 130(12), 3116S–3121S. https://doi.org/10.1093/jn/130.12.3116S.

Mottram, D. S. 1998. Flavour formation in meat and meat products: a review. *Food Chem.* 62, 415–424. https://doi.org/10.1016/S0308-8146(98)00076-4.

Murray, T. R. 1980. *Studies on the Differentiation of 3T3-L Preadipocytes into Adipose Cells (Ph.D.).* University of Miami, Florida.

Nath, S. C., Nagamori, E., Horie, M. and Kino-oka, M. 2017. Culture medium refinement by dialysis for the expansion of human induced pluripotent stem cells in suspension culture. *Bioprocess Biosyst. Eng.* 40(1), 123–131. https://doi.org/10.1007/s00449-016-1680-z.

National Center for Biotechnology Information 2022. *PubChem Compound Summary for CID* (vol. 3758). National Center for Biotechnology Information, p. 3-Isobutyl-1-methylxanthine.

Nawrocka, D., Krzyscik, M. A., Opaliński, Ł., Zakrzewska, M. and Otlewski, J. 2020. Stable fibroblast growth factor 2 dimers with high pro-survival and mitogenic potential. *Int. J. Mol. Sci.* 21(11), 4108. https://doi.org/10.3390/ijms21114108.

Nie, Z. 1992. Fetuin: its enigmatic property of growth promotion. *Am. J. Physiol.* 263(3 Pt 1), C551–C562. https://doi.org/10.1152/ajpcell.1992.263.3.C551.

Nie, Z., Jellinek, D. and Ham, R. G. 1991. Separation of growth-promoting activity for human muscle cells from fetuin. *Biochem. Biophys. Res. Commun.* 178(3), 959–966. https://doi.org/10.1016/0006-291x(91)90985-g.

Nilausen, K. 1978. Role of fatty acids in growth-promoting effect of serum albumin on hamster cells in vitro. *J. Cell. Physiol.* 96(1), 1–14. https://doi.org/10.1002/jcp.1040960102.

Noh, S. M., Park, J. H., Lim, M. S., Kim, J. W. and Lee, G. M. 2017. Reduction of ammonia and lactate through the coupling of glutamine synthetase selection and downregulation of lactate dehydrogenase-A in CHO cells. *Appl. Microbiol. Biotechnol.* 101(3), 1035–1045. https://doi.org/10.1007/s00253-016-7876-y.

Okamoto, Y., Haraguchi, Y., Sawamura, N., Asahi, T. and Shimizu, T. 2020. Mammalian cell cultivation using nutrients extracted from microalgae. *Biotechnol. Prog.* 36(2), e2941. https://doi.org/10.1002/btpr.2941.

Okamoto, Y., Haraguchi, Y., Yoshida, A., Takahashi, H., Yamanaka, K., Sawamura, N., Asahi, T. and Shimizu, T. 2022. Proliferation and differentiation of primary bovine myoblasts using Chlorella vulgaris extract for sustainable production of cultured meat. *Biotechnol. Prog.* 38(3), e3239. https://doi.org/10.1002/btpr.3239.

Olson, E. N., Sternberg, E., Hu, J. S., Spizz, G. and Wilcox, C. 1986. Regulation of myogenic differentiation by type beta transforming growth factor. *J. Cell Biol.* 103(5), 1799–1805. https://doi.org/10.1083/jcb.103.5.1799.

O'Neill, B. T., Lauritzen, H. P. M. M., Hirshman, M. F., Smyth, G., Goodyear, L. J. and Kahn, C. R. 2015. Differential role of insulin/IGF-1 receptor signaling in muscle growth and glucose homeostasis. *Cell Rep.* 11(8), 1220–1235. https://doi.org/10.1016/j.celrep.2015.04.037.

O'Neill, E. N., Ansel, J. C., Kwong, G. A., Plastino, M. E., Nelson, J., Baar, K. and Block, D. E. 2022. Spent media analysis suggests cultivated meat media will require species and cell type optimization. *npj Sci. Food* 6(1), 46. https://doi.org/10.1038/s41538-022-00157-z.

O'Neill, E. N., Cosenza, Z. A., Baar, K. and Block, D. E. 2021. Considerations for the development of cost-effective cell culture media for cultivated meat production. *Compr. Rev. Food Sci. Food Saf.* 20(1), 686–709. https://doi.org/10.1111/1541-4337.12678.

Ong, K. J., Johnston, J., Datar, I., Sewalt, V., Holmes, D. and Shatkin, J. A. 2021. Food safety considerations and research priorities for the cultured meat and seafood industry. *Compr. Rev. Food Sci. Food Saf.* 20(6), 5421–5448. https://doi.org/10.1111/1541-4337.12853.

Onuma, Y., Higuchi, K., Aiki, Y., Shu, Y., Asada, M., Asashima, M., Suzuki, M., Imamura, T. and Ito, Y. 2015. A stable chimeric fibroblast growth factor (FGF) can successfully replace basic FGF in human pluripotent stem cell culture. *PLoS ONE* 10(4), e0118931. https://doi.org/10.1371/journal.pone.0118931.

Osherov, N. and Levitzki, A. 1994. Epidermal-growth-factor-dependent activation of the src-family kinases. *Eur. J. Biochem.* 225(3), 1047–1053. https://doi.org/10.1111/j.1432-1033.1994.1047b.x.

Oshima, M., Miyamoto, I., Takaoka, T., Goto, K. and Kagawa, Y. 1987. Unsaturated fatty acids in mammalian cell strains cultured for more than 10 years in serum-free synthetic media. *Jpn. J. Exp. Med.* 57(6), 351–354.

Pak, S. C. O., Hunt, S. M. N., Bridges, M. W., Sleigh, M. J. and Gray, P. P. 1996. Super-CHO—A cell line capable of autocrine growth under fully defined protein-free conditions. *Cytotechnology* 22(1–3), 139–146. https://doi.org/10.1007/BF00353933.

Panella, S., Muoio, F., Jossen, V., Harder, Y., Eibl-Schindler, R. and Tallone, T. 2021. Chemically defined Xeno- and serum-free cell culture medium to grow human adipose stem cells. *Cells* 10(2), 466. https://doi.org/10.3390/cells10020466.

Paula, P. C., Oliveira, J. T. A., Sousa, D. O. B., Alves, B. G. T., Carvalho, A. F. U., Franco, O. L. and Vasconcelos, I. M. 2017. Insulin-like plant proteins as potential innovative drugs to treat diabetes—the Moringa oleifera case study. *New Biotechnol.* 39(A), 99–109. https://doi.org/10.1016/j.nbt.2016.10.005.

Pawlowski, M., Ortmann, D., Bertero, A., Tavares, J. M., Pedersen, R. A., Vallier, L. and Kotter, M. R. N. 2017. Inducible and deterministic forward programming of human pluripotent stem cells into neurons, skeletal myocytes, and oligodendrocytes. *Stem Cell Rep.* 8(4), 803–812. https://doi.org/10.1016/j.stemcr.2017.02.016.

Pérez-Fernández, B. A., Fernandez-de-Cossio-Diaz, J., Boggiano, T., León, K. and Mulet, R. 2021. In-silico media optimization for continuous cultures using genome scale metabolic networks: the case of CHO-K1. *Biotechnol. Bioeng.* 118(5), 1884–1897. https://doi.org/10.1002/bit.27704.

Pierson Jr., R. W. and Temin, H. M. 1972. The partial purification from calf serum of a fraction with multiplication-stimulating activity for chicken fibroblasts in cell culture and with non-suppressible insulin-like activity. *J. Cell. Physiol.* 79(3), 319–330. https://doi.org/10.1002/jcp.1040790302.

Pietrzkowski, Z., Lammers, R., Carpenter, G., Soderquist, A. M., Limardo, M., Phillips, P. D., Ullrich, A. and Baserga, R. 1992. Constitutive expression of insulin-like growth factor 1 and insulin-like growth factor 1 receptor abrogates all requirements for exogenous growth factors. *Cell Growth Differ.* 3(4), 199–205.

Potter, G., Smith, A. S. T., Vo, N. T. K., Muster, J., Weston, W., Bertero, A., Maves, L., Mack, D. L. and Rostain, A. 2020. A more open approach is needed to develop cell-based fish technology: it starts with zebrafish. *One Earth* 3(1), 54–64. https://doi.org/10.1016/j.oneear.2020.06.005.

Poudel, S. B., Min, C. K., Lee, J. H., Shin, Y. J., Kwon, T. H., Jeon, Y. M. and Lee, J. C. 2019. Local supplementation with plant-derived recombinant human FGF2 protein enhances bone formation in critical-sized calvarial defects. *J. Bone Miner. Metab.* 37(5), 900–912. https://doi.org/10.1007/s00774-019-00993-2.

Powers, M. L. and Florini, J. R. 1975. A direct effect of testosterone on muscle cells in tissue Culture1. *Endocrinology* 97(4), 1043–1047. https://doi.org/10.1210/endo-97-4-1043.

Puig, S., Ramos-Alonso, L., Romero, A. M. and Martínez-Pastor, M. T. 2017. The elemental role of iron in DNA synthesis and repair. *Metallomics* 9(11), 1483–1500. https://doi.org/10.1039/c7mt00116a.

Quinlan, G. J., Martin, G. S. and Evans, T. W. 2005. Albumin: biochemical properties and therapeutic potential. *Hepatology* 41(6), 1211–1219. https://doi.org/10.1002/hep.20720.

Rabinowitz, R., Rachmilewitz, B., Rachmilewitz, M. and Schlesinger, M. 1982. Production of transcobalamin II by various murine and human cells in culture. *Isr. J. Med. Sci.* 18(7), 740–745.

Radošević, K., Dukić, B., Andlar, M., Slivac, I. and Gaurina Srček, V. 2016. Adaptation and cultivation of permanent fish cell line CCO in serum-free medium and influence of protein hydrolysates on growth performance. *Cytotechnology* 68(1), 115–121. https://doi.org/10.1007/s10616-014-9760-x.

Ragozzino, E., Brancaccio, M., Di Costanzo, A., Scalabrì, F., Andolfi, G., Wanderlingh, L. G., Patriarca, E. J., Minchiotti, G., Altamura, S., Summa, V. and Varrone, F. 2019. 6-Bromoindirubin-3′-oxime intercepts GSK3 signaling to promote and enhance skeletal muscle differentiation affecting miR-206 expression in mice. *Sci. Rep.* 9(1), 18091. https://doi.org/10.1038/s41598-019-54574-4.

Rajala, K., Lindroos, B., Hussein, S. M., Lappalainen, R. S., Pekkanen-Mattila, M., Inzunza, J., Rozell, B., Miettinen, S., Narkilahti, S., Kerkelä, E., Aalto-Setälä, K., Otonkoski, T., Suuronen, R., Hovatta, O. and Skottman, H. 2010. A defined and xeno-free culture method enabling the establishment of clinical-grade human embryonic, induced pluripotent and adipose stem cells. *PLoS ONE* 5(4), e10246. https://doi.org/10.1371/journal.pone.0010246.

Raoufinia, R., Mota, A., Keyhanvar, N., Safari, F., Shamekhi, S. and Abdolalizadeh, J. 2016. Overview of albumin and its purification methods. *Adv. Pharm. Bull.* 6(4), 495–507. https://doi.org/10.15171/apb.2016.063.

Reed, B. C., Kaufmann, S. H., Mackall, J. C., Student, A. K. and Lane, M. D. 1977. Alterations in insulin binding accompanying differentiation of 3T3-L1 preadipocytes. *Proc. Natl. Acad. Sci. U. S. A.* 74(11), 4876–4880. https://doi.org/10.1073/pnas.74.11.4876.

Reed, B. C. and Lane, M. D. 1980. Expression of insulin receptors during preadipocyte differentiation. *Adv. Enzyme Regul.* 18, 97–117. https://doi.org/10.1016/0065-2571(80)90011-4.

Reig, M. and Toldrá, F. 2008. Veterinary drug residues in meat: concerns and rapid methods for detection. Symposium on Meat safety: From Abattoir to Consumer 78. *Meat Sci.* 78(1–2), 60–67. https://doi.org/10.1016/j.meatsci.2007.07.029.

Relaix, F., Bencze, M. and Borok, M. J., Der Vartanian, A., Gattazzo, F., Mademtzoglou, D., Perez-Diaz, S., Prola, A., Reyes-Fernandez, P.C., Rotini, A. and Taglietti. 2021. Perspectives on skeletal muscle stem cells. *Nat. Commun.* 12, 692. https://doi.org/10.1038/s41467-020-20760-6.

Ren, X., Zhao, M., Lash, B., Martino, M. M. and Julier, Z. 2019. Growth factor engineering strategies for regenerative medicine applications. *Front. Bioeng. Biotechnol.* 7, 469.

Reyne, Y., Nougès, J. and Dulor, J. P. 1989. Differentiation of rabbit adipocyte precursor cells in a serum-free medium. *In Vitro Cell. Dev. Biol.* 25(8), 747–752. https://doi.org/10.1007/BF02623728.

Rinderknecht, E. and Humbel, R. E. 1976. Polypeptides with nonsuppressible insulin-like and cell-growth promoting activities in human serum: isolation, chemical characterization, and some biological properties of forms I and II. *Proc. Natl. Acad. Sci. U. S. A.* 73(7), 2365–2369. https://doi.org/10.1073/pnas.73.7.2365.

Ringer, S. 1883. A further contribution regarding the influence of the different constituents of the blood on the contraction of the heart. *J. Physiol.* 4(1), 29–42.3.

Ringer, S. 1882. Concerning the Influence exerted by each of the Constituents of the Blood on the Contraction of the ventricle. *J. Physiol.* 3(5–6), 380–393.

Risner, D., Li, F., Fell, J. S., Pace, S. A., Siegel, J. B., Tagkopoulos, I. and Spang, E. S. 2020. Preliminary techno-economic assessment of animal cell-based meat. *Foods* 10(1), 3. https://doi.org/10.3390/foods10010003.

Rizzino, A. and Sato, G. 1978. Growth of embryonal carcinoma cells in serum-free medium. *Proc. Natl. Acad. Sci. U. S. A.* 75(4), 1844–1848.

Roberts, A. B., Lamb, L. C., Newton, D. L., Sporn, M. B., De Larco, J. E. and Todaro, G. J. 1980. Transforming growth factors: isolation of polypeptides from virally and chemically transformed cells by acid/ethanol extraction. *Proc. Natl. Acad. Sci. U. S. A.* 77(6), 3494–3498. https://doi.org/10.1073/pnas.77.6.3494.

Roberts, P. J. and Der, C. J. 2007. Targeting the Raf-MEK-ERK mitogen-activated protein kinase cascade for the treatment of cancer. *Oncogene* 26(22), 3291–3310. https://doi.org/10.1038/sj.onc.1210422.

Roche, M., Rondeau, P., Singh, N. R., Tarnus, E. and Bourdon, E. 2008. The antioxidant properties of serum albumin. *FEBS Lett.* 582(13), 1783–1787. https://doi.org/10.1016/j.febslet.2008.04.057.

Rockwell, G. A., Sato, G. H. and McClure, D. B. 1980. The growth requirements of SV40 virus transformed Balb/C-3T3 cells in serum-free monolayer culture. *J. Cell. Physiol.* 103(2), 323–331. https://doi.org/10.1002/jcp.1041030218.

Rodriguez, A. M., Elabd, C., Delteil, F., Astier, J., Vernochet, C., Saint-Marc, P., Guesnet, J., Guezennec, A., Amri, E. Z., Dani, C. and Ailhaud, G. 2004. Adipocyte differentiation of multipotent cells established from human adipose tissue. *Biochem. Biophys. Res. Commun.* 315(2), 255–263. https://doi.org/10.1016/j.bbrc.2004.01.053.

Rommel, C., Bodine, S. C., Clarke, B. A., Rossman, R., Nunez, L., Stitt, T. N., Yancopoulos, G. D. and Glass, D. J. 2001. Mediation of IGF-1-induced skeletal myotube hypertrophy by PI(3)K/Akt/mTOR and PI(3)K/Akt/GSK3 pathways. *Nat. Cell Biol.* 3(11), 1009–1013. https://doi.org/10.1038/ncb1101-1009.

Rosen, O. M., Smith, C. J., Hirsch, A., Lai, E. and Rubin, C. S. 1979. Recent Studies of the 3T3-L1 Adipocyte-Like Cell Line11This. In: Proceedings of the 1978 Laurentian Hormone Conference. *Recent Prog. Horm. Res.* Greep, R. O. (Ed.). Academic Press, Boston 35, 477–499. https://doi.org/10.1016/B978-0-12-571135-7.50015-1.

Rubio, N., Datar, I., Stachura, D., Kaplan, D. and Krueger, K. 2019. Cell-based fish: A novel approach to seafood production and an opportunity for cellular agriculture. *Front. Sustain. Food Syst.* 3.

Rubio, N. R., Fish, K. D., Trimmer, B. A. and Kaplan, D. L. 2019. Possibilities for engineered insect tissue as a food source. *Front. Sustain. Food Syst.* 3.

Russell, T. R. and Ho, R. 1976. Conversion of 3T3 fibroblasts into adipose cells: triggering of differentiation by prostaglandin F2alpha and 1-methyl-3-isobutyl xanthine. *Proc. Natl. Acad. Sci. U. S. A.* 73(12), 4516–4520. https://doi.org/10.1073/pnas.73.12.4516.

Ryall, J. G. 2013. Metabolic reprogramming as a novel regulator of skeletal muscle development and regeneration. *FEBS Journal* 280(17), 4004–4013. https://doi.org/10.1111/febs.12189.

Sakwe, A. M., Koumangoye, R., Goodwin, S. J. and Ochieng, J. 2010. Fetuin-A (α2HS-glycoprotein) is a major serum adhesive protein that mediates growth signaling in breast tumor cells. *J. Biol. Chem.* 285(53), 41827–41835. https://doi.org/10.1074/jbc.M110.128926.

Sanford, K. K., Earle, W. R. and Likely, G. D. 1948. The growth in vitro of single isolated tissue cells. *J. Natl. Cancer Inst.* 9(3), 229–246.

Sargent, P. J., Farnaud, S., Cammack, R., Zoller, H. M. P. and Evans, R. W. 2006. Characterisation of recombinant unglycosylated human serum transferrin purified from Saccharomyces cerevisiae. *Biometals* 19(5), 513–519. https://doi.org/10.1007/s10534-005-5532-6.

Schenzle, L., Egger, K., Fuchs, A. and Pichler, H. 2022. Never let me down: Optimizing performance of serum free culture medium for bovine satellite cells. https://doi.org/10.1101/2022.11.13.516330.

Schwingshackl, L., Zähringer, J., Beyerbach, J., Werner, S. S., Heseker, H., Koletzko, B. and Meerpohl, J. J. 2021. Total dietary fat intake, fat quality, and health outcomes: A scoping review of systematic reviews of prospective studies. *Ann. Nutr. Metab.* 77(1), 4–15. https://doi.org/10.1159/000515058.

Scott, M. A., Nguyen, V. T., Levi, B. and James, A. W. 2011. Current methods of adipogenic differentiation of mesenchymal stem cells. *Stem Cells Dev.* 20(10), 1793–1804. https://doi.org/10.1089/scd.2011.0040.

Serrano, A. L., Baeza-Raja, B., Perdiguero, E., Jardí, M. and Muñoz-Cánoves, P. 2008. Interleukin-6 is an essential regulator of satellite cell-mediated skeletal muscle hypertrophy. *Cell Metab.* 7(1), 33–44. https://doi.org/10.1016/j.cmet.2007.11.011.

Serrero, G. and Khoo, J. C. 1982. An in vitro model to study adipose differentiation in serum-free medium. *Anal. Biochem.* 120(2), 351–359. https://doi.org/10.1016/0003-2697(82)90357-8.

Serrero, G. and Mills, D. 1987. Differentiation of newborn rat adipocyte precursors in defined serum-free medium. *In Vitro Cell. Dev. Biol.* 23(1), 63–66. https://doi.org/10.1007/BF02623495.

Shah, B. H., Neithardt, A., Chu, D. B., Shah, F. B. and Catt, K. J. 2006. Role of EGF receptor transactivation in phosphoinositide 3-kinase-dependent activation of MAP kinase by GPCRs. *J. Cell. Physiol.* 206(1), 47–57. https://doi.org/10.1002/jcp.20423.

Sheehan, S. M. and Allen, R. E. 1999. Skeletal muscle satellite cell proliferation in response to members of the fibroblast growth factor family and hepatocyte growth factor. *J. Cell. Physiol.* 181(3), 499–506. https://doi.org/10.1002/(SICI)1097-4652(199912)181:3<499::AID-JCP14>3.0.CO;2-1.

Sheikhpour, R. 2017. A survey on the role of fetuin-A in breast cancer: a review. *Basic Clin. Cancer Res.* 9, 21–25.

Siddle, K. 2011. Signalling by insulin and IGF receptors: supporting acts and new players. *J. Mol. Endocrinol.* 47(1), R1–R10. https://doi.org/10.1530/JME-11-0022.

Sikdar, S. K. and Sawant, S. B. 1994. Ammonia removal from mammalian cell culture medium by ion-exchange membranes*. *Sep. Sci. Technol.* 29(12), 1579–1591. https://doi.org/10.1080/01496399408007375.

Simsa, R., Yuen, J., Stout, A., Rubio, N., Fogelstrand, P. and Kaplan, D. L. 2019. Extracellular heme proteins influence bovine myosatellite cell proliferation and the color of cell-based meat. *Foods* 8(10), E521. https://doi.org/10.3390/foods8100521.

Sinke, P. and Odegard, I. 2021. *LCA of Cultivated Meat. Future Projections for Different Scenarios*. council of Europe, Delft.

Skrivergaard, S., Young, J. F., Sahebekhtiari, N., Semper, C., Venkatesan, M., Savchenko, A., Stogios, P. J., Therkildsen, M. and Rasmussen, M. K. 2023. A simple and robust serum-free media for the proliferation of muscle cells. *Food Res. Int.* 172, 113194. https://doi.org/10.1016/j.foodres.2023.113194.

Slaaby, R., Schäffer, L., Lautrup-Larsen, I., Andersen, A. S., Shaw, A. C., Mathiasen, I. S. and Brandt, J. 2006. Hybrid receptors formed by insulin receptor (IR) and insulin-like growth factor I receptor (IGF-IR) have low insulin and high IGF-1 affinity irrespective of the IR splice variant. *J. Biol. Chem.* 281(36), 25869–25874. https://doi.org/10.1074/jbc.M605189200.

Song, J. A., Koo, B. K., Chong, S. H., Kwak, J., Ryu, H. B., Nguyen, M. T., Vu, T. T. T., Jeong, B., Kim, S. W. and Choe, H. 2013. Expression and Purification of Biologically Active Human FGF2 Containing the b'a' Domains of Human PDI in Escherichia coli. *Appl. Biochem. Biotechnol.* 170(1), 67–80. https://doi.org/10.1007/s12010-013-0140-3.

Sotiropoulos, A., Ohanna, M., Kedzia, C., Menon, R. K., Kopchick, J. J., Kelly, P. A. and Pende, M. 2006. Growth hormone promotes skeletal muscle cell fusion independent of insulin-like growth factor 1 up-regulation. *Proc. Natl. Acad. Sci. U. S. A.* 103(19), 7315–7320. https://doi.org/10.1073/pnas.0510033103.

Spangenburg, E. E. and Booth, F. W. 2002. Multiple signaling pathways mediate LIF-induced skeletal muscle satellite cell proliferation. *Am. J. Physiol. Cell Physiol.* 283(1), C204–C211. https://doi.org/10.1152/ajpcell.00574.2001.

Spiro, R. G. 1960. Studies on fetuin, a glycoprotein of fetal serum: I. *J. Biol. Chem.* 235(10), 2860–2869. https://doi.org/10.1016/S0021-9258(18)64553-3.

St. Clair, J. A., Meyer-Demarest, S. D. and Ham, R. G. 1992. Improved medium with EGF and BSA for differentiated human skeletal muscle cells. *Muscle Nerve* 15(7), 774–779. https://doi.org/10.1002/mus.880150705.

Staiger, H. and Loffler, G. 1998. The role of PDGF-dependent suppression of apoptosis in differentiating 3T3-L1 preadipocytes. *Eur. J. Cell Biol.* 77(3), 220–227. https://doi.org/10.1016/S0171-9335(98)80110-6.

Steinlein, L. M. and Ikeda, R. A. 1993. Production of N-terminal and C-terminal human serum transferrin in Escherichia coli. *Enzyme Microb. Technol.* 15(3), 193–199. https://doi.org/10.1016/0141-0229(93)90137-Q.

Stout, A. J., Kaplan, D. L. and Flack, J. E. 2023. Cultured meat: Creative solutions for a cell biological problem. *Trends Cell Biol.* 33(1). https://doi.org/10.1016/j.tcb.2022.10.002.

Stout, A. J., Mirliani, A. B., Rittenberg, M. L., Shub, M., White, E. C., Yuen, J. S. K. and Kaplan, D. L. 2022. Simple and effective serum-free medium for sustained expansion of bovine satellite cells for cell cultured meat. *Commun. Biol.* 5(1), 466. https://doi.org/10.1038/s42003-022-03423-8.

Stout, A. J., Mirliani, A. B., Soule-Albridge, E. L., Cohen, J. M. and Kaplan, D. L. 2020. Engineering carotenoid production in mammalian cells for nutritionally enhanced cell-cultured foods. *Metab. Eng.* 62, 126–137. https://doi.org/10.1016/j.ymben.2020.07.011.

Stout, A. J., Rittenberg, M. L., Shub, M., Saad, M. K., Mirliani, A. B., Dolgin, J. and Kaplan, D. L. 2023. A Beefy-R culture medium: Replacing albumin with rapeseed protein isolates. *Biomaterials* 296, 122092. https://doi.org/10.1016/j.biomaterials.2023.122092.

Stout, A. J., Zhang, X., Letcher, S. M., Rittenberg, M. L., Shub, M., Chai, K. M., Kaul, M. and Kaplan, D. L. 2023. Engineered autocrine signaling eliminates muscle cell FGF2 requirements for cultured meat production. *bioRxiv*. https://doi.org/10.1101/2023.04.17.537163.

Syverud, B. C., VanDusen, K. W. and Larkin, L. M. 2016. Effects of dexamethasone on satellite cells and tissue engineered skeletal muscle units. *Tissue Eng. Part A* 22(5–6), 480–489. https://doi.org/10.1089/ten.TEA.2015.0545.

Tierney, M. T., Aydogdu, T., Sala, D., Malecova, B., Gatto, S., Puri, P. L., Latella, L. and Sacco, A. 2014. STAT3 signaling controls satellite cell expansion and skeletal muscle repair. *Nat. Med.* 20(10), 1182–1186. https://doi.org/10.1038/nm.3656.

Tuomisto, H. L. and Teixeira de Mattos, M. J. 2011. Environmental impacts of cultured meat production. *Environ. Sci. Technol.* 45(14), 6117–6123. https://doi.org/10.1021/es200130u.

Urban, R. J., Bodenburg, Y. H., Nagamani, M. and Peirce, J. 1994. Dexamethasone potentiates IGF-I actions in porcine granulosa cells. *Am. J. Physiol.* 267(1 Pt 1), E115–E123. https://doi.org/10.1152/ajpendo.1994.267.1.E115.

U.S. Food and Drug Administration. 2023. Human food made with cultured animal cells inventory. Available at: https://www.cfsanappsexternal.fda.gov/scripts/fdcc/?set=AnimalCellCultureFoods.

Venâncio, T. M., Oliveira, A. E. A., Silva, L. B., Machado, O. L. T., Fernandes, K. V. S. and Xavier-Filho, J. 2003. A protein with amino acid sequence homology to bovine insulin is present in the legume Vigna unguiculata (cowpea). *Braz. J. Med. Biol. Res.* 36(9), 1167–1173. https://doi.org/10.1590/S0100-879X2003000900004.

Venkatesan, M., Semper, C., DiLeo, R., Mesa, N., Stogios, P. J. and Savchenko, A. 2022. Recombinant production of growth factors for application in cell culture. https://doi.org/10.1101/2022.02.15.480596.

Vierck, J. L., Icenoggle, D. L., Bucci, L. and Dodson, M. V. 2003. The effects of ergogenic compounds on myogenic satellite cells. *Med. Sci. Sports Exerc.* 35(5), 769–776. https://doi.org/10.1249/01.MSS.0000065005.96298.01.

Vijayasankaran, N., Li, J., Shawley, R., Chen, A., Shiratori, M., Gawlitzek, M., Li, F., Kiss, R. and Amanullah, A. 2010. *Animal Cell Culture Media*. https://doi.org/10.1002/9780470054581.eib030.

Volz, A. C. and Kluger, P. J. 2018. Completely serum-free and chemically defined adipocyte development and maintenance. *Cytotherapy* 20(4), 576–588. https://doi.org/10.1016/j.jcyt.2018.01.004.

Wabitsch, M., Hauner, H., Heinze, E. and Teller, W. M. 1995. The role of growth hormone/insulin-like growth factors in adipocyte differentiation. Metabolism, international symposium on glucose metabolism and growth factors 44, 45–49. https://doi.org/10.1016/0026-0495(95)90220-1.

Warner, R. D. 2017. Chapter 14. The eating quality of meat. In: *Lawrie´s Meat Science*, Woodhead Publishing Series in Food Science, Technology and Nutrition (8th edn.) (vol. IV) Water-Holding Capacity and Juiciness, Toldra, F. (Ed.). Woodhead Publishing, pp. 419–459. https://doi.org/10.1016/B978-0-08-100694-8.00014-5.

Watanabe, F. and Bito, T. 2018. Vitamin B12 sources and microbial interaction. *Exp. Biol. Med. (Maywood)* 243(2), 148–158. https://doi.org/10.1177/1535370217746612.

Watson, K., Koumangoye, R., Thompson, P., Sakwe, A. M., Patel, T., Pratap, S. and Ochieng, J. 2012. Fetuin-A triggers the secretion of a novel set of exosomes in detached tumor cells that mediate their adhesion and spreading. *FEBS Lett.* 586(19), 3458–3463. https://doi.org/10.1016/j.febslet.2012.07.071.

Waymouth, C. 1959. Rapid proliferation of sublines of NCTC Clone 929 (Strain L) mouse cells in a simple chemically defined medium (MB 752/1)2. *J. Natl. Cancer Inst.* 22(5), 1003–1017. https://doi.org/10.1093/jnci/22.5.1003.

Werner, H., Weinstein, D. and Bentov, I. 2008. Similarities and differences between insulin and IGF-I: structures, receptors, and signalling pathways. *Arch. Physiol. Biochem.* 114(1), 17–22. https://doi.org/10.1080/13813450801900694.

Westall, F. C., Lennon, V. A. and Gospodarowicz, D. 1978. Brain-derived fibroblast growth factor: identity with a fragment of the basic protein of myelin. *Proc. Natl. Acad. Sci. U. S. A.* 75(10), 4675–4678.

Will, K., Schering, L., Albrecht, E., Kalbe, C. and Maak, S. 2015. Differentiation of bovine satellite cell-derived myoblasts under different culture conditions. *In Vitro Cell. Dev. Biol. Anim.* 51(9), 885–889.

Wroblewski, O. M., Vega-Soto, E. E., Nguyen, M. H., Cederna, P. S. and Larkin, L. M. 2021. Impact of human epidermal growth factor on tissue-engineered skeletal muscle structure and function. *Tissue Eng. Part A* 27(17–18), 1151–1159. https://doi.org/10.1089/ten.TEA.2020.0255.

Wu, Z., Bucher, N. L. and Farmer, S. R. 1996. Induction of peroxisome proliferator-activated receptor gamma during the conversion of 3T3 fibroblasts into adipocytes is mediated by C/EBPbeta, C/EBPdelta, and glucocorticoids. *Mol. Cell. Biol.* 16(8), 4128–4136. https://doi.org/10.1128/MCB.16.8.4128.

Xie, Y., Su, N., Yang, J., Tan, Q., Huang, S., Jin, M., Ni, Z., Zhang, B., Zhang, D., Luo, F., Chen, H., Sun, X., Feng, J. Q., Qi, H. and Chen, L. 2020. FGF/FGFR signaling in health and disease. *Signal Transduct. Target. Ther.* 5(1), 181. https://doi.org/10.1038/s41392-020-00222-7.

Xing, Z., Kenty, B., Koyrakh, I., Borys, M., Pan, S.-H. and Li, Z. J. 2011. Optimizing amino acid composition of CHO cell culture media for a fusion protein production. *Process Biochem.* 46(7), 1423–1429. https://doi.org/10.1016/j.procbio.2011.03.014.

Xu, M., Chen, X., Chen, D., Yu, B. and Huang, Z. 2017. FoxO1: a novel insight into its molecular mechanisms in the regulation of skeletal muscle differentiation and fiber type specification. *Oncotarget* 8(6), 10662–10674. https://doi.org/10.18632/oncotarget.12891.

Yablonka-Reuveni, Z., Danoviz, M. E., Phelps, M. and Stuelsatz, P. 2015. Myogenic-specific ablation of FGFR1 impairs FGF2-mediated proliferation of satellite cells at the myofiber niche but does not abolish the capacity for muscle regeneration. *Front. Aging Neurosci.* 7 (85), 1–16.

Yamane, I., Kan, M., Hoshi, H. and Minamoto, Y. 1981. Primary culture of human diploid cells and its long-term transfer in a serum-free medium. *Exp. Cell Res.* 134(2), 470–474. https://doi.org/10.1016/0014-4827(81)90448-1.

Yao, T. and Asayama, Y. 2017. Animal-cell culture media: history, characteristics, and current issues. *Reprod. Med. Biol.* 16(2), 99–117. https://doi.org/10.1002/rmb2.12024.

Yasin, R. and van Beers, G. 1983. Growth and differentiation of human muscle cells in defined medium. In: Fischer, G. and Wieser, R. J. (Eds). *Hormonally Defined Media.*

Springer, Berlin, Heidelberg, pp. 407–410. https://doi.org/10.1007/978-3-642-69290-1_62.

Yoshiko, Y., Hirao, K., Sakabe, K., Seiki, K., Takezawa, J. and Maeda, N. 1996. Autonomous control of expression of genes for insulin-like growth factors during the proliferation and differentiation of C2C12 mouse myoblasts in serum-free culture. *Life Sci.* 59(23), 1961–1968. https://doi.org/10.1016/S0024-3205(96)00547-4.

Yoshioka, G., Imaeda, N., Ohtani, T. and Hayashi, K. 2005. Effects of cortisol on muscle proteolysis and meat quality in piglets. *Meat Sci.* 71(3), 590–593. https://doi.org/10.1016/j.meatsci.2005.05.015.

Yuen Jr, J. S. K., Saad, M. K., Xiang, N., Barrick, B. M., DiCindio, H., Li, C., Zhang, S. W., Rittenberg, M., Lew, E. T., Zhang, K. L., Leung, G., Pietropinto, J. A. and Kaplan, D. L. 2022. Aggregating in vitro-grown adipocytes to produce macroscale cell-cultured fat tissue with tunable lipid compositions for food applications. *eLife* 12, e82120. https://doi.org/10.7554/eLife.82120.

Yuen, J. S. K., Stout, A. J., Kawecki, N. S., Letcher, S. M., Theodossiou, S. K., Cohen, J. M., Barrick, B. M., Saad, M. K., Rubio, N. R., Pietropinto, J. A., DiCindio, H., Zhang, S. W., Rowat, A. C. and Kaplan, D. L. 2022. Perspectives on scaling production of adipose tissue for food applications. *Biomaterials* 280, 121273. https://doi.org/10.1016/j.biomaterials.2021.121273.

Yun, Z. Y., Zhang, H. P., Cai, X. Z., Wang, A. P. and Zhang, L. B. 2007. Kinetic and thermodynamic studies on the thermal denaturation of bovine milk insulin-like growth factor-I in model systems. *Lait* 87(2), 139–148. https://doi.org/10.1051/lait:2007005.

Zhang, D., Nandi, S., Bryan, P., Pettit, S., Nguyen, D., Santos, M. A. and Huang, N. 2010. Expression, purification, and characterization of recombinant human transferrin from rice (Oryza sativa L.). *Protein Expr. Purif.* 74(1), 69–79. https://doi.org/10.1016/j.pep.2010.04.019.

Zheng, X., Baker, H., Hancock, W. S., Fawaz, F., McCaman, M. and Pungor Jr., E. 2006. Proteomic analysis for the assessment of different lots of fetal bovine serum as a raw material for cell culture. Part IV. Application of proteomics to the manufacture of biological drugs. *Biotechnol. Prog.* 22(5), 1294–1300. https://doi.org/10.1021/bp060121o.

Zizola, C. F., Balañá, M. E., Sandoval, M. and Calvo, J. C. 2002. Changes in IGF-I receptor and IGF-I mRNA during differentiation of 3T3-L1 preadipocytes. *Biochimie* 84(10), 975–980. https://doi.org/10.1016/S0300-9084(02)00009-3.

Chapter 7

Bioreactors for cell multiplication in cultured meat product development

Stefan Schlößer and Philipp Nold, Eppendorf SE, Germany

1 Introduction

Cultured meat holds great promise for the future with its potential to satisfy the growing demand for meat while reducing harm to animals and the negative environmental impact of livestock husbandry at the same time (Post, 2014). However, the innovative idea of isolating livestock muscle and fat stem cells and growing them into muscle tissue poses challenges that need to be overcome.

High cell numbers are required for sufficient meat production; estimates suggest 10^{12}–10^{13} cells are needed to produce 10–100 kg of cultured meat (Bodiou et al., 2020). Consequently, high cell-density cultures are essential in order to satisfy the anticipated market. Thus, while the first cultured meat burger was literally created on a lab bench scale (Post, 2014), an increase in scale is inevitable to bring this novel food alternative to the general public.

Furthermore, the burger created by Post and colleagues utilized static two-dimensional (2D) cell culturing methods in which thousands of muscle fibers were grown in culture dishes, a strategy sufficient for proof-of-concept but one that does not offer optimal growth conditions and efficiency (Post, 2014; Bellani

http://dx.doi.org/10.19103/AS.2023.0130.09

et al., 2020). In fact, the potential cell numbers that could be obtained in this system are limited by a low growth surface-to-volume ratio, as well as a lack of control over important growth parameters, such as gas and nutrient supply or pH (Bellani et al., 2020).

To that end, a suitable environment needs to be created and maintained to facilitate efficient cell proliferation and sufficient cell numbers, ideally through precise parameter control and growth surface increase by switching from 2D to three-dimensional (3D) cultures. Furthermore, all this needs to be achieved at a price acceptable to consumers.

A promising way to confront these challenges is by switching from conventional 2D cultures in flasks placed in CO_2 incubators to 3D cultures in bioreactors.

In this chapter, we explore bioreactors and their working principles, as well as their role as a vital tool to tackle the challenges of creating suitable growth conditions for anchorage-dependent stem cells and process scale-up. Furthermore, we evaluate measures with the potential to reduce the production costs of cultured meat.

2 The bioreactor: principles and structure

Even though bioreactors come in many shapes and forms, their basic principle is straightforward: a vessel providing a 3D growth environment mimicking the natural conditions for prokaryotic and eukaryotic cells due to culture parameter control (Fig. 1). Yet, in order to achieve this simple goal, a whole industry has developed providing innovative solutions for all bioprocessing needs.

As a result, the bioreactor field is quite diverse. Bioreactors can be made from stainless steel, glass, or plastic. Steel and glass reactors are re-usable, with steel favored for large vessels. Contemporary plastic reactors, on the other hand, are smaller than their steel counterparts but offer the advantages of a

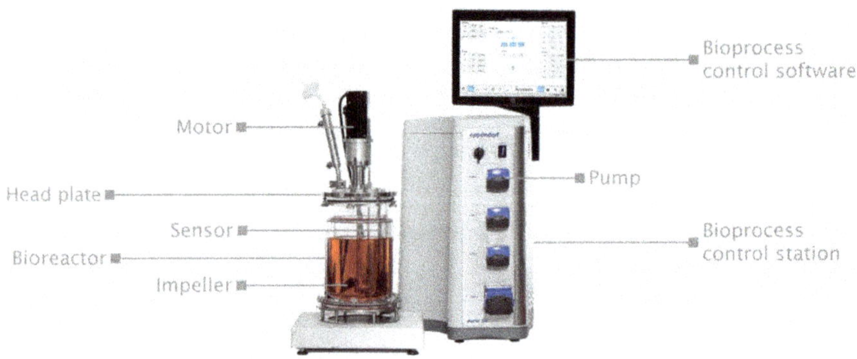

Figure 1 Profile of a bioreactor. Components of a stirred tank-bioreactor (left) and a connected control station (right).

Published by Burleigh Dodds Science Publishing Limited, 2023.

single-use product, such as less maintenance effort and reduced downtime between batches.

Different bioreactor working principles provide a variety of options to choose from. Out of these options, we will have a look at four examples which are commonly employed for cell culture, namely hollow fiber, rocking bed, bubble column, and stirred-tank bioreactors.

The hollow fiber reactor uses the principle of mimicking blood vessel structures to provide a cellular growth environment. Here, a cylindrical vessel is stacked with semipermeable hollow fibers with varying pore sizes able, for example, to let pass medium components or cellular products while retaining the cells at the same time (Bellani et al., 2020). The fibers are commonly made of cellulose and polyethersulfone (Pajčin et al., 2022). Both adherent and suspension cells can be cultivated inside or outside of the fibers while being perfused by medium and medium components using the principle of diffusion (Frank et al., 2019; Bellani et al., 2020).

The rocking-bed or wave bioreactor utilizes a large, disposable polymeric bag as a culture vessel that is placed on a rocking platform (Pajčin et al., 2022). The bag is inflated through an air circulation system and the rocking movement creates a wave motion that enables mixing of medium components and cells, as well as oxygenation of the culture. Thus, this bioreactor type does not utilize internal stirring parts to mix the culture.

This is also true for the bubble column bioreactor type, such as air-lift bioreactors (Guan et al., 2021). Here, culture mixing within the bioreactor tank is achieved by an air bubble stream inducing a circular medium flow resulting in the lifting and sinking of the cells.

Lastly, the stirred-tank bioreactor utilizes a motorized impeller to swirl the cell suspension, thereby supporting gas exchange, nutrient distribution, and culture homogeneity (Fig. 1).

Each bioreactor type has its pros and cons. For example, hollow fiber, wave, and air-lift bioreactors are known to cause low shear stress (Bellani et al., 2020; Al-Mashhadani et al., 2015; Pajčin et al., 2022), a property especially important for shear-sensitive cells like those used for cultured meat production. However, due to the gentle nature of their working principles, mixing and aeration efficiency is lower compared to stirred tank bioreactors (Bartczak et al., 2022; Al-Mashhadani et al., 2015). Furthermore, the structure of hollow fiber bioreactors with cells growing around tubular structures can lead to gradients where cells growing closer to the fiber surface receive optimal nutrient, oxygen, and pH conditions, whereas cells growing in the periphery are not supplied adequately (Pajčin et al., 2022). This limits its use for operation at a larger scale (Rodrigues et al., 2011), for example, when compared to airlift or stirred-tank bioreactors. In fact, the stirred-tank bioreactor is one of the most widely utilized systems for mammalian cell culture on a larger scale (Bellani et al., 2020).

However, its more efficient and homogenous culture conditions come with the need to balance shear forces by choosing the right impeller type and adjusting the agitation speed properly.

Within this chapter, the stirred-tank bioreactor will be used as an example to convey the principles of bioprocessing and its role in culturing anchorage-dependent stem cells.

Culture volumes in bioreactors can range from a few milliliters to tens of thousands of liters; all these proportions have a place in process development. The initial growth conditions for a given cell type are typically determined in small-scale approaches before scale-up, the process of stepwise increasing the dimensions of the bioprocess, is initiated (Fig. 2).

The aim of this strategy is to ensure that yields and critical quality features obtained at small working volumes during process development can be reproduced at larger cell culture volumes. A common method to achieve this goal is to identify the most relevant cultivation parameters and try to replicate them at the different culture sizes. Stirred-tank bioreactors are well suited in this context, as the vessel dimensions are widely scalable and reproducibility of relevant vessel properties (for example, oxygen transfer capabilities or mixing time) is predictable at different scales.

Constant cultivation parameters are usually achieved by a control station connected to the bioreactor. By utilizing a sensor-feedback loop, it constantly monitors, adjusts, and maintains previously selected growth parameters, including temperature, gas concentrations such as dissolved oxygen (DO), pH, and culture mixing (Fig. 3). Sensor readings are transferred to the bioprocess software and compared to a setpoint. Setpoint deviations are then equalized by actuators in automated feedback control loops. Actuators can be heating

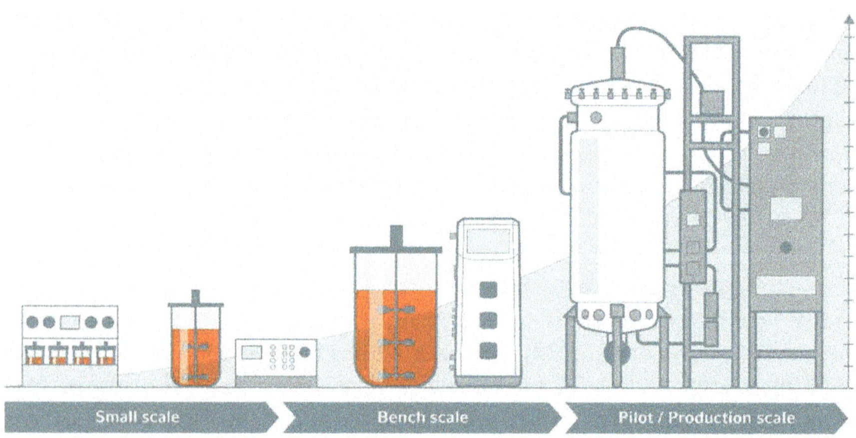

Figure 2 Bioreactor scales during process development.

Published by Burleigh Dodds Science Publishing Limited, 2023.

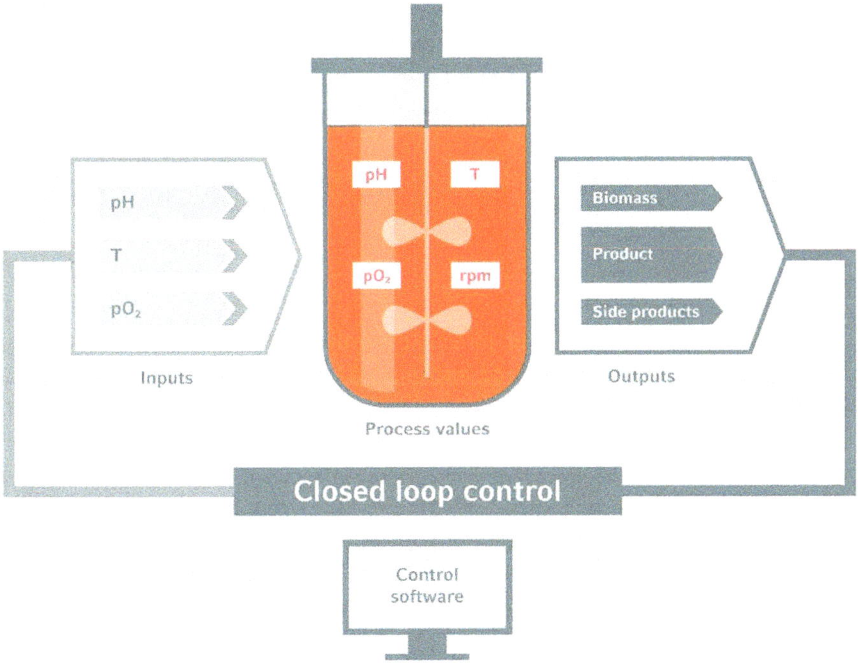

Figure 3 Working principle of a bioreactor. Regulation of parameters, such as pH, temperature (T), or oxygen (O_2) with a control station and specific software creates a suitable growth environment for cells.

devices to adjust the temperature, pumps to deliver liquids for pH control, or gassing devices to provide air, oxygen, nitrogen, or carbon dioxide to the culture.

Being able to adjust the parameters instantaneously according to the culture's needs is one of the biggest advantages of a bioreactor over static cell culture in incubators. It creates and maintains a stable and optimized environment for the cells to thrive in which ensures reproducibility between batches at different scales, thus enabling higher cell densities.

In cell culture, the availability of nutrients and the concentration of toxic by-products are crucial for the achievable cell density. Both can be influenced by the operation mode of the bioreactor. If high cell numbers need to be generated, as in the production of cultured meat, the bioreactor operation mode is, therefore, a critical parameter.

3 Bioreactor operation modes

In general, all cells, no matter if prokaryotic or eukaryotic, undergo the same growth kinetic during cultivation: an initial lag phase after inoculation of the bioreactor in which the cells adapt to the newly established environment, followed by an exponential phase with increasing cell division and cell density, then a stationary phase in which nutrient limitations halt growth and, in case of sustained nutrient limitation, a death phase. Even though these growth phases are inevitable, their properties can be modulated, depending on the bioreactor operation mode.

Three operation modes are commonly used for bioprocessing, namely batch, fed-batch, and continuous cultivation (Fig. 4).

3.1 Batch cultivation

Batch cultivation is carried out in a fixed volume. No medium or feeding solutions are added, removed, or exchanged during the process (Yang and Sha, 2019; Reiss et al., 2021). The cells continuously consume the nutrients and undergo the typical growth phases. Once the nutrients are depleted, cellular growth comes to a halt and the process is terminated. As there are almost no external interventions, batch cultivation bears a low risk of contamination. However, the cells are presented with a constantly changing environment and accumulation of potentially toxic by-products within the medium (Yang and Sha, 2019). This may affect the growth efficiency of the culture and by-products might cause problems in downstream processes.

3.2 Fed-batch cultivation

Fed-batch cultivation builds upon the batch mode to address its disadvantages. Here, nutrients are replenished during the incubation period, thus establishing a more stable environment for the cells. Feeding of carbon and nitrogen sources to the cells is usually initiated after the initial lag phase. Thus, fed-batch cultivation prolongs the availability of nutrients, giving the cells more time to grow before entering the stationary phase. However, the external nutrient

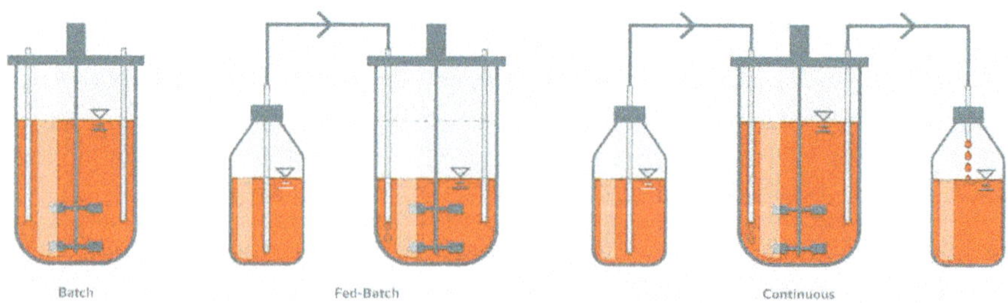

Figure 4 Bioreactor operation modes. Batch cultivation carried out in a fixed volume, fed-batch where medium and nutrients are added after the initial lag phase until the end of the process, and continuous cultivation where constant medium/nutrient replenishment is employed to ensure a stable growth environment.

supply also bears a higher contamination risk and leads to a volume increase over time (Reiss et al., 2021).

3.3 Continuous cultivation

Continuous cultivation is the most complex operation mode. However, once it is thoroughly adjusted it offers the most stable cell environment and the longest possible running times. This is obtained by the continuous removal of the used medium and cells combined with a constant flow of fresh medium and nutrients (Yang and Sha, 2019). The result is a steady state that ideally keeps the volume, nutrient concentrations, and biomass constant while removing toxic by-products at the same time (Reiss et al., 2021). The introduction of a cell retention device can be used to exchange the medium while the cells are maintained within the bioreactor. This subtype of continuous cultivation, known as perfusion (Allan et al., 2019), might be favorable for cultured meat production to obtain even higher cell densities than in a fed-batch process. However, some drawbacks of the fed-batch mode also occur here. Multiple external connections and the possibility of long-term incubation make maintaining sterile and stable growth conditions more challenging.

Nevertheless, fed-batch and continuous cultivation are the methods of choice for large-scale applications, such as cultured meat production, as they allow for easier automation and culture medium recycling (Reiss et al., 2021).

4 Creating a stem cell niche in a bioreactor: important bioprocess parameters

Even though pluripotent stem cells are an attractive option for cultured meat production, multipotent (also called adult) stem cells are the more common cell sources. Among those, muscle satellite cells, mesenchymal stem/stromal cells, and fibro/adipogenic progenitors are the cell types of choice in cultured meat production, as they can be isolated relatively easily by tissue and muscle biopsy (Reiss et al., 2021). In contrast to their pluripotent counterparts, these progenitor cells can only differentiate into specific cell types associated with the relevant organs (Reiss et al., 2021).

However, regardless of the stem cell's origin, the basic demands concerning their biological environment are similar. In this section, we will have a look at some of the important stem cell niche parameters and how they can be maintained in a bioreactor.

4.1 Temperature

A suitable temperature is the most basic parameter of any cell culture and yet one of the most important ones in order to ensure cell viability and growth. Most mammalian cells are cultivated at a temperature of 37°C (Rubio et al., 2019), even though higher temperatures were described to be adequate for myoblast development, such as using 37–39°C in a pig cell model (Metzger et al., 2021). Other possible cell sources for cultured meat production include fish or birds. Fish cells are usually cultured at 15–30°C (Rubio et al., 2019). Avian muscle cells, on the other hand, are commonly incubated at around 38°C, but studies with deviant incubation temperatures were carried out as well (Clark et al., 2016; Reed et al., 2017; Reed et al., 2017).

Also, these studies suggest an influence of temperature differences on muscle cell proliferation and differentiation (Reed et al., 2017; Clark et al., 2016; Reed et al., 2017). Such indications of temperature affecting the cell fate highlight the importance of precise temperature control during the culturing process.

Within a bioreactor, the temperature can be measured and monitored by different sensor types. The most common ones are thermocouple and resistance sensors (Djisalov et al., 2021). In both cases, the sensor readings act as feedback for the control station to maintain a previously set temperature within the bioreactor. This is achieved by electrical heating jackets (Fig. 5)

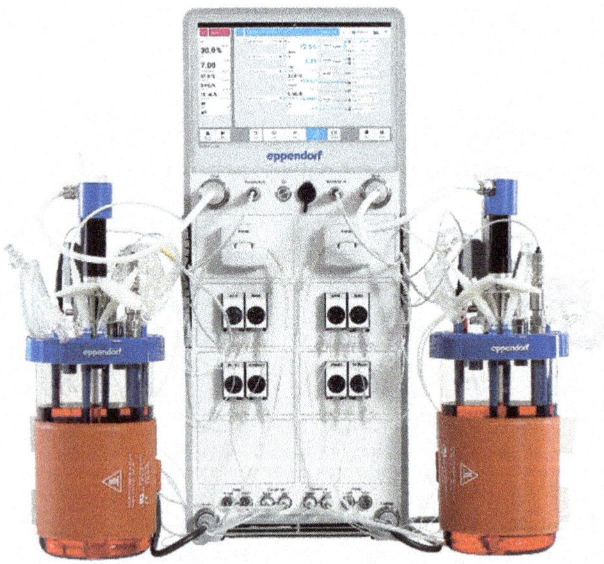

Figure 5 Two bioreactors wrapped in heating jackets and attached to a bioprocess control station. Besides electrical heating jackets, water jackets can be an option for temperature control in a bioreactor. The optimal growth temperature is regulated and maintained by the control station.

Published by Burleigh Dodds Science Publishing Limited, 2023.

wrapped around the bioreactor. Furthermore, certain glass bioreactors are enclosed by a water jacket, a surrounding glass cylinder in which heating or cooling water circulates.

4.2 Oxygen

Traditional *in vitro* cell cultures provide a static environment of ca. 5% CO_2 and 20% O_2 (Liu et al., 2013). However, the *in vivo* situation is much more variable with oxygen levels typically ranging from 3% to 9% in tissues and even 1.3% to 2.5% within a cell (Thomas, 2016). Such low oxygen tensions are termed 'hypoxic' and described to influence the stem cell fate.

For instance, oxygen levels of 1% maintained human stem cells at a primitive progenitor state and promoted self-renewal (Ivanovic et al., 2000; Ivanovic et al., 2002). Similar observations were made for human mesenchymal stem cells cultured at 2% oxygen which expressed more stem cell-related genes compared to cells incubated with 20% oxygen (Grayson et al., 2006). Additional effects of low oxygen tension on mesenchymal stem cells include reduced deoxyribonucleic acid damage (Estrada et al., 2012), better engraftment (Hung et al., 2007), and inhibition of senescence (Tsai et al., 2011). Also, the cultivation of old muscle satellite cells in a 3% oxygen atmosphere greatly increased their proliferation potential *in vitro* (Chakravarthy et al., 2001). Moreover, different oxygen tensions can affect the differentiation behavior of stem cells. Incubation of human mesenchymal stem cells at 5% O_2 resulted in chondrocyte development whereas incubation at 20% O_2 induced adipogenic differentiation (Lovett et al., 2010).

Those examples illustrate the impact of oxygen, as certain concentrations induce cell proliferation while others shift the cells toward differentiation. As the process of cultured meat production incorporates both proliferation and differentiation, being able to precisely control DO levels is of high importance. Using a bioreactor not only provides this control but also enables mid-process adjustments, for example, if a switch from pure proliferation toward differentiation is required.

The DO concentrations in a bioreactor are usually measured by one of three sensor types: electrochemical, optical, or paramagnetic (Djisalov et al., 2021). Regardless of the measuring method, the DO levels are processed by the control station which constantly compares the actual values to a setpoint. Deviations between the setpoint and measured value can be resolved through various countermeasures. The control station can use a cascade to adjust the DO levels. Depending on the requirements of the specific culture, DO levels can be upregulated, for example, by increasing the airflow rate, adding pure oxygen, or increasing the agitation of the impeller, or downregulated by mixing the four gases used in most bioreactors (air, oxygen, nitrogen, and carbon dioxide) accordingly, for example, by blending nitrogen into the air to lower the oxygen concentration.

4.3 pH

Another critical part of successful cell culturing is the correct pH value. The pH optimum for animal cells lies around 7.4 (Djisalov et al., 2021). However, uncontrolled pH in a bioprocess usually leads to medium pH changes due to the increase of by-products such as lactic acid (Lawson et al., 2017; Djisalov et al., 2021) and small deviations can affect cell survival and growth (Djisalov et al., 2021).

Indeed, there are indications of pH value fluctuations altering stem cell fate determination. For instance, short-term acidic pre-treatment of pH 6.8 was shown to enhance the viability and proliferation of murine bone marrow-derived mesenchymal stem cells, as well as an increased expression of stem cell markers (Hazehara-Kunitomo et al., 2019). Another study found that acidic conditions (pH 6.8) maintain the pluripotency of murine embryonic stem cells (Teo et al., 2014). Even though few studies on the direct impact of pH value exist, it is hypothesized that modulations affect stem cells fate-related phenomena, such as alternative mRNA splicing or mitochondrial activity (Kim, 2021). Thus, providing a stable pH is important to maintain optimal growth conditions.

In a bioreactor, this is typically monitored either by electrochemical or optical sensors (Djisalov et al., 2021). The feedback provided by the sensors enables the control station to adjust the pH value accordingly by either adding base or acid.

4.4 Lactic acid

Bioreactors enable dynamic feeding with nutrients such as glucose-containing solutions. However, metabolizing glucose results in increasing lactic acid levels during the culturing process with the potential to alter stem cell properties. Even though no direct mechanism has been proposed yet (Nalbandian et al., 2020), several studies suggest an impact on muscle-related stem cells and their progeny *in vitro*.

For example, treating human mesenchymal stem cells with low sodium lactate concentrations of 1 mM promoted stemness properties (Sun et al., 2020). Experiments with a murine myoblast cell line, the downstream differentiation step of muscle satellite cells, demonstrated that a daily 2 h exposure to lactic acid (10 mM and 20 mM) induces early but delays late differentiation marker expression in a time- and dose-dependent manner (Willkomm et al., 2014). Moreover, myotubes differentiated from this cell line showed reduced myogenesis after long-term treatment with 8 mM of lactic acid for 24 h (Oh et al., 2019). Furthermore, increased lactic acid levels can alter the pH of the medium (Djisalov et al., 2021), leading to the consequences discussed earlier.

Thus, monitoring lactic acid and other compounds such as glucose in a bioreactor and controlling them accordingly by removing used and adding fresh medium is an important tool to maintain optimal culture conditions (Fig. 6).

For instance, knowledge of the medium's nutrient and by-product composition enables the determination of medium feeding time points or the evaluation of medium quality in the process of medium recycling (Djisalov et al., 2021). For that, process analytic technology can be employed to analyze critical compound levels in the medium, either by an online or offline approach. Online means the analysis device is integrated into the feedback loop of the bioreactor control station software which then adjusts medium addition and removal automatically. Offline means removing the medium manually through specific sampling ports without disturbing the sterile conditions of the culture and analyzing it in an external device. The obtained values can then be used to adjust feeding and removal rates manually. Both online and offline metabolite monitoring was shown to be viable tools for process optimization of mammalian cell culture in stirred-tank bioreactors (Craven, 2015; Escobar Ivirico and Sha, 2020).

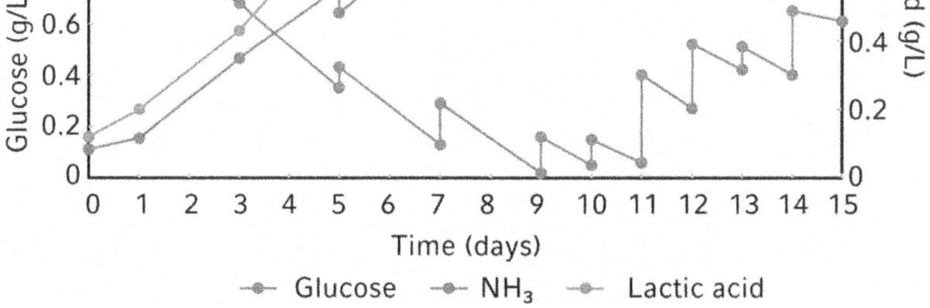

Figure 6 Example of controlling glucose, lactic acid, and ammonia levels during mesenchymal stem cell culture in a bioreactor (Escobar Ivirico and Sha, 2020). Lactic acid, a by-product of glycolysis, increases when glucose is consumed by the cells. Adjusting the glucose feeding rates and regular medium exchanges can help to keep the lactic acid levels from exceeding a critical value that can be harmful to the cells. In addition, medium exchanges can also be a measure to regulate the levels of other potentially toxic by-products, such as ammonia (NH_3).

Figure 6 illustrates how metabolite monitoring and feeding rate adjustment can maintain a balance of nutrients such as glucose and potentially harmful by-products, including lactic acid and also ammonia, a by-product of glutamine metabolism known for its toxic effects on mammalian cells (Escobar Ivirico and Sha, 2020; Mirabet et al., 1997; Moritz et al., 2015).

4.5 Culture mixing

Efficient nutrient distribution, gas-liquid mass transfer, and culture homogeneity are the key factors for successful bioprocessing. A stirred-tank bioreactor achieves this through the use of rotating impellers. However, mammalian cells are known to be shear-sensitive (Allan et al., 2019) and factors such as wrong impeller design, increasing agitation speeds, and bursting bubbles from an oxygen sparger can induce cell damage (Martens et al., 1996; Fan et al., 2015).

Furthermore, it was observed that certain shear forces can affect cell differentiation. For instance, fluid flow-induced shear stress was shown to drive human and goat mesenchymal stem cells toward osteogenic differentiation (Yourek et al., 2010; Knippenberg et al., 2005). While this is a noteworthy observation, it bears the risk of directing the cells to an unwanted cell fate determination.

Thus, shear stress regulation is crucial to steer the cells toward the optimal growth and differentiation path. In a stirred-tank bioreactor, shear forces can be regulated by choosing the right impeller type and adjusting the agitation speed. In our studies, pitched blade impellers (Fig. 7) combined with low agitation speed have emerged as a proven tool to efficiently mix shear-sensitive mammalian cells in a stirred-tank bioreactor, including mesenchymal and pluripotent stem cells (Olmer et al., 2015; Dufey et al., 2016; Siddiquee and Sha, 2014; Roldão et al., 2018; Escobar Ivirico and Sha, 2020).

Figure 7 A pitched eight-blade impeller. This impeller type with blade angles of 60°C is suitable for stem cell cultivation within a stirred-tank bioreactor.

5 Cultivating anchorage-dependent cells in stirred-tank bioreactors

Like most other tissue-derived cells, stem cells employed for cultured meat production are anchorage-dependent, meaning attachment to a surface is needed for normal proliferation (Merten, 2015). Therefore, scaffolds enabling cell adherence are required for efficient culturing. This aspect can be tackled by employing microcarriers. Microcarriers are beads with diameters of 60–300 μm (Bellani et al., 2020) that can be added to the culture medium, thereby providing a growth surface to the cells (Fig. 8) and increasing the surface-to-volume ratio within the bioreactor.

Microcarriers are available with varying porosity from a range of materials, including polystyrene, gelatin, dextran, and collagen (McKee and Chaudhry, 2017). Besides those materials, research is performed on biodegradable and edible microcarriers (Bodiou et al., 2020). These would be a valuable addition to cultured meat production as they have the potential to facilitate or even eliminate downstream microcarrier removal and cell-detachment steps which are laborious and can affect cell integrity (Reiss et al., 2021).

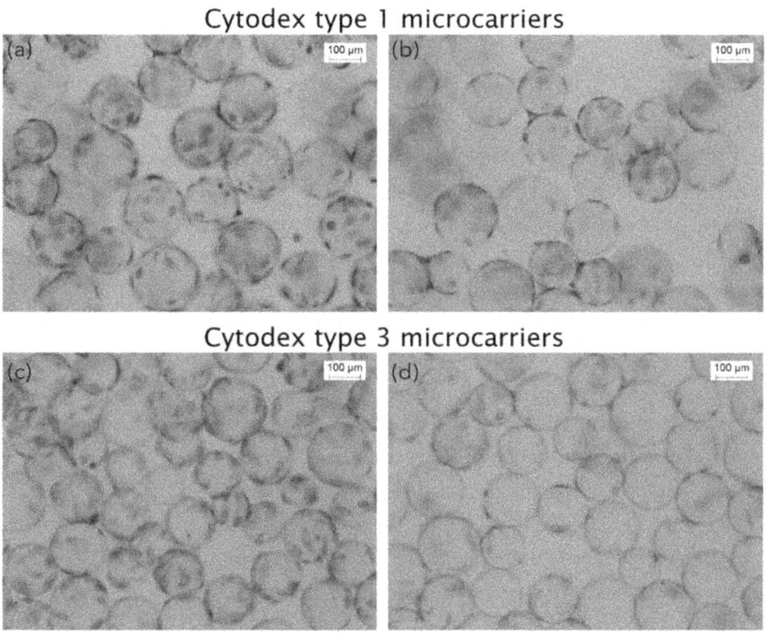

Figure 8 Viable mesenchymal stem cells (blue) on different microcarrier types (transparent spheres) (Dufey et al., 2016). Cells were stained with (3-(4,5-dimethylthiazol-2-yl)-2,5-diphenyltetrazolium bromide (MTT). Light microscopy pictures of cells (a) (c) before and (b) (d) 1 day after microcarrier addition to the culture demonstrate the quick transfer of cells to the new beads.

Moreover, microcarriers can be used in many bioreactor types and are a vital tool to facilitate scale-up. This is because the available growth surface during bioprocessing can be increased by simply adding fresh microcarriers to the bioreactor culture, a phenomenon facilitated by bead-to-bead transfer (Bodiou et al., 2020). Bead-to-bead transfer is the exchange of cells from one bead to another, probably either by direct bead-to-bead contact or pick-up of detached floating cells by microcarriers moving in the cell suspension (Verbruggen et al., 2018). Several studies, including our own, demonstrated the feasibility of microcarriers in culturing stem cells and their progeny in bioreactors. The cell types in these studies also include myoblasts, mesenchymal stem cells, and pluripotent stem cells (Verbruggen et al., 2018; Dufey et al., 2016; Siddiquee and Sha, 2014; Escobar Ivirico and Sha, 2020; Tejerina et al., 2021). Besides that, the differentiation of microcarrier-attached satellite cells and cells of a myoblast cell line into myotubes was observed in different bioreactor models (Molnar et al., 1997; Torgan et al., 2000; Moritz et al., 2015).

Another study explored the possibility of bead-to-bead transfer as a scale-up strategy with human mesenchymal stem cells. The study successfully employed the bead-to-bead transfer to scale up from 4 L to 50 L in stirred-tank bioreactors. Moreover, the work demonstrated that the addition of fresh microcarriers to a culture of cell-populated microcarriers followed by intermediate agitation increased proliferation compared to continuous agitation (Chen et al., 2021).

An alternative to culturing anchorage-dependent cells on microcarriers is letting them grow as cell aggregates. Aggregates are free-floating 3D spheroid structures consisting of agglomerated cells which act as growth surfaces to each other and are formed at high initial cell density (Moritz et al., 2015). In theory, this organization as a united cell structure more closely resembles the natural *in vivo* environment of stem cells than microcarrier cultures (Bodiou et al., 2020). Furthermore, with the possibility of protecting individual cells from shear forces, aggregate culture bears the potential to reach high cell densities (Allan et al., 2019).

However, the structure of aggregates can lead to spatial gradients of nutrients and especially oxygen (Allan et al., 2019; Reiss et al., 2021) resulting in heterogeneity concerning viability, proliferation, and differentiation (Fan et al., 2015; Wu et al., 2014). Understandably, the effects of nutrient diffusion seem to be dependent on the aggregate size (Wu et al., 2014). Therefore, it is important to control aggregate dimensions during cultivation, a process that is difficult to handle (Allan et al., 2019; Bodiou et al., 2020). Despite that, it was shown that step-by-step optimization of process conditions in a stirred-tank bioreactor can enable high numbers of viable induced pluripotent stem cells (iPSC) while maintaining their differentiation capability and controlling aggregate size. The study demonstrated an almost 70-fold increase of human iPSC numbers from initially 5 × 10^5 cells/mL to 33 × 10^6 cells/mL after 7 days of culture by systematically optimizing pH, culture feeding, and agitation rates in combination with *in silico*-directed process modelling (Manstein et al., 2021). The aggregate size was controlled by adjusting the agitation speed while using shear-protective substances to reduce cell stress.

At present, aggregates seem to be more relevant to pluripotent stem cell culture (Reiss et al., 2021). However, studies on other cell types are gaining momentum. Contemporary reports investigated the potential of mesenchymal stem cell aggregate culture in terms of their therapeutic use (Sart et al., 2014; Follin et al., 2016; Schäfer et al., 2016; Petrenko et al., 2017), yet an extensive evaluation in terms of cultured meat production is still pending. Also, a few reports on myogenic cell aggregates exist (Westerman et al., 2010; Wei et al., 2011; Hosoyama et al., 2013), even though cultivation was performed under static conditions. Suspension culture studies utilizing the myogenic C2C12 cell line resulted in aggregate formation. However, these aggregates expressed markers of quiescent satellite cells (Aguanno et al., 2019) which do not meet proliferation requirements (Bodiou et al., 2020). Future studies might shed light on the role of aggregate cultures in cultured meat production.

In this context, it is worth mentioning that microcarriers in cell culture can also form aggregates associated with cell proliferation inhibition (Bodiou et al., 2020). On the other hand, positive aspects of microcarrier aggregates are reported as well. A study suggests that inoculating a new culture with myoblast-loaded microcarrier aggregates might decrease the initial lag phase and promote scaling-up due to the protection of the cells from mechanical stress (Boudreault et al., 2001).

It is proposed that microcarrier aggregates might be a problem of mesenchymal stem cells rather than satellite cell culture, as the latter produces less extracellular matrix and therefore bears a reduced aggregation potential (Bodiou et al., 2020). Furthermore, other than cell aggregates the size control of microcarrier aggregates can be achieved more easily by simply adding more microcarriers to the culture (Verbruggen et al., 2018).

6 Scale-up of a bioprocess in bioreactors

As mentioned previously, scale-up is the increase of the bioprocess dimensions to meet the demand for larger working volumes and higher cell numbers. While all the described cell niche parameters are applicable for both small- and larger-scale approaches, scaling up is not just a one-step process of simply enlarging bioreactors and volumes.

Increased bioreactor geometries and working volumes can aggravate establishing homogeneous cell cultures by prolonged mixing times and induced temperature, nutrient, or oxygen gradients. To minimize the effects specified in the previous sentence and thus ensure reproducibility for different bioreactor scales, it is critical to maintain similar properties, such as geometry, but also special physiochemical scale-up parameters that were established over time (Suttle et al., 2021).

Among the commonly employed scale-up parameters are power input/volume ratio (P/V), impeller tip speed, constant volumetric mass transfer, and constant oxygen transfer rate (OTR).

A constant (impeller) P/V is one of the most widely used scale-up strategies; it can be calculated from the formula:

$$P/V = (Np \times \rho \times N^3 \times d^5)/V$$

where Np is the dimensionless impeller power or Newton number, ρ is the density of water (1000 kg/m³), N is the agitation speed (rps), d is the impeller's outer diameter (m), and V is the bioreactor's working volume (m³) (Suttle et al., 2021). The power number is experimentally determined for a given impeller by measuring its torque but can also be obtained from standard impeller references or vendors (Li et al., 2016; Yang, 2014).

The tip speed of an impeller influences oxygen transfer and mixing times but is also linked to shear forces, an important point to be kept in mind when working with sensitive cells. It can be calculated from the formula:

$$\text{tip speed} = \pi \times d \times N$$

where π is the ratio of a circle's circumference to its diameter (~3.14159), d is the impeller's outer diameter (m), and N is the agitation speed (rps) (Suttle et al., 2021).

The OTR describes the oxygen-to-liquid transfer in the bioreactor and is dependent on the volumetric mass transfer coefficient (k_La). k_La is a measure of the oxygen-to-bioreactor culture medium delivery efficiency for a given combination of parameter conditions (Suttle et al., 2021).

Ideally, all these parameters should be kept constant between different scale sizes. However, due to the physical properties of a bioreactor, this is not possible (Yang, 2014; Suttle et al., 2021). Therefore, the common practice is to select one or more parameters to be kept constant during the scale-up process (Suttle et al., 2021). As the choice depends on the process approach, it is important to test and identify critical culture parameters early in small-scale experiments before moving to larger dimensions.

Such small-scale bioreactors able to accommodate working volumes as small as 60 mL have been successfully used by us and others for efficient stem cell culture establishment (Olmer et al., 2015; Manstein et al., 2021; Dufey et al., 2016; Escobar Ivirico and Sha, 2020). Figure 9 showcases how the growth of a

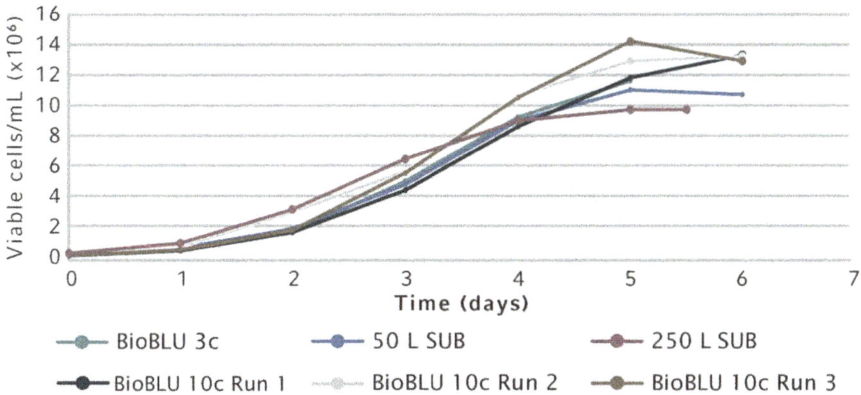

Figure 9 Examples of a scale-up experiment using a mammalian Chinese hamster ovary (CHO) cell line (Suttle et al., 2021). Scale-up was performed using four different single-use bioreactor (SUB) models with working volumes of 3 L (BioBLU 3c), 10 L (BioBLU 10c), 40 L (50 L SUB), and 150 L (250 L SUB). Applying a constant impeller power input per unit volume (P/V) resulted in comparable cell growth between all volumes over the course of the bioprocess run.

mammalian cell line can be kept consistent between different bioreactor sizes with working volumes of 3 L, 10 L, 40 L, and 150 L by maintaining a constant impeller power input per unit volume (P/V) (Suttle et al., 2021).

7 Considerations for cost reduction and process optimization in bioreactors

In order to create general acceptance in society for alternative meat options, it is important to offer the product at a competitive price. While production costs for cultured meat have tremendously decreased over the last years, they still surpass those of conventionally produced meat (Bodiou et al., 2020; Guan et al., 2021). During bioreactor cell cultivation and upscaling, resource efficiency is crucial to maintain the costs at reasonable levels. This means generating maximized yields with minimal use of resources. In this section, we will consider three possibilities with the potential to reduce production costs.

7.1 Medium recycling

In cultured meat production, the most impactful cost driver is the culture medium (Treich, 2021). One possibility to reduce its cost is changing the medium's composition and replace costly ingredients, such as the ethically questionable animal-derived fetal bovine serum, an issue that is reviewed elsewhere in this book. Another option in a bioreactor could be conserving the medium as efficiently as possible. This can be achieved by creating a loop to re-introduce the used medium back into the bioreactor multiple times.

Of course, for this recycling process, pretreatment of the medium is needed before its re-introduction to the cell culture. Utilized nutrients, like glucose and glutamine, need to be replenished and unwanted by-products, such as lactic acid or ammonia, need to be removed. In relation to by-product removal, this could be facilitated, for instance, by chromatographic methods (Moritz et al., 2015). For this reconditioning process, close monitoring of the medium properties is necessary to enable proper adjustments accordingly (Guan et al., 2021). In a bioreactor, critical parameters can already be monitored routinely. Others may be determined by online or offline sampling.

This optimization effort could result in a more resource- and cost-efficient production process. Medium recycling could be, for example, implemented in the continuous cultivation operation mode. Furthermore, it has been suggested that the used medium might be enriched with growth factors produced by the cultured cells which, in turn, could promote subsequent cell proliferation upon medium re-introduction (Moritz et al., 2015). Thus, medium recycling bears the potential to reduce process cost and enhance cell growth.

7.2 Combining cell proliferation and differentiation in one bioreactor

Another consideration for cost reduction is to establish an approach that combines stem cell proliferation and differentiation within the same bioreactor. The usual practice of growing myotubes from muscle stem cells is to expand the cells in one bioreactor before transferring them into another reactor for differentiation (Moritz et al., 2015). This process is laborious and includes a relatively sudden change in the stem cell growth environment.

Thus, combining the two steps into one bioreactor would reduce work and open the possibility to gradually change the environment from proliferation to differentiation, for example, by altering growth factor concentrations within the medium during continuous cultivation. Molnar and Torgan identified precursor differentiation into myotubes taking place on microcarriers (Molnar et al., 1997; Torgan et al., 2000). While microcarriers are mainly used for the cell expansion step, this may provide a model for combining proliferation and differentiation steps in one bioreactor by altering the culture conditions. It is an interesting observation, yet it remains to be determined if such structures are able to develop into mature muscle fibers with the same physical properties as the ones grown by the established process (Moritz et al., 2015). Also, by having differentiated myotubes attached to the surface, detachment for downstream processing bears the risk of damaging the structures. Thus, the availability of edible microcarriers would gain importance in this setting.

Further investigation is still needed to clarify this possibility. However, if it could be realized, combining the proliferation and differentiation process in one vessel would save time, material, and costs during cultured meat production.

7.3 Single-use technology

It might appear counterintuitive at first, but another cost-saving and environmentally friendly strategy could be found in disposable products. When it comes to choosing a bioreactor, two main options exist: reusable units, such as traditional stainless steel bioreactors, or single-use bioreactors, such as bags or vessels usually made from plastic compounds (Fig. 10). Re- or multi-usable bioreactors are an established and reliable tool for bioprocessing, but single-use technology also has its advantages.

As cultured meat technology is still in the beginning stages of establishing large-scale production processes and facilities, an evaluation of established industries might be helpful for decision-making.

A possible model is the pharmaceutical industry, where single-use technology gained momentum over the years, for instance, in the production of monoclonal antibodies. Here, reduced downtime between batches, minimal

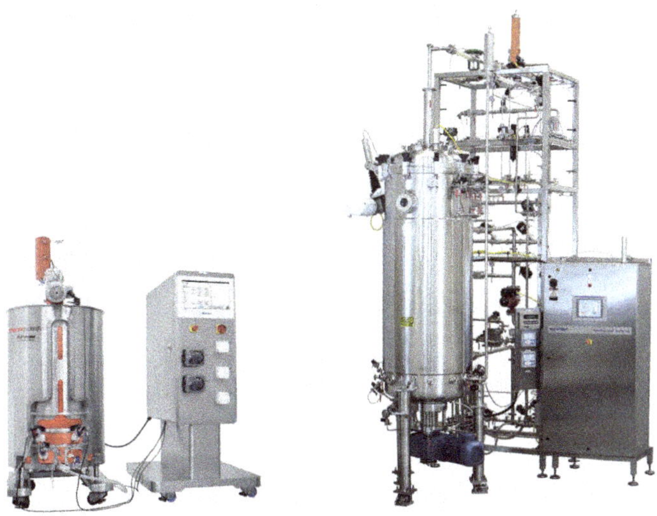

Figure 10 Examples of a single-use stirred-tank bioreactor using a disposable plastic bag for cell culture (left) and a multi-use stainless steel stirred-tank bioreactor (right), both attached to a bioreactor control system.

cleaning effort and decreased contamination risks are strong arguments for single-use products (Allan et al., 2019). Admittedly, even though available working volumes are increasing, the current maximum working volumes of single-use bioreactors are typically smaller than their stainless steel counterparts, ranging from several thousand liters in single-use to tens of thousands of liters in stainless steel reactors. However, scale-out, the process of using multiple smaller units in parallel, might be an alternative to the classical scale-up approach and could compensate for reduced volumetric capacities. Furthermore, operating multiple bioreactors in parallel potentially reduces the risk of contamination-induced batch loss. In general, scale-out offers a more flexible process operation and performance improvement, while scale-up enables better reproducibility at the expense of handling simplicity (Bellani et al., 2020).

Single-use technology also bears the potential for cost reduction at the bioreactor facility level. Compared to a stainless steel bioreactor facility for monoclonal antibody production, the use of disposable bioreactors was estimated to reduce costs significantly, mainly due to less labor, maintenance, and waste (Jacquemart et al., 2016). Another study concluded that despite increased production of plastic waste, a single-use facility would have an overall reduced environmental impact compared to a traditional stainless steel bioreactor facility. The authors explain this by reduced water and energy consumption, less space requirements, and an overall smaller carbon footprint (Sinclair et al., 2008). However, in order to keep a single-use facility productive, consistent quality, reproducibility, and availability of single-use components need to be ensured by the supplier. Additionally, the plastic material needs to comply with threshold values for extractable and leachable compounds (Jacquemart et al., 2016), a criterion with special significance for food products like cultured meat.

Thus, future cost analyses will be essential to determine if single-use or reusable bioreactors are the right approach for cultured meat production (Allan et al., 2019).

8 Conclusion and future trends

The world's population is projected to reach 8 billion in late 2022 and further growth is expected in the years to come (United Nations Department of Economic and Social Affairs; Population Division, 2022). Such an expansion of humanity will increase the demand for meat, a trend that will be harder and harder to satisfy with conventional meat production methods. Moreover, conventional meat production is increasingly tied to negative impacts on animal welfare, environment, and climate. The concept of cultured meat offers an exciting new avenue to address the social, ethical, and environmental challenges.

However, tacking up challenges like those mentioned in the previous sentences entails its own trails. Introducing cultured meat to the general public requires a step-up in cell numbers and productivity, as well as ensuring a stable and maintainable growth environment for the cells to provide consistent quality; all at a price that customers are willing to pay.

To this end, bioreactors are a powerful tool to help reach this goal. Established in other industries for decades and with an ever-expanding application repertoire, bioreactors are both a proven and future-proof technology that can also aid cultured meat production.

Their ability to precisely control a diversity of growth parameters facilitates the creation of a 3D environment for various cell types to thrive. Several bioreactor types, operation modes, and accessories enable process structures tailored to one's specific needs. Scale-up and scale-out options, as well as cell density-increasing measures, including the use of microcarriers, can help to augment cell yields and efficiency. Furthermore, process adaptation can be used to reduce costs and the environmental footprint.

Constant improvement in bioreactor and process control design, material options, and culturing methods will continue to push the boundaries of what is possible and what is feasible in cell culture technology.

In conclusion, bioreactors are a vital tool to overcome current and future challenges that the relatively new field of novel food is facing these days.

9 Where to look for further information

A useful introduction to stirred-tank bioreactors, bioreactor accessories, incubation parameters, and a typical bioprocess run can be found in this white paper by Eppendorf:

- Rasche, U. (2019). Bioreactors and fermentors-powerful tools for resolving cultivation bottlenecks. *Eppendorf White Paper* 21.

A detailed explanation of the different bioreactor operation modes along with a comparative cost analysis are available in the following application note by Eppendorf:

- Yang, Y. and Sha, M. (2019). A beginner's guide to bioprocess modes-batch, fed-batch, and continuous fermentation. *Eppendorf Application Note* 408.

Published by Burleigh Dodds Science Publishing Limited, 2023.

Information about important scale-up parameters can be accessed by reading the following publications:

- Li, B., Becken, U. and Sha, M. (2016). Fermentation systems for bioprocess scale-up from small to pilot/production. *Genetic Engineering and Biotechnology News* 36(9).
- Suttle, A., Rasche, U. and Sha, M. (2021). From R&D to production: How the bioreactor choice can streamline biologics scale-up. *International Pharmaceutical Industry* 13(3), 60–62.

10 References

Aguanno, S., Petrelli, C., di Siena, S., de Angelis, L., Pellegrini, M. and Naro, F. (2019). A three-dimensional culture model of reversibly quiescent myogenic cells. *Stem Cells International* 2019, 7548160.

Allan, S. J., de Bank, P. A. and Ellis, M. J. (2019). Bioprocess design considerations for cultured meat production with a focus on the expansion bioreactor. *Frontiers in Sustainable Food Systems* 3, 44.

Al-Mashhadani, M. K. H., Wilkinson, S. J. and Zimmerman, W. B. (2015). Airlift bioreactor for biological applications with microbubble mediated transport processes. *Chemical Engineering Science* 137, 243–253.

Bartczak, M., Wierzchowski, K. and Pilarek, M. (2022). Mixing performance in a litre-scale rocking disposable bioreactor: DoE-based investigation of mixing time dependence on operational parameters. *Chemical Engineering Journal* 431, 133288.

Bellani, C. F., Ajeian, J., Duffy, L., Miotto, M., Groenewegen, L. and Connon, C. J. (2020). Scale-up technologies for the manufacture of adherent cells. *Frontiers in Nutrition* 7, 575146.

Bodiou, V., Moutsatsou, P. and Post, M. J. (2020). Microcarriers for upscaling cultured meat production. *Frontiers in Nutrition* 7, 10.

Boudreault, P., Tremblay, J. P., Pépin, M. F. and Garnier, A. (2001). Scale-up of a myoblast culture process. *Journal of Biotechnology* 91(1), 63–74.

Chakravarthy, M. V., Spangenburg, E. E. and Booth, F. W. (2001). Culture in low levels of oxygen enhances in vitro proliferation potential of satellite cells from old skeletal muscles. *Cellular and Molecular Life Sciences: CMLS* 58(8), 1150–1158.

Chen, S., Sato, Y., Tada, Y., Suzuki, Y., Takahashi, R., Okanojo, M. and Nakashima, K. (2021). Facile bead-to-bead cell-transfer method for serial subculture and large-scale expansion of human mesenchymal stem cells in bioreactors. *Stem Cells Translational Medicine* 10(9), 1329–1342.

Clark, D. L., Coy, C. S., Strasburg, G. M., Reed, K. M. and Velleman, S. G. (2016). Temperature effect on proliferation and differentiation of satellite cells from Turkeys with different growth rates. *Poultry Science* 95(4), 934–947.

Craven, S. (2015). A quality-by-design approach to upstream bioprocess interrogation and intensification. *Eppendorf Application Note* 300.

Djisalov, M., Knežić, T., Podunavac, I., Živojević, K., Radonic, V., Knežević, N. Ž., Bobrinetskiy, I. and Gadjanski, I. (2021). Cultivating multidisciplinarity: Manufacturing and sensing challenges in cultured meat production. *Biology* 10(3), 204.

Dufey, V., Tacheny, A., Art, M., Becken, U. and de Longueville, F. (2016). Expansion of human bone marrow-derived mesenchymal stem cells in BioBLU® 0.3c single-use bioreactors. *Eppendorf Application Note* 305, 1-8.

Escobar Ivirico, J. L. and Sha, M. (2020). Stem cell exosome production on the SciVario® twin, a flexible controller for your bioprocess needs. *Eppendorf Application Note* 435.

Estrada, J. C., Albo, C., Benguría, A., Dopazo, A., López-Romero, P., Carrera-Quintanar, L., Roche, E., Clemente, E. P., Enríquez, J. A., Bernad, A. and Samper, E. (2012). Culture of human mesenchymal stem cells at low oxygen tension improves growth and genetic stability by activating glycolysis. *Cell Death and Differentiation* 19(5), 743–755.

Fan, Y., Jing, D. and Tzanakakis, E. S. (2015). Bioreactors and the design of the stem cell niche. In: *Biology in Stem Cell Niche*, pp. 107–128.

Follin, B., Juhl, M., Cohen, S., Perdersen, A. E., Kastrup, J. and Ekblond, A. (2016). Increased paracrine immunomodulatory potential of mesenchymal stromal cells in three-dimensional culture. *Tissue Engineering: Part B, Reviews* 22(4), 322–329.

Frank, N. D., Jones, M. E., Vang, B. and Coeshott, C. (2019). Evaluation of reagents used to coat the hollow-fiber bioreactor membrane of the Quantum® cell expansion system for the culture of human mesenchymal stem cells. *Materials Science and Engineering C, Materials for Biological Applications* 96, 77–85.

Grayson, W. L., Zhao, F., Izadpanah, R., Bunnell, B. and Ma, T. (2006). Effects of hypoxia on human mesenchymal stem cell expansion and plasticity in 3D constructs. *Journal of Cellular Physiology* 207(2), 331–339.

Guan, X., Lei, Q., Yan, Q., Li, X., Zhou, J., Du, G. and Chen, J. (2021). Trends and ideas in technology, regulation and public acceptance of cultured meat. *Future Foods* 3, 100032.

Hazehara-Kunitomo, Y., Hara, E. S., Ono, M., Aung, K. T., Komi, K., Pham, H. T., Akiyama, K., Okada, M., Oohashi, T., Matsumoto, T. and Kuboki, T. (2019). Acidic pre-conditioning enhances the stem cell phenotype of human bone marrow stem/progenitor cells. *International Journal of Molecular Sciences* 20(5), 1097.

Hosoyama, T., Meyer, M. G., Krakora, D. and Suzuki, M. (2013). Isolation and in vitro propagation of human skeletal muscle progenitor cells from fetal muscle. *Cell Biology International* 37(2), 191–196.

Hung, S. C., Pochampally, R. R., Hsu, S. C., Sanchez, C. C., Chen, S. C., Spees, J. and Prockop, D. J. (2007). Short-term exposure of multipotent stromal cells to low oxygen increases their expression of CX3CR1 and CXCR4 and their engraftment in vivo. *PLoS ONE* 2(5), e416.

Ivanovic, Z., Belloc, F., Faucher, J. L., Cipolleschi, M. G., Praloran, V. and Dello Sbarba, P. (2002). Hypoxia maintains and interleukin-3 reduces the pre-colony-forming cell potential of dividing CD34 murine bone marrow cells. *Experimental Hematology* 30(1), 67–73.

Ivanovic, Z., Dello Sbarba, P., Trimoreau, F., Faucher, J. L. and Praloran, V. (2000). Primitive human HPCs are better maintained and expanded in vitro at 1 percent oxygen than at 20 percent. *Transfusion* 40(12), 1482–1488.

Jacquemart, R., Vandersluis, M., Zhao, M., Sukhija, K., Sidhu, N. and Stout, J. (2016). A single-use strategy to enable manufacturing of affordable biologics. *Computational and Structural Biotechnology Journal* 14, 309–318.

Kim, N. (2021). pH variation impacts molecular pathways associated with somatic cell reprogramming and differentiation of pluripotent stem cells. *Reproductive Medicine and Biology* 20(1), 20–26.

Knippenberg, M., Helder, M. N., Zandieh Doulabi, B. Z., Semeins, C. M., Wuisman, P. I. J. M. and Klein-Nulend, J. (2005). Adipose tissue-derived mesenchymal stem cells acquire bone cell-like responsiveness to fluid shear stress on osteogenic stimulation. *Tissue Engineering* 11(11–12), 1780–1788.

Lawson, T., Kehoe, D. E., Schnitzler, A. C., Rapiejko, P. J., Der, K. A., Philbrick, K., Punreddy, S., Rigby, S., Smith, R., Feng, Q., Murrell, J. R. and Rook, M. S. (2017). Process development for expansion of human mesenchymal stromal cells in a 50 L single-use stirred tank bioreactor. *Biochemical Engineering Journal* 120, 49–62.

Li, B., Becken, U. and Sha, M. (2016). Tackling the challenge of scalability: Fermentation systems for bioprocess scale-up from small to pilot/production. *Genetic Engineering and Biotechnology News* 36(9), 28–29. Available at: www.eppendorf.com.

Liu, M., Liu, N., Zang, R., Li, Y. and Yang, S. T. (2013). Engineering stem cell niches in bioreactors. *World Journal of Stem Cells* 5(4), 124–135.

Lovett, M., Rockwood, D., Baryshyan, A. and Kaplan, D. L. (2010). Simple modular bioreactors for tissue engineering: A system for characterization of oxygen gradients, human mesenchymal stem cell differentiation, and prevascularization. *Tissue Engineering. Part C, Methods* 16(6), 1565–1573.

Manstein, F., Ullmann, K., Kropp, C., Halloin, C., Triebert, W., Franke, A., Farr, C. M., Sahabian, A., Haase, A., Breitkreuz, Y., Peitz, M., Brüstle, O., Kalies, S., Martin, U., Olmer, R. and Zweigerdt, R. (2021). High density bioprocessing of human pluripotent stem cells by metabolic control and in silico modeling. *Stem Cells Translational Medicine* 10(7), 1063–1080.

Martens, D. E., Nollen, E. A. A., Hardeveld, M., van der Velden-de Groot, C. A. M., de Gooijer, C. D., Beuvery, E. C. and Tramper, J. (1996). Death rate in a small air-lift loop reactor of vero cells grown on solid microcarriers and in macroporous microcarriers. *Cytotechnology* 21(1), 45–59.

McKee, C. and Chaudhry, G. R. (2017). Advances and challenges in stem cell culture. *Colloids and Surfaces. B, Biointerfaces* 159, 62–77.

Merten, O. W. (2015). Advances in cell culture: Anchorage dependence. *Philosophical Transactions of the Royal Society of London: Series B, Biological Sciences* 370(1661), 20140040.

Metzger, K., Dannenberger, D., Tuchscherer, A., Ponsuksili, S. and Kalbe, C. (2021). Effects of temperature on proliferation of myoblasts from donor piglets with different thermoregulatory maturities. *BMC Molecular and Cell Biology* 22(1), 36.

Mirabet, M., Navarro, A., Lopez, A., Canela, E. I., Mallol, J., Lluis, C. and Franco, R. (1997). Ammonium toxicity in different cell lines. *Biotechnology and Bioengineering* 56(5), 530–537.

Molnar, G., Schroedl, N. A., Gonda, S. R. and Hartzell, C. R. (1997). Skeletal muscle satellite cells cultured in simulated microgravity. *In Vitro Cellular and Developmental Biology Animal* 33(5), 386–391.

Moritz, M. S. M., Verbruggen, S. E. L. and Post, M. J. (2015). Alternatives for large-scale production of cultured beef: A review. *Journal of Integrative Agriculture* 14(2), 208–216.

Nalbandian, M., Radak, Z. and Takeda, M. (2020). Lactate metabolism and satellite cell fate. *Frontiers in Physiology* 11, 610983.

Oh, S. S. J., Kim, S., Moon, S., Park, D. H. and Kang, J. H. (2019). Lactate overload inhibits myogenic activity in C2C12 myotubes. *Open Life Sciences* 14(1), 29–37.

Olmer, R., Kropp, C., Huether-Franken, C., Schlottbom, C. and Zweigerdt, R. (2015). Scalable expansion of human pluripotent stem cells in Eppendorf BioBLU® 0.3 single-use bioreactors. *Eppendorf Application Note* 292, 1–4.

Pajčin, I., Knežić, T., Azoulay, I. S., Vlajkov, V., Djisalov, M., Janjušević, L., Grahovac, J. and Gadjanski, I. (2022). Bioengineering outlook on cultivated meat production. *Micromachines* 13(3), 402.

Petrenko, Y., Syková, E. and Kubinová, Š. (2017). The therapeutic potential of three-dimensional multipotent mesenchymal stromal cell spheroids. *Stem Cell Research and Therapy* 8(1), 94.

Post, M. J. (2014). Cultured beef: Medical technology to produce food. *Journal of the Science of Food and Agriculture* 94(6), 1039–1041.

Rasche, U. (2019). Bioreactors and fermentors-powerful tools for resolving cultivation bottlenecks. *Eppendorf White Paper* 21.

Reed, K. M., Mendoza, K. M., Abrahante, J. E., Barnes, N. E., Velleman, S. G. and Strasburg, G. M. (2017). Response of turkey muscle satellite cells to thermal challenge. I. Transcriptome effects in proliferating cells. *BMC Genomics* 18(1), 352.

Reed, K. M., Mendoza, K. M., Strasburg, G. M. and Velleman, S. G. (2017). Response of Turkey muscle satellite cells to thermal challenge. II. Transcriptome effects in differentiating cells. *Frontiers in Physiology* 8(V), 948.

Reiss, J., Robertson, S. and Suzuki, M. (2021). Cell sources for cultivated meat: Applications and considerations throughout the production workflow. *International Journal of Molecular Sciences* 22(14), 7513.

Rodrigues, C. A. V., Fernandes, T. G., Diogo, M. M., da Silva, C. L. and Cabral, J. M. S. (2011). Stem cell cultivation in bioreactors. *Biotechnology Advances* 29(6), 815–829.

Roldão, A., Sousa, M., Santo, V. E., Becken, U. and Serra, M. (2018). Culture of 3D cell aggregates in perfusion in a DASbox® mini bioreactor system. *Eppendorf Application Note* 360.

Rubio, N., Datar, I., Stachura, D., Kaplan, D. and Krueger, K. (2019). Cell-based fish: A novel approach to seafood production and an opportunity for cellular agriculture. *Frontiers in Sustainable Food Systems* 3, 43.

Sart, S., Tsai, A. C., Li, Y. and Ma, T. (2014). Three-dimensional aggregates of mesenchymal stem cells: Cellular mechanisms, biological properties, and applications. *Tissue Engineering: Part B, Reviews* 20(5), 365–380.

Schäfer, R., Spohn, G. and Baer, P. C. (2016). Mesenchymal stem/stromal cells in regenerative medicine: Can preconditioning strategies improve therapeutic efficacy? *Transfusion Medicine and Hemotherapy: Offizielles Organ Der Deutschen Gesellschaft Fur Transfusionsmedizin und Immunhamatologie* 43(4), 256–267.

Siddiquee, K. and Sha, M. (2014). Large-scale production of human mesenchymal stem cells in BioBLU ® 5c single-use bioreactors. *Eppendorf Application Note* 334.

Sinclair, A., Leveen, L., Monge, M., Lim, J. and Cox, S. (2008). The environmental impact of disposable technologies can disposables reduce your facility's environmental footprint? *Biopharm International* 2008(7).

Sun, H., Zhang, X., Dai, J., Pan, Z., Wu, Y., Yu, D., Zhu, S., Chen, Y., Qin, T. and Ouyang, H. (2020). Sodium lactate promotes stemness of human mesenchymal stem cells through KDM6B mediated glycolytic metabolism. *Biochemical and Biophysical Research Communications* 532(3), 433–439.

Suttle, A., Longsworth, J. and Sha, M. (2021). From shaker to pilot/production bioreactor: How scale up assist using the BioFlo® 720 bioreactor control system can help your antibody production workflow. *Eppendorf Application Note*, p. 436.

Suttle, A., Rasche, U. and Sha, M. (2021). From R&D to production: How the bioreactor choice can streamline biologics scale-up. *International Pharmaceutical Industry* 13(3), 60–62. Available at: www.international-pharma.com.

Tejerina, S., Dufey, V., Hoet, J.-F., Tacheny, A. and de Longueville, F. (2021). The DASbox® mini bioreactor system as a tool for process development and stem-cell derived exosome production in standardized culture conditions. *Eppendorf Application Note* 440.

Teo, A., Mantalaris, A. and Lim, M. (2014). Influence of culture pH on proliferation and cardiac differentiation of murine embryonic stem cells. *Biochemical Engineering Journal* 90, 8–15.

Thomas, K. A. (2016). Angiogenesis. In: *Encyclopedia of Cell Biology*. Waltham: Elsevier, pp. 102–116.

Torgan, C. E., Burge, S. S., Collinsworth, A. M., Truskey, G. A. and Kraus, W. E. (2000). Differentiation of mammalian skeletal muscle cells cultured on microcarrier beads in a rotating cell culture system. *Medical and Biological Engineering and Computing* 38(5), 583–590.

Treich, N. (2021). Cultured meat: Promises and challenges. *Environmental and Resource Economics* 79(1), 33–61.

Tsai, C. C., Chen, Y. J., Yew, T. L., Chen, L. L., Wang, J. Y., Chiu, C. H. and Hung, S. C. (2011). Hypoxia inhibits senescence and maintains mesenchymal stem cell properties through down-regulation of E2A-p21 by HIF-TWIST. *Blood* 117(2), 459–469.

United Nations Department of Economic and Social Affairs. (2022). *World Population Prospects 2022: Summary of Results*. Population Division 2022. Available at: UN DESA/POP/2022/TR/NO. 3.

Verbruggen, S., Luining, D., van Essen, A. and Post, M. J. (2018). Bovine myoblast cell production in a microcarriers-based system. *Cytotechnology* 70(2), 503–512.

Wei, Y., Li, Y., Chen, C., Stoelzel, K., Kaufmann, A. M. and Albers, A. E. (2011). Human skeletal muscle-derived stem cells retain stem cell properties after expansion in myosphere culture. *Experimental Cell Research* 317(7), 1016–1027.

Westerman, K. A., Penvose, A., Yang, Z., Allen, P. D. and Vacanti, C. A. (2010). Adult muscle 'stem' cells can be sustained in culture as free-floating myospheres. *Experimental Cell Research* 316(12), 1966–1976.

Willkomm, L., Schubert, S., Jung, R., Elsen, M., Borde, J., Gehlert, S., Suhr, F. and Bloch, W. (2014). Lactate regulates myogenesis in C2C12 myoblasts in vitro. *Stem Cell Research* 12(3), 742–753.

Wu, J., Rostami, M. R., Cadavid Olaya, D. P. and Tzanakakis, E. S. (2014). Oxygen transport and stem cell aggregation in stirred-suspension bioreactor cultures. *PLoS ONE* 9(7), e102486.

Yang, X. (2014). Scale-up of microbial fermentation process. In: *Manual of Industrial Microbiology and Biotechnology*. Washington, DC: ASM Press, pp. 669–675.

Yang, Y. and Sha, M. (2019). A beginner's guide to bioprocess modes-batch, fed-batch, and continuous fermentation. *Eppendorf Application Note* 408.

Yourek, G., McCormick, S. M., Mao, J. J. and Reilly, G. C. (2010). Shear stress induces osteogenic differentiation of human mesenchymal stem cells. *Regenerative Medicine* 5(5), 713–724.

Published by Burleigh Dodds Science Publishing Limited, 2023.

Chapter 8

Tissue engineering using scaffolds in cultured meat product development

Iris Ianovici, Idit Goldfracht, Yana Shaulov Dvir, Hagit Shoyhat and Shulamit Levenberg, Technion - Israel Institute of Technology, Israel

1 Introduction

To address the environmental, health, and ethical issues associated with the large-scale animal slaughter-based meat industry, significant efforts are being made to develop cultured meat (also known as cultivated meat or cell-based meat; CM). The main principle of CM is the production of meat products via cellular agriculture (CA), which implements tissue engineering (TE) techniques to fabricate edible muscle and fat tissues. Although extensive research has been conducted in TE in the context of regenerative medicine, its application toward CM production only began approximately 20 years ago. To make significant progress in the field, both existing and novel TE technologies must be harnessed, supported by the combined efforts of TE, food engineering, and biomaterial engineering experts.

In TE, cells are often grown on scaffolds in controlled environments within bioreactors, with the use of cultivation media and chemical, mechanical, or electrical cues - all orchestrated into mimicking the natural microenvironment as closely as possible. The scaffolding materials are typically used as three-dimensional (3D) frameworks, which must meet several key criteria for adequate tissue formation and function. However, apart from the functional requirements for muscle and fat TE, scaffolds for CM purposes must meet additional demands, such as nutritional and sensorial attributes, toxicity, sustainability, and scale-up possibilities. The feasibility of creating such scaffolds leans on the

http://dx.doi.org/10.19103/AS.2023.0130.10

ability to tailor the used biomaterial composition and fabrication technique for the variety of desired CM products. Generally, there are three potential material origins that could be considered for CM:

- Plant-/algae-/fungus-/bacteria-derived biomaterials;
- Recombinant ECM molecules; and
- Synthetic polymers.

In terms of material processing into 3D cell cultivation platforms, several types of constructs and respective fabrication techniques can be used:

- Microcarriers;
- Porous constructs;
- Fibrous scaffolds;
- Hydrogels; and
- 3D-printed scaffolds.

Thus, the key criteria that must be met for this application, possible scaffolding compositions, and their processing methods into 3D platforms for cell culture will be elaborated in this chapter, along with the latest innovations in the field.

2 Key scaffold properties as three-dimensional meat cultivation platforms

2.1 Tissue engineering-related criteria

In general, CM engineering is essentially a sub-discipline of TE, with the ultimate aim of producing edible meat products. In both cases, the engineered tissues are achieved through the controlled proliferation and subsequent differentiation of progenitor cells into their desired mature forms of interest. In the context of CM, once the cells reach a certain maturation stage, they can be used to create the final desired products, which will preferably be composed of skeletal muscle, connective tissue, and adipose tissue (Bomkamp et al., 2022). Such overall methodology is relevant to fabricate both structured (steak/fillet) and unstructured (sausages/ground meat, etc.) meat. *In vitro* tissue formation is usually achieved via cell cultivation on 3D scaffolds which facilitate the process by mimicking the natural cellular microenvironment, i.e. the extracellular matrix (ECM). *In vivo*, the composition of the ECM, including secreted proteins and proteoglycans, and its subsequent physical and bio-chemical attributes has a substantial influence on cell behavior. Thus, all scaffolds used for TE, and more specifically CM, must feature several key characteristics to resemble the native ECM and allow for adequate muscle and adipose tissue formation. Scaffolds designed for CM must possess suitable physical properties, such as cell type-tailored porosity, architecture, and stiffness, while supporting cell adhesion, without imparting cytotoxic effects.

2.1.1 Scaffold porosity and architecture

Among the most important features of cell cultivation platforms are their porosity and inner architecture (Lutzweiler et al., 2020; Reddy et al., 2021). Currently, CM does not require the development of cell-based vasculature for organoleptic or nutritional purposes. However, to produce 3D CM resembling conventional meat and to support the survival of thick tissues, an alternative to the natural vasculature present *in vivo*, such as the use of highly porous scaffolds with an interconnected structure, is crucial (Chen et al., 2022). Such architecture can promote cell growth and uniform distribution of cells, as sufficient material transfer and inner voids enable the proliferation, migration, and overall organization of cells throughout the construct. For this purpose, the recommended overall porosity of non-vascularized cultivation platforms should reach ~90%, although several works demonstrated successful cell cultivation on scaffolds with lower porosities (Song et al., 2017; Bomkamp et al., 2022). Moreover, although the exact architecture (shape/channel distribution) of natural vascular networks should not necessarily be recapitulated for CM, channel-structured pores could be used for sufficient material and oxygen diffusion to every cell in the construct (e.g. should be ~200 μm away from any given cell). Importantly, while higher porosities improve nutrient diffusion and hydraulic permeability, they generally inversely correlate with the scaffold's mechanical properties, which can impact cell behavior (Lutzweiler et al., 2020). In addition to their role in material transfer and uniform cell distribution, pore size and pore shape are key determinants of cell behavior. More specifically, previous studies found that pores of 50–150 μm or larger are suitable for myogenic cells, while 40–400 μm pores (with emphasis on 170 μm) are advised for adipogenic cells (Bomkamp et al., 2022). In both cases, an optimized pore size can allow for adequate cell maturation; myogenic progenitors will be able to fuse into multinucleated myofibers, while adipogenic progenitors will mature into large lipid droplet-containing adipocytes. Moreover, numerous studies demonstrated that highly anisotropic cultivation platforms promoted the unidirectional alignment and fusion of muscle cells. Such scaffolds were already shown to be fabricated by various methods such as electrospinning (Choi et al., 2008), unidirectional freeze-drying (Pot et al., 2015), and 3D printing (Kang et al., 2016). Alternatively, anisotropic structures of decellularized plants were also found as promising CM scaffolds in that aspect. Taken together, the use CM 3D scaffolds with suitable porosity and architecture can orchestrate myogenesis and adipogenesis - which are essential to recapitulate the flavor, texture, and nutritional value of conventional meat.

2.1.2 Scaffold mechanical properties

In addition to the inner structure and porosity of the scaffold, cells are sensitive to the stiffness of their surroundings. This process termed 'mechanotransduction' occurs naturally *in vivo* and is mediated by integrins on the cell membrane and the ECM. Mechanosensing triggers cell type-specific effects on cell proliferation, migration, and differentiation. For example, studies have shown that while satellite cells (muscle progenitors; SC) proliferate well on constructs with muscle-like stiffness (2–12 kPa), their differentiation is further enhanced on stiffer ones (Gilbert et al., 2010; Bodiou et al., 2020). Therefore, Bodiou et al. concluded that constructs with 11–21 kPa stiffness could apply to both stages of cell cultivation (Bodiou et al., 2020). In other works, improved myoblast spreading and proliferation were noted on viscoelastic compared to elastic hydrogels of the same stiffness, underscoring the potential to adjust scaffold viscoelasticity to optimize cell expansion (Ng and Kurisawa, 2021). In the context of adipose tissue development, mesenchymal stem cells (MSCs) or adipose-derived stem cells (ASCs) are often used. While MSCs were shown to differentiate into adipocytes on a wide range of 3D scaffold matrices, soft substrates mimicking natural adipose tissue (~2–3 kPa) enhanced MSC adipogenesis (Xie et al., 2018; Jain et al., 2021; Bomkamp et al., 2022; Zagury et al., 2022). Culturing ASC on gels with such stiffness (2 kPa) supported adipogenesis even in the absence of growth factors or small molecules (Young et al., 2013). Thus, proper design of scaffold mechanical properties is not only essential to support adequate cell behavior but might also minimize the use of other chemical compounds, potentially simplifying the tissue fabrication process. The growth factors can be physically integrated and immobilized within the scaffold to support cell differentiation with high spatial resolution.

2.1.3 Scaffold cytocompatibility

When designing the composition and fabrication process of scaffolds intended to promote tissue formation, it is key to both promote cell adhesion and minimize any cytotoxic effects. The increased interest in more sustainable and low-cost sources of scaffolds for TE and especially for CM has spurred a pursuit after 'green', potentially edible ECM-mimicking scaffolding biopolymers. As such materials are essentially not animal derived, it is crucial to evaluate whether they are suitable for mammalian/avian/fish cell culture in the most basic terms of cytotoxicity and cell adhesiveness. For example, undesirable cell death could stem from potential microbial contamination (i.e. endotoxins), non-cytocompatible cross-linking mechanisms during scaffold formation (especially relevant when encapsulating cells in the material), and more (Xi et al., 2009; Sze et al., 2016; Ng and Kurisawa, 2021). Therefore, cell viability and metabolic activity are crucial assessments to determine the possible cytotoxic effects of such materials and cross-linking mechanisms.

2.1.4 Promoting cell adherence

Apart from avoiding any toxicity to the developing cells, the scaffolding biomaterials should also support cellular attachment through adequate surface chemistry. This is especially crucial for anchorage-dependent cells, such as muscle cells. Although cell attachment is ultimately dictated by multiple factors, the net charge, hydrophilicity, and presence/absence of anchoring motifs on the biomaterial's surface are among the main determinants. Positively charged surfaces generally increase cell attachment, as cells possess a slightly negative charge on their surface. Moderately hydrophilic surfaces can indirectly increase cell attachment by increasing surface protein adsorption (Ng and Kurisawa, 2021). Interestingly, it was demonstrated that while one surface chemistry element might be unsuitable for adhesion, others can compensate for it and suffice for cellular attachment. For example, similar cell attachment to net-negatively charged Synthemax microcarriers and to net-positively charged microcarriers was documented, as the enhanced protein adsorption caused by the hydrophilicity of the negatively charged microcarriers outweighed the effect of the net charge on cell attachment (Verbruggen et al., 2018). Such interplay between the various attributes of the scaffold composition opens the opportunity to screen natural components to find cheaper sources of cell-adhesive materials for CM. Finally, while some materials are functional and do not require further alterations for cell attachment, others can be modified for this purpose. Cell adhesion motifs at optimal densities signal cell survival and proliferation and can be introduced via coating or covalent modification of the scaffold biomaterials with recombinant proteins (e.g. recombinant gelatin and recombinant collagen) or ECM cell adhesion motifs (e.g. RGD, PHSRN, and YIGSR). Recombinant proteins with cell adhesion domains can also be incorporated into the scaffolding composition itself to create composite scaffolds (Ng and Kurisawa, 2021; Bomkamp et al., 2022).

2.2 Additional scaffold constraints for cultivated meat production

For CM to become a genuine alternative to the meat industry, the choice of materials and scaffold fabrication techniques must also consider aspects regarding sustainability, nutrition, food safety, organoleptic attributes, cost, and scale-up possibilities.

2.2.1 Sustainability

The combination of the rapid global population growth and improving economies in developing countries has resulted in a tremendous boost in meat consumption, which is expected to at least double by 2050 (Lee et al., 2020). Novel methods to meet such high demands are warranted, particularly in light of the current meat industry's reliance on large-scale animal slaughter, which has raised global environmental, health, and ethical concerns. To ensure the sustainability of CM, fabrication processes must be environmentally conscious throughout all stages. Essentially, all used materials, including those used for scaffold fabrication, must be derived from sources other than animals. Moreover, the entire process should reach maximum production efficiency while exploiting minimal resources. To this end, extensive research conducted in pursuit of suitable materials for CM scaffolds has already identified plant-/algae-/fungus-/bacteria-derived candidates, as well as relevant synthetic polymers and recombinant ECM protein variants, all of which will be further elaborated on in this chapter (Campuzano and Pelling, 2019). Notably, the integration of food industry by-products as novel scaffolding materials or cross-linking agents for construct fabrication purposes would be preferable in terms of promoting circular economy (CE). For example, cellulose and pectin, both commonly used for TE scaffolds, can be extracted from waste leaves and fruit peels (Zamri et al., 2021). Another example is Genipin, a promising understudied cross-linker extracted from *Gardenia jasminoides* Ellis fruit and flowers (Dimida et al., 2017; Zamri et al., 2021). Genipin was found to possess lower cytotoxicity compared to other amino-side group cross-linkers and was successfully used in the fabrication of chitosan scaffolds (Dimida et al., 2017). Minimizing the environmental impacts of CM fabrication would also warrant biomaterials recycling, where possible, thereby further improving cost- and resource efficiency. Recyclable scaffolds are beneficial if used as temporary cultivation platforms from which cells are later harvested and not as integral parts of the final product. For unstructured CM, microcarriers are especially relevant recyclable materials and are being investigated for cell detachment yields and separation via chemical, mechanical, or thermal mechanisms (Bodiou et al., 2020). Some microcarriers, such as Cytodex-3, are already considered reusable for cell culture of mammalian cells (Bodiou et al., 2020; Reddy et al., 2021).

2.2.2 Nutrition and food safety

All materials included in the CM production process must meet essential food safety standards, dictated by regulatory organizations (e.g. Food and Drug Administration (FDA), European Food Safety Authority (EFSA), and local authorities) (Bodiou et al., 2020; Bomkamp et al., 2022; Levi et al., 2022). Thus, the used materials must be suitable for human consumption or otherwise removed completely from the final product (Seah et al., 2022). For instance, the former criterion is enforced on the scaffolding materials when those are intended to be part of the final product. In such cases, it is preferred that the constructs will be made of materials that bear nutritional benefits of their own. Edible plant-/algae-/fungus-/bacteria-derived proteins and polysaccharides have already been assessed for cell-cultivation purposes (Gong et al., 2006; Dhingra et al., 2012; Xu et al., 2014; Wongkanya et al., 2017; Yang et al., 2017; Ahn et al., 2018; Jahangirian et al., 2019; Ben-Arye et al., 2020; Pérez-Guzmán and Castro-Muñoz, 2020; Qi et al., 2020; Cheng et al., 2020; Mohammadinejad et al., 2020; Jin et al., 2021; Xiang et al., 2022; Xiang et al., 2022; Zheng et al., 2022; Dutta et al., 2022; Ianovici et al., 2022). For example, plant proteins, such as soy, zein, and pea, have been integrated as scaffolding components as they are abundant, inexpensive, and protein rich and can be processed into scaffolds and hydrogels via food-grade cross-linking mechanisms (Ianovici et al., 2022; Levi et al., 2022; Seah et al., 2022; Wollschlaeger et al., 2022). Moreover, scaffolding components of indigestible polysaccharides such as pectin, cellulose, or alginate can serve as dietary fibers in the final product (Dhingra et al., 2012; Ng and Kurisawa, 2021). While some materials might require chemical modifications to become cell adherent, such modified biomaterials and the clearance of potentially toxic residues formed in the process must be evaluated to confirm such options as food grade. While RGD modification is a commonly used chemical alteration to enhance cell attachment, where the cell recognition motif R of arginine, G for glycine, and D for aspartic acid is covalently linked to another material, it is yet to be approved for food-grade purposes as it is mostly based on carbodiimide chemistry (Bodiou et al., 2020; Bomkamp et al., 2022). One noteworthy patent describes the possible use of edible pectin-cardosin A (RGD-containing polypeptide) microcarriers for CM (Marga et al., 2017). Thus, opting for food-grade materials already possessing surface chemistry suitable for cell attachment could minimize potential food safety issues. Notably, the inclusion of food allergens, such as soy protein or gluten, in scaffolding components, will hinder CM consumption by certain consumers and would require suitable labeling (Levi et al., 2022). If the scaffolding materials serve as temporary cell cultivation platforms, a cell-harvesting or biomaterial-degradation step is necessary to ensure their complete removal. Inedible, non-degradable, or slowly degrading synthetic materials can be considered as temporary scaffolds, especially microcarriers, as the cells can be efficiently harvested at the end of cultivation. Some groups have suggested that by applying precise design of synthetic materials, scaffolds with faster/controlled-degradation rates could be created, rendering them potential candidates for tissue-embedded structures. Plant-/algae-/fungus-/bacteria-derived materials subjected to chemical, enzymatic, or pH-dependent degradation may also be suitable for fast and stimulus-specific scaffold degradation. For example, ionically cross-linked alginate may degrade in response to bivalent cation chelators such as ethylenediaminetetraacetic acid (EDTA). Moreover, both alginate and other biomaterials (e.g. dextran and pectin) could be considered for enzymatic degradation in case recombinant enzymes that function under cell-culture conditions are available (Ng and Kurisawa, 2021).

2.2.3 Organoleptic attributes

To increase consumer acceptance, CM should mimic natural meat-based products as closely as possible. Both scaffolding materials aimed to be used as integral parts of the final products and scaffold fabrication techniques can considerably contribute to the organoleptic attributes of CM (Ng and Kurisawa, 2021). For example, the texture of structured meat (steak/fillet) differs greatly from that of unstructured meat (e.g. ground meat and sausages), with the former displaying the highly ordered fibrous hierarchical architecture of natural meat. Such structure is relatively similar in mammalian, avian, and fish skeletal muscle tissues, where the smallest cellular unit is the myofiber (multinucleated muscle cell), surrounded by connective tissue, vasculature, nerves, and intra-muscular fat. In mammalian/avian muscle, fibers are organized into fascicles and whole muscles via three connective tissue layers (endomysium, perimysium, and epimysium), while in fish muscle, the equivalent repeating muscle tissue units are shorter and w-shaped and are also surrounded by a connective tissue (Bomkamp et al., 2022). Therefore, to achieve structural resemblance, the hierarchical organization of muscle fibers, intra- and inter-muscular fat, and connective tissue must be tailored in a product-specific fashion. This might be achieved using various scaffold fabrication techniques that enable control over the 3D construct architecture, such as processing methods to create fibrous structures (Dekkers et al., 2018). Additionally, some groups suggest that repurposing of meat quality augmenting and food-grade biopolymers into scaffolding components can bring the post-cultivation organoleptic traits of CM products closer to those of conventional meat (Ng and Kurisawa, 2021). For example, water-binding agents, such as cold-swelling gums or starches, can improve water retention during cooking, and thereby influence the final product's juiciness. Alternatively, organoleptic or nutritional aspects can be improved by designing scaffolds to degrade/release molecules such as flavor-generating essential vitamins or aromatic compounds (Ben-Arye and Levenberg, 2019; Ng and Kurisawa, 2021). Similarly to plant-based meat alternatives which often contain added vitamin B12, the same methodology could potentially be harnessed for CM to enhance its nutritional value (Fraeye et al., 2020). Cooking meat includes a series of reactions resulting in a specific aroma and taste, such as Maillard reactions, lipid oxidation, interactions between the products of the two reactions, and thiamine degradation (Kaczmarska et al., 2021). Therefore, tailoring scaffold compositions to promote the onset of such reactions under heat treatment, by including relevant amino acids, carbohydrates, or other contributing components, may also enhance the flavor, smell, and color of the cooked meat. For example, the widely studied hemoproteins myoglobin and hemoglobin were shown to catalyze the production of various flavor- and aroma-contributing compounds during cooking (Han et al., 1994; Devaere et al., 2022). Moreover, when hemoproteins were added to the culturing media of bovine muscle satellite cells (BSCs), cell proliferation and tissue coloration were improved (Simsa et al., 2019). Thus, heme proteins from non-animal origins such as leghemoglobin, a plant-extracted structural ortholog of myoglobin already harnessed for commercially available meat analogs (Fraser et al., 2018; Kyriakopoulou et al., 2018), may also be investigated as components in the culturing media or scaffolding composition to optimize the resultant attributes of CM.

2.2.4 Scalability and cost

For CM to enter the market and become available to a wide range of consumers, it must compete with conventional meat product prices (~US$2/kg) (Risner, et al., 2020). Currently, CM manufacturing is significantly more expensive than conventional meat due to costly bioreactor equipment and media ingredients (Levi et al., 2022). Moreover, it is estimated that the production cost of structured CM products will be higher compared to unstructured CM, as the former will most likely integrate 3D scaffolds to better mimic the traits of whole-cut meat. Therefore, the development of low-cost, scalable biomaterial supply chains and fabrication techniques for CM scaffold production is essential.

As discussed earlier, animal-derived scaffolding biomaterials are not suitable for CM fabrication. While plant-/algae-/fungus-/bacteria-derived, synthetic, or recombinant ECM biomaterials can be considered, each must be optimized before their widespread integration into CM production processes. For example, although inedible synthetic microcarriers are beneficial for unstructured CM fabrication in terms of recyclability, sustainability, batch-to-batch consistency, and production costs, their use will require a cell dissociation step which could increase costs and raise safety concerns. In addition, if a CM scaffolding material must be chemically modified with recombinant ECM motifs, further improvements in recombinant-protein fabrication processes are warranted to reach cost-effective and large-scale production. To put things into perspective, food-grade alginate is more than ten times cheaper than food-grade microbial-derived hyaluronic acid (HA) (~US$10/kg vs. ~US$130/kg, respectively) (Ng and Kurisawa, 2021). Notably, cost reduction attempts can include the evaluation of combinations of recombinant ECM motifs mixed with other lower-cost materials. However, any introduction of novel chemically modified or synthetic materials to the food system will also require additional regulatory evaluations, resulting in longer development times and higher costs. Such limitations serve as strong incentives to take a more cost-effective approach, such as examination of plant-/algae-/fungus-derived scaffolding compositions which may be adequate for cell cultivation and are already considered food safe and even used in the current meat industry (Ng and Kurisawa, 2021; Bomkamp et al., 2022).

Apart from the search for suitable, cost-effective biomaterials, new scaffold design, and fabrication methodologies must also be developed and optimized for cost-effective, large-scale CM production. Such technologies will be essential to create desirable scaffold properties in a controlled, consistent fashion, tailored to the desired structured/unstructured meat products. For example, as recombinant proteins and growth factors constitute the majority (~99%) of total culture media costs (Chen et al., 2022), designing scaffolds with growth factors incorporated in specific regions may significantly reduce CM expenses. Several works already demonstrated the potential use of scaffolding materials as slow-release platforms of growth factors or other molecules, with some also demonstrating the ability to design such constructs with morphogenic concentration gradients (Tzezana et al., 2012; Lev and Seliktar, 2018). For example, alginate hydrogels were successfully incorporated with insulin-like growth factor-1 (Lev and Seliktar, 2018). The growth factors can be physically integrated and immobilized within the scaffold to support cell differentiation with high spatial resolution, resulting in lower amounts of used growth factors. Notably, such an approach can be of special benefit toward the creation of structured CM products involving multiple cell types (Chen et al., 2022). The fabrication and cultivation of individual constructs as building blocks later combined into a single heterogenous structured product has also been suggested to simplify the process. In this manner, building blocks can be fabricated using a range of techniques (e.g. electro-spinning, 3D-printing, and mold-casting), each of which must be further adapted for scale-up production. Innovations in assembling techniques of various construct types already show promising results in creating multi-tissue constructs (Kang et al., 2021; Li et al., 2022; Zagury et al., 2022) and will be further elaborated in later sections. For unstructured CM, separate growth of fat and muscle tissues and their later combination in a final processing step may be the simplest and most cost-effective solution. Of note, the current research time and cost could also be reduced. For example, computational modeling can be applied to investigate various aspects in scaffold design. However, highly complex multi-cell-type scaffolds with nonuniform geometries will be computationally intensive and will pose new challenges (Bomkamp et al., 2022).

3 Potential materials for cultivated meat scaffolds

3.1 Plant-/algae-/fungus/bacteria-derived biomaterials

Plant-, algae-, and microbial-derived biomaterials bear highly promising scaffolding potential for CM purposes, as they are of low cost, edible, and potentially nutritious, are associated with high consumer acceptance, and can benefit from readily available processing techniques. As many of these materials are already considered food safe and are already being used in the current food industry, they are of high relevance as integral scaffolding components in final CM products (Ng and Kurisawa, 2021). For example, plant protein isolated from corn (Gong et al., 2006; Yang et al., 2017; Jahangirian et al., 2019; Pérez-Guzmán and Castro-Muñoz, 2020; Ng and Kurisawa, 2021; Xiang et al., 2022), soy (Wongkanya et al., 2017; Ahn et al., 2018; Ben-Arye et al., 2020; Ng and Kurisawa, 2021; Bomkamp et al., 2022; Dutta et al., 2022; Ianovici et al., 2022; Wollschlaeger et al., 2022; Xiang et al., 2022), pea (Dutta et al., 2022; Ianovici et al., 2022; Wollschlaeger et al., 2022), wheat (Xu et al., 2014; Jahangirian et al., 2019; Dutta et al., 2022; Xiang et al., 2022; Xiang, Yuen, et al., 2022), and peanut (Zheng et al., 2022) have already been investigated as scaffolding components in cell cultivation platforms. While protein isolates are widely available, nutritious, low-cost, and already applied in vegan alternatives mimicking conventional meat products, the broad usage of allergenic components should be minimized. Several studies demonstrated the fabrication of glutenin films and porous 3D sponges supportive of BSC proliferation and differentiation (Xiang et al., 2022; Xiang et al., 2022). However, as wheat gluten consists of glutenin and gliadin, a protein associated with celiac disease and cytotoxicity for cell culture, glutenin must be separated from the gluten for further use as a scaffold. It should be noted that although separation is possible, immunogenic epitopes are also found in glutenin and further study is needed to determine whether suitable processing could reduce its immunogenicity. Ianovici et al. (Ianovici et al., 2022) compared two formulations for the creation of 3D-printed plant protein-enriched scaffolds, where either soy or pea protein isolates were integrated as nutritious components. Both compositions either as mold-casted or as 3D-printed scaffolds supported BSC growth and differentiation. As pea protein is nutritious yet non-allergenic, its incorporation as a scaffolding component in the context of CM production could be favorable.

Various polysaccharides, such as those derived from plants (e.g. cellulose, starch, pectin, and guar gum), algae (e.g. alginate, agarose, and carrageenan), fungi (e.g. chitosan and Hyaluronic acid), or bacteria (e.g. bacterial-cellulose, xanthan gum, and gellan gum), can also serve as CM scaffolding ingredients. Many have already been proven suitable for TE applications and are regarded as non-toxic, food-safe materials (Mohammadinejad et al., 2020; Qi et al., 2020; Jin et al., 2021; Ng and Kurisawa, 2021; Bomkamp et al., 2022; Levi et al., 2022). Some might also be regarded as dietary fibers (Dhingra et al., 2012). For example, alginate, an edible algal polysaccharide widely used in the food industry, constitutes a promising material for CM scaffolds, as demonstrated by its suitability in MSC culturing and adipogenesis (Mehta et al., no date; Zagury et al., 2022). Moreover, as alginate possesses reversible ionic gelation, tunable mechanical properties, and is easily processable, it has been successfully applied in various fabrication techniques. For example, while alginate can easily form soft hydrogels potentially suited for adipose tissue (Chandler et al., 2011; Freeman and Kelly, 2017), stiffer hydrogels can also be produced to support myogenic differentiation (Ansari et al., 2016). It was also shown as a promising stabilizing bioink component for creating 3D-printed constructs for adipocytes (Freeman and Kelly, 2017; Ianovici et al., 2022) and myoblast cultivation (Distler et al., 2020; Bolívar-Monsalve et al., 2021; Ianovici et al., 2022). Other studies examined the use of alginate as a suitable material for microbeads fabrication for TE purposes (Yao et al., 2012; de Lucena-Thomas et al., 2020). However, alginate does not contribute nutritional value, lacks cell adhesiveness in its nonmodified state, and is nondegradable by mammalian cells. Partial oxidation and biofunctionalization of the alginate can accelerate degradation and cell attachment, respectively (Ansari et al., 2016). To enhance its nutritional value, alginate must be combined with other materials, a course of action generally applied in the design of composite scaffolds that compensate for component shortcomings by integrating compensatory materials. In the previously mentioned work by Ianovici et al. (Ianovici et al., 2022), the two nutritious protein isolates were mixed with an RGD-modified alginate, which served as a non-animal-derived, stable, cell-adhesive scaffolding component, leading to scaffolds that were both nutritious and supportive of BSC growth and maturation. Another recent study by Wollschlaeger et al. investigated agarose, gellan, and xanthan-locust bean gum blend as tasteless mechanical support materials, mixed with nutritious pea and soy proteins, and showed all blends could be potential candidates for CM scaffolds (Wollschlaeger et al., 2022). Others have applied different kinds of cellulose, one of the most abundant polysaccharides on the earth and widely used for TE, such as cellulose nanofiber (CNF) and bacterial cellulose (BC) (Khan et al., 2016; Campodoni et al., 2020; Mohammadinejad et al., 2020; Jin et al., 2021; Ng and Kurisawa, 2021; Bomkamp et al., 2022). CNF is obtained from renewable sources and has been investigated for its potential as a fat replacer, while BC is a CNF produced with high purity from aerobic bacteria and has been applied in food production (Xiang et al., 2022). Although a recent study by Xiang et al. (Xiang et al., 2022) showed both CNF and BC did not adequately support cell attachment on their own, other studies successfully harnessed their mechanical attributes and allowed cell-adhesiveness by creating composite scaffolds of CNF or BC mixed with gelatin(Khan et al., 2016; Campodoni et al., 2020). Cellulose in the form of decellularized plants/algae can also be used for TE – a scaffold fabrication method that will be further elaborated on this chapter. Although cellulose alone is inert, several decellularized plant tissues (e.g. celery, green onion bulbs, spinach leaves, and grass) were shown to promote muscle-cell attachment and alignment (Hickey et al., 2018; Campuzano et al., 2020; Cheng et al., 2020; Allan et al., 2021; Jones et al., 2021). Such capabilities stem from their naturally heterogeneous cellulose-rich compositions which could include lignin, a hydrophobic protein hypothesized to alter overall hydrophilicity and promote cell attachment (Campuzano et al., 2020). Chitosan, alone or in combination with other materials, presents another promising edible polysaccharide supportive of cell adhesion and cultivation (Jana et al., 2013; Rubio et al., 2019; Kolasa et al., 2022; Zernov et al., 2022). In conclusion, readily available natural materials can be harnessed in different fashions to fabricate tunable, sustainable, low-cost, food-safe, and potentially nutritious 3D scaffolds for CM.

3.2 Recombinant extracellular matrix molecules

One of the main goals of CM scaffolds is to recapitulate the complex mechanical and biological properties of the ECM that influence both cell behavior and tissue formation, as well as the texture and nutritional value of the final product. In natural skeletal muscle, the ECM is primarily comprised of collagen, glycosaminoglycans, and fibrous proteins, such as elastin, fibronectin, and laminin, which are known to play key roles in cell attachment and behavior (Ahmad et al., 2021). As CM production cannot rely on animal-derived ECM proteins, substantial research is being conducted to produce its main components through microbial fermentation or plant molecular farming. Thus far, only some of the ECM proteins, such as collagen, gelatin, elastin, elastin-like proteins (ELPs), and fibronectin fragments, have been successfully produced using alternative methods (Olsen et al., 2003; Staunton et al., 2009; Miranda-Nieves and Chaikof, 2017; Peng et al., 2020; Ahmad et al., 2021). These might be sufficient when incorporated as scaffolding materials alone or in composite constructs to promote secretion of ECM components by embedded cells and to bring matrix complexity closer to that of the natural ECM. When considering the final CM product, recombinant ECM components are likely to be cleared as food safe, as their natural counterparts are present in natural meat. Moreover, their degradation is optimally controlled by the enzymes and inhibitors secreted by seeded developing cells during matrix turnover (Bomkamp et al., 2022), thanks to their naturally suited composition for such mechanisms. Therefore, this category of materials may be mainly relevant as integral scaffolding components in the final CM product, although other uses, such as sacrificial patterning or coating of other biomaterials for enhanced cell attachment, could also be considered. Hereafter will be presented the ECM proteins assessed as TE scaffolds, with emphasis on those for which non-animal-derived production feasibility has already been demonstrated. Importantly, further research is currently needed to promote more cost-effective and scalable recombinant expression methodologies, which will be a crucial step toward turning such biomaterials into a viable option for CM scaffolds.

Collagen, the most abundant structural protein in skeletal muscle ECM, has been widely investigated as a scaffolding biomaterial for skeletal muscle TE (Ahmad et al., 2021). As collagen promotes cell adherence, proliferation, and differentiation, plays a role in the mechano-transduction of forces from tissues to cells, and is mechanically suitable for tissue formation, it was used in various fabrication techniques to successfully create TE cultivation platforms. For example, 3D freeze-dried collagen scaffolds were shown to support myoblast alignment and myogenesis (Chen et al., 2015). It has also been applied to coat various microcarrier types, such as those comprised of alginate or Cytodex 3, to enhance cell attachment (Gröhn et al., 1997; Verbruggen et al., 2018). Moreover, collagen is known to contribute to the local storage and release of growth factors and cytokines during tissue repair processes. In a work conducted by Kanematsu et al., an FGF-loaded collagen hydrogel successfully allowed for the slow release of the growth factor, expanding its potential as a biomaterial suitable for more complex scaffold designs (Kanematsu et al., 2004) and more cost-efficient CM production. Hydrolyzed collagen, i.e. gelatin, which is generally formed in regular meat during cooking and denaturation of collagen, has also been applied in muscle cultivation platforms. Composite scaffolds comprised of gelatin, alginate, agarose, and glycerol displayed promising porosities and biocompatibility and supported myoblast attachment and growth (Enrione et al., 2017). In another work, fibrous gelatin scaffolds were successfully produced to enable the growth of rabbit myoblasts and bovine smooth muscle cells, in a well-aligned structure (MacQueen et al., 2019). Contessi Negrini et al. showed the fabrication of micro-patterned gelatin scaffolds by using sacrificial alginate-based microbeads and filaments, which later supported MSC adipogenesis (Contessi Negrini et al., 2019). Due to its non-cytotoxic thermo-reversible gelation, gelatin may also be used as a sacrificial material to enhance porosity in fabricated constructs, i.e. as a porogen (Liu et al., 2022). For example, gelatin beads were used as a sacrificial porogen for generating pores within cell-laden alginate hydrogels, resulting in improved cell viability and proliferation (Hwang et al., 2010). Modified versions of both collagen and gelatin, such as methacrylated collagen (CollMa) and gelatin (GelMa), were both used for TE scaffold fabrication (Aubin et al., 2010; Gaudet and Shreiber, 2012; Colle et al., 2020; Yang et al., 2020; Bolívar-Monsalve et al., 2021; Albrecht, et al., 2022; Hwangbo et al., 2022; Ronzoni et al., 2022). However, these semi-synthetic materials remain to be approved for human consumption. Although TE-related research mostly uses animal-derived collagen and gelatin, recombinant versions of collagen have been successfully produced in plants, yeast, and bacteria, some of which have already been proven to be biocompatible *in vitro* and *in vivo* (Campuzano and Pelling, 2019; Peng et al., 2020; Szklanny et al., 2021). Yet, while genetic engineering platforms have supported the expression of COL1A1 and COL1A2 genes and the fabrication of procollagen through several recombinant platforms, its post-translational modifications are crucial to the achievement of mechanical and biochemical properties similar to those of native collagen. Unlike other production platforms, tobacco plants can express both COL1A1 and COL1A2 proteins together with the necessary post-translational modifying proteins. Such a platform is currently the most promising in terms of scalability and production yield (Campuzano and Pelling, 2019).

ECM components, such as fibronectin, elastin, vitronectin, and laminin, were also found to promote cell attachment, and their recombinant production was examined. Fibronectin and laminin, two oligosaccharide-containing glycoproteins known to directly link the muscle fiber cell membrane to the basement membrane, have been assessed for their contribution to skeletal muscle TE. Indeed, satellite cells of porcine origin exhibited improved attachment to fibronectin and laminin as compared to collagen or gelatin surface coatings (Wilschut et al., 2010). Others demonstrated improved adhesion of C2C12 myoblast and human skeletal myoblasts to fibronectin-coated decellularized plant tissues (Cheng et al., 2020). Roy et al. successfully produced several recombinant fibronectin fragments in *Escherichia coli*. After their purification, the materials were later used to successfully fabricate ECM-mimetic cell attachment surfaces (Roy and Hocking, 2013). Various laminin isoforms were also investigated for their potential to promote cell attachment, proliferation, and differentiation (Garg, 2017). Penton et al. showed that laminin isoforms 111, 521, and 511 promoted mouse myoblast proliferation and differentiation (Penton et al., 2016). Another study demonstrated that 2D-patterned laminin substrates provided significantly enhanced support for myotube formation by murine C2C12 and human satellite cells, as compared to fibronectin and collagen substrates (Duffy et al., 2016). Some laminin isoforms were successfully produced in recombinant expression systems (Doi et al., 2002; Phan et al., 2009; Kinsella and Monk, 2012). Elastin, an ECM protein elastomer with considerable mechanical strain, is found mainly in connective tissues and basement membranes and is in great demand for regenerative TE applications (Peng et al., 2020). As a cell-adhesive component, elastin was shown to successfully enhance the biocompatibility of composite scaffolds made of polycaprolactone (PCL) and elastin, intended for muscle regeneration (Perez-puyana et al., 2021). Recombinant production of the precursor protein tropoelastin has already been attempted in several bacterial systems, and the production of full-length tropoelastin has recently been explored in tobacco plants but still requires further optimization (Miranda-Nieves and Chaikof, 2017; Peng et al., 1622020). Recombinant ELPs have been successfully generated by D'Andrea et al., who later applied them to create adhesive substrates supportive of myoblast attachment, growth, and maturation (D'Andrea et al., 2017). Vitronectin has also been demonstrated as a promising material for cell adhesion and culture (Molnar et al., 2007; Clevenger et al., 2016; Fan et al., 2017). Interestingly, Clevenger et al. (Clevenger et al., 2016) demonstrated the positive adipogenic effect of a short vitronectin-derived RGD-containing peptide incorporated into a polyethylene glycol (PEG) hydrogel. Production of a recombinant vitronectin version has been successfully demonstrated in bacteria (Halford et al., 2014).

3.3 Synthetic polymers

As mentioned earlier, CM scaffolds should optimally recapitulate the attributes of natural ECM. Although synthetic polymers are not present in nature, are mostly inedible, lack adequate surface chemistry for cell adhesion, and degrade slowly, they possess other beneficial qualities for CM cultivation platforms. Synthetic materials can be reproducibly generated with tunable properties and at low production costs, rendering them well-suited components for temporary CM cultivation platforms. As in the pharmaceutical industry, a likely configuration for scaled-up CM production would be suspension cultures in bioreactors. In such a context, these materials are mostly assessed as microcarriers from which cells are later separated to create an unstructured meat product. Although the overall cost will rise due to such additional processing steps, the potential recyclability and sustainability of synthetic polymers might improve overall efficiency and lower overall expenses. Publications report on the use of such materials for various scaffold fabrication techniques in the TE field, e.g. porous or fibrous scaffold production and 3D printing. However, for synthetic materials to be considered for scaffolds embedded within formed tissues, additional investigations will be required to assess controlled degradation possibilities, and the effect of remaining materials on tissue function, edibility, and organoleptic properties.

To date, mostly inedible materials such as polylactic acid (PLA), polylactic acid-glycolic acid (PLGA), PCL, and PEG have been assessed for microcarrier production in various contexts (Huang et al., 2020; Wang et al., 2023). For example, PLA, one of the most extensively used synthetic polymers in TE (Seah et al., 2022), was processed into collagen-coated PLA microspheres which enabled cell attachment (Hong et al., 2005). A recent study by Shi et al. also demonstrated that application by using aminolysis and hydrolysis - two simple and economical degradation methods - effectively modified porous PLA spheres into cell adherent microcarriers (Shi et al., 2019). PLGA, another widely investigated polyester, was harnessed to create porous microcarriers by microfluidic processing, with sacrificial gelatin serving as a porogen, which was found to promote C2C12 cell adhesion, proliferation, and differentiation (Kankala et al., 2019). Porous PLGA microspheres were also used as cell-adhesive microcarriers for the attachment and adipogenesis of mouse preadipocyte aggregates (Chung and Park, 2009). Another study demonstrated the modification of the porosity and inner geometry of synthetic PCL microspheres, by adjusting the PCL concentration and the ratio between the porogen and PCL (Zhou et al., 2019). Interestingly, both PLA and PLGA polyesters are produced by lactic and glycolic acid fermentation from sugars, rendering them more ecologically friendly than other polyesters that are synthesized from petrochemical sources (Elmowafy et al., 2019). Moreover, they were recently flagged as promising scaffolding materials for circular economy-based CM production (Bomkamp et al., 2022). As lactic acid is a major metabolite produced by growing cells, the potentially substantial amount of lactic acid produced in large-scale CM fabrication can be harnessed to create additional scaffolding material.

Significant knowledge on PLA, PLGA, and PCL in other scaffolding configurations applied for TE is available (Lee et al., 2007; Levy-mishali et al., 2009; Badrossamay et al., 2010; Kim and Kim, 2013; Shin et al., 2015; Kim et al., 2018; Luo et al., 2018; Chen et al., 2019; Gruen et al., 2020; Perez-puyana et al., 2021; Perez-puyana et al., 2021). However, further research to determine their potential use as materials embedded within CM tissue is crucial. For example, while PLA or PLGA removal via enhanced degradation rates might promote the edibility of the final product, complete degradation might be necessary to ensure food-grade consumption. However, such manipulations might be detrimental to the developing tissue, as elevated lactic acid concentrations in cell culture media can adversely affect cellular activity. For example, when cells were grown on either PCL spheres or degradable PLGA-based microcarriers, high lactic acid release from the latter resulted in slower cell growth and longer duplication time (Muoio et al., 2021). Thus, further assessments must be made to find the delicate, balanced interplay between the different aspects for such a version of CM production.

Other synthetic polymers are already approved for consumption. For example, PEG is widely used for TE applications, including skeletal muscle and adipose tissue development (Teßmar et al., 2010; Fuoco et al., 2014; Clevenger et al., 2016), yet its edibility is limited to small amounts only (Bomkamp et al., 2022). Partially synthetic materials can be generated by various chemical modifications to the original polymers (e.g. methacrylation and conjugation of adhesive components), such as RGD-modified alginate, gelatin methacrylate (GelMa), and collagen methacrylate (CollMa) (Aubin et al., 2010; Chandler et al., 2011; Gaudet and Shreiber, 2012; Colle et al., 2020; Yang et al., 2020; Bolívar-Monsalve et al., 2021; Albrecht et al., 2022; Hwangbo et al., 2022; Ianovici et al., 2022; Ronzoni et al., 2022). For example, a cell-laden GelMa bioink was successfully used for 3D-printing muscle constructs with a high degree of uniaxial alignment (Hwangbo et al., 2022). However, although PEG and such semi-synthetic polymers make promising scaffolding TE materials, every modification requires reevaluation of the product's suitability for food applications in case they are to remain in the final product.

4 Construct types and fabrication techniques

A variety of scaffold fabrication techniques are available for the creation of structured and unstructured CM products. For each CM type, the selected technique must support the fabrication of constructs with the desired biological, organoleptic, and nutritional properties, in a scalable, sustainable, low-cost manner. For example, for unstructured meat products, such as sausages and ground meat, microcarriers might be suitable. Porous and fibrous prefabricated scaffolds can be harnessed to allow the maturation of seeded cells while enabling medium perfusion. Cell-containing hydrogels which come in a multitude of configurations can also serve as cell cultivation platforms. 3D bioprinting, which uses hydrogel inks to create complex structures in well-defined geometries, is a more advanced technique and may be suitable for the creation of structured CM. Hereafter, the various scaffold-based cultivation platforms and fabrication methodologies will be elaborated. Recent reviews detail other structuring approaches such as scaffold-free methodologies that may also be applied for CM production (Srutee et al., 2021; Bomkamp et al., 2022; Levi et al., 2022; Seah et al., 2022).

4.1 Microcarriers

Microcarriers (MCs) are small polymeric particles (100-200 μM in diameter) (Bodiou et al., 2020) used in biomedicine for the purpose of culturing or delivering cells. The concept of growing adherent cells on MCs in suspension was first introduced in 1967 (Wezel, 1967). Live cells can be cultured within the porous structure of MCs or on their surfaces. Alternatively, cells can be encapsulated within the MCs, enabling the controlled release of medicinal agents produced by the cells, through their semipermeable barrier (Acarregui et al., 2012). MCs boast a large surface area-to-volume ratio, rendering them good 3D scaffold candidates (Cahn, 1990; Chen et al., 2011; Tavassoli et al., 2018). Lithography, micromolding, emulsion drops, imprinting, and microfluidic techniques are a partial list of the methods used to control MC synthesis and encapsulation efficiency (Choi et al., 2012; Shim et al., 2013; Zhang et al., 2016).

As commercially relevant MCs for scalable cell cultivation were developed over the years, such a concept has become an attractive approach for CM. Like other scaffolding platforms, several MC properties specific to muscle and adipose tissues must be tailored, such as porosity, mechanical support, surface chemistry for cell attachment, high surface area, and more. For example, as skeletal muscle tissue forms upon differentiation of satellite cells into multinucleated myotubes, such tissue requires sufficient mechanical support against the strains applied on the surface by the cells, sufficient porosity to allow their spreading and communication, and a preferably fibrous microstructure which is known to induce a unidirectional organization. On the other hand, adipose tissue formation requires different conditions for its growth and differentiation, such as softer materials. Although such constraints pose challenges for MC synthesis for the purpose of advanced tissue formation, research examining their use for such applications has already been reported and may be harnessed for CM production. Three scenarios are possible for MCs to be applicable in the CM industry: MCs can serve as temporary (possibly recyclable) carriers after which the cells are removed, as temporary carriers after which they are degraded or dissolved, or as edible carriers embedded in the final product (Bodiou et al., 2020).

MCs can be fully fabricated from synthetic materials, which are often non-edible and non-degradable. Thus, such MCs can be used to culture cells, such as satellite cells and adipocytes or MSCs, mainly for expansion purposes. The most common materials used to generate MCs offering large surface area and significant mechanical support are PLGA microspheres and glass beads (Kim et al., 2006; Kankala et al., 2019; Kong et al., 2022). For example, Kankala et al. synthesized PLGA microspheres with an inner structure of interconnected pores achieved by using sacrificial gelatin during particle fabrication. The highly porous structure combined with the stiffness of the PLGA allowed skeletal muscle cell culturing (Kankala et al., 2019) (Fig. 1). After the culturing period, the cells can be harvested from the MCs, which may then be recycled for further use. Cell disassociation from the micro-scaffolds is an important step and can be achieved in several ways. The most common method includes mechanical stress, e.g. agitation and sonication. Enzymatic disassociation is also possible but is more challenging in such systems. Nienow et al. presented a promising disassociation method based on both agitation for short time periods and enzymatic detachment resulting in increased efficiency and cell viability, which could be beneficial for large-scale settings (Nienow et al., 2014).

Alternatively, MCs can be fabricated using materials that can be degraded under mild conditions after microtissue formation, allowing the separation and collection of the formed tissues for CM purposes. Some commercially available MC materials are essentially safe and may be adapted to food-grade levels (Andreassen et al., 2022; Liu et al., 2022), such as alginate which can be degraded using ion chelators. Other stimuli-responsive polymers can undergo reversible gelation processes and can be dissolved by changing conditions such as temperature or pH (Choi and Park, 2002; Custódio et al., 2015; Guarino et al., 2016; Parmar et al., 2017). Custódio et al. designed a chitosan microgel that could bind specific cell types via proteins anchored to the particle surface. Chitosan gels are pH sensitive and can dissolve under mild acidic conditions and

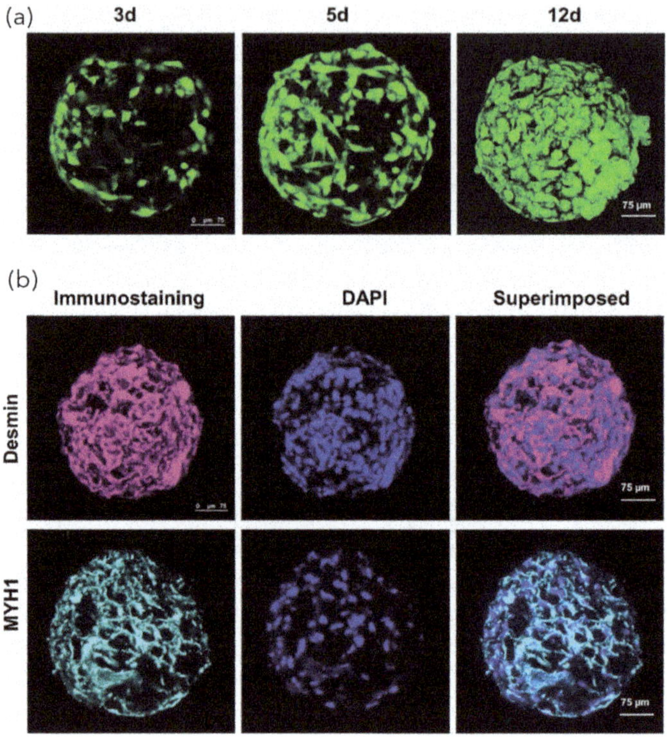

Figure 1 Microcarriers harnessed for myoblast cultivation. Highly open porous PLGA microspheres allowed muscle cell growth and myogenesis. (a) Confocal imaging confirmed myoblast proliferation on the PLGA MCs for 12 days. (b) Imaging of stained myoblasts after 7 days of cultivation revealed the cells differentiated into myotubes on the PLGA-based MCs (desmin = magenta, MYH1 = light blue, DAPI-stained nuclei = blue). Source: Adapted with permission from Kankala et al. (2019).

thus may be well suited to serve as temporary scaffolding platforms (Custódio et al., 2015; Guarino et al., 2016). Such types of cell-friendly degradable MCs can protect cell aggregates which are sensitive to mechanical stress, thus allowing their safe collection for subsequent usage.

MCs can also be composed of edible materials suitable for incorporation in the final product, which may reduce and possibly eliminate costly cell harvesting steps and corresponding loss of cellular biomass (Reiss et al., 2021). Thus, the composition and fabrication process of such MCs must comply with food regulations. Andreassen et al. proposed using edible MCs made of sustainable raw materials from food industry by-products that do not require dissociation or degradation steps and can be included in final meat products (Andreassen et al., 2022). Such MCs could potentially further contribute to the organoleptic and nutritional values of the CM product. Relevant edible MC polymers can be classified into several groups: polysaccharides such as alginate (Gröhn et al.,

1997; Perteghella et al., 2017), chitosan (Chui et al., 2019), and gellan gum (Wang et al., 2008); polypeptides such as collagen, zein (Li et al., 2016), and gelatin (Radaei et al., 2017); and lipids such as shellac (Yuan et al., 2021). Chitosan, cellulose, and alginate are excellent edible MC candidates, as they are widely available and known to be biocompatible and biodegradable (Shit and Shah, 2014). To date, Matrix Meats, Tantti Laboratory, and SingCell are among the few companies that offer edible MCs specified for CM (Bomkamp et al., 2022). Norris et al. recently demonstrated the expansion and differentiation of C2C12 and BSCs on edible gelatin MCs prepared using water-in-oil emulsion templates (Norris et al., 2022). In terms of the MCs' production process, physical cross-linking of polymers seems more promising for CM when compared to chemical methods, as it better maintains low toxicity (Ali and Ahmed, 2018). To improve cell adherence, MC surface chemistry can be optimized by coating or modifying surfaces with ECM proteins, such as collagen, laminin, fibronectin, or vitronectin (Chen et al., 2011), which contain an RGD domain (Ruoslahti and Pierschbacher, 1986; Derakhti et al., 2019). For example, RGD-modified chitosan MCs developed by Yang et al. were shown to enhance cytocompatibility, resulting in improved cell adhesion and proliferation compared to nonfunctionalized MCs (Yang et al., 2014). Despite the great potential of modified edible materials, any chemical modification requires careful reevaluation, both of their edibility and of the potential toxicity of residues.

4.2 Porous constructs

Porous scaffolding platforms include those which are produced via polymer solution processing of porosity-modifying or texturizing techniques or those harnessed from natural plant/algae tissues via decellularization. Such sponge-like scaffolds are produced relatively easily and at low costs and feature micro-scale porosity and inner structures that can facilitate post-fabrication cellular seeding, growth and differentiation, and enhance material transfer. Construct fabrication via polymer solution processing enables a reproducible, large-scale design of the exact desired scaffold composition and allows a delicate interplay between conflicting factors, such as high porosity together with sufficient mechanical properties. However, optimization of scaffolding design using this approach will require further research, and the final edibility of such scaffolds as integrated product components will need to be individually evaluated for each material combination. On the other hand, while plant/algae tissues are naturally readily available, edible structured constructs, the decellularization process must be non-toxic, and additional steps of biofunctionalization or coating with functional surface proteins might be necessary (Seah et al., 2022). In both configurations, the presence of larger pores suitable for media perfusion through the constructs will likely support tissue maintenance, positioning them as candidates for structured CM fabrication. Thus, both promising approaches are currently being explored and developed to meet the various requirements for CM production.

Porous scaffolds fabricated using polymer processing techniques can generally be prepared by harnessing methods either from the TE or food-engineering fields. In TE, common scaffold fabrication techniques, such as freeze-drying, particle-leaching, and gas foaming, can be applied on a variety of materials to generate cultivation platforms and have been shown to be advantageous with regards to control over the construct composition, inner structure, and mechanical properties. For example, freeze-dried porous composite scaffolds made from gelatin, alginate, agarose, and glycerol had relatively large pore sizes and permitted myoblast adherence and growth, while exhibiting slow degradation profiles, making them potentially suitable integrated constructs in CM (Enrione et al., 2017). Mechanical and microstructural properties of gelatin-chitosan scaffolds were shown to be highly dependent on the proportion of the raw materials used, as well as on modifications introduced during freeze-dry processing (Perez-Puyana et al., 2020). Interestingly, several works demonstrated the use of unidirectional freeze-drying to create scaffolds with microtubular pores (Kroehne et al., 2008; Pot et al., 2015). When Kroehne et al. applied this technique on collagen hydrogels, the resultant constructs possessed inner parallel pores which guided C2C12 alignment and muscle tissue formation (Kroehne et al., 2008). Moreover, a study showed that manipulation of the biomaterials ratio in alginate-bacterial-nanocellulose-based scaffolds and the freezing rate, optimized pore size, and resultant MSC adipogenesis (Krontiras et al., 2015). Interestingly, porous scaffolds can be used to fabricate tissue 'building blocks' to later be applied for complex tissue fabrication. For example, in a recent work by Zaguri et al. (Zagury et al., 2022), engineered bovine muscle prepared on freeze-dried scaffolds was combined with adipocyte-containing hydrogels into structures replicating conventional meat texture in several geometrical configurations. The gentle stitching process used to produce such multi-tissue constructs enabled co-culturing while preserving the mature myotubes and adipocytes.

Texturized constructs made of vegetable proteins (TVPs) have been fabricated by extrusion-cooking in the food industry to produce meat analogs. As such commercially available constructs are already fabricated using a large-scale and low-cost texturing technology and existing food-relevant materials, recent works with promising early results suggest examining their potential repurposing as CM scaffolds. Ben-Arye et al. demonstrated that textured soy protein constructs (TSP) had porous structures capable of supporting the attachment, proliferation, and differentiation of BSCs, resulting in 3D muscle tissues (Ben-Arye et al., 2020) (Fig. 2). While the TSP had inter-connected porous structures and enhanced nutritional values, additional adaptations are needed

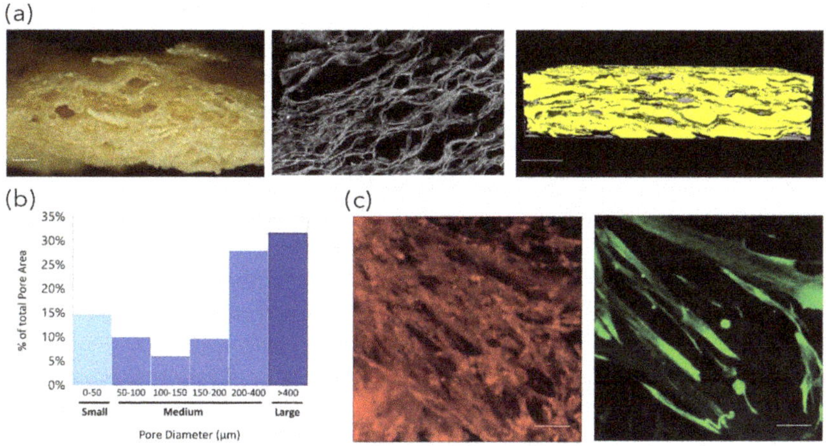

Figure 2 Porous constructs for engineering muscle tissue. Porous textured soy protein (TSP) constructs harnessed as scaffolds for BSC cultivation into muscle. (a) Representative images of empty TSP using regular (left), confocal (middle), and micro-CT 3D reconstruction (right) imaging (yellow = largest interconnected pores). Scale bars: left and middle = 500 μm, right = 1 mm. (b) Pore area percentage as a function of pore diameter. (c) BSC spreading and differentiation after 4 days of cultivation on TSP scaffolds are shown. Confocal imaging revealed elongated cell morphologies according to membranal Dil (red; left), and successful myogenesis according to immuno-staining against desmin (green; right). Scale bars: left = 50 μm, right = 30 μm. Source: Adapted with permission from Ben-Arye et al. (2020).

to avoid the dependency on animal-derived materials such as fibrinogen, which was used as a hydrogel for cell seeding. The surface chemistry of scaffold materials can be carefully adjusted to enhance cell attachment. For example, lunasin, a polypeptide commonly found in soy protein isolate, was found to possess an RGD-like cell adhesion motif (Chatterjee et al., 2018). Moreover, both thermal and mechanical forces applied in the extrusion cooking of meat analogs were shown to induce the reorganization of chemical bonds and structural orientation within protein ingredients, including the exposure of hydrophobic residues (Schmid et al., 2022). In addition, post-processing of extruded constructs, such as enzymatic hydrolysis, can affect TVP surface hydrophobicity (Ma et al., 2018). Therefore, further research regarding the impacts of processing and post-processing conditions on the surface chemistry of TVPs will be essential to guide the tailoring of candidate CM scaffold properties.

In addition to polymer-processing-generated scaffolds, porous constructs can be fabricated by decellularizing natural tissues of plants, fungi, and algae. Decellularization technology removes cells from a tissue while preserving its ECM and inner structure and was initially developed to create scaffolds for medical use. However, in recent years, researchers have been seeking the most low-cost, readily available scaffolds from non-animal origin (Gershlak et al., 2017). Constructs of various non-animal origins were shown to possess patterns supportive of both muscle cell alignment and branched vasculature which mimics the one found in animal tissues, rendering them promising scaffolding platforms for CM. For example, spinach leaves enabled the perfusion of media through their inner vasculature, as well as the attachment of MSCs on their surface (Gershlak et al., 2017). A later work found spinach scaffolds to support the cultivation of BSC, which resulted in directional BSC alignment and differentiation (Jones et al., 2021). Other plants with preferable anisotropic inner structures were examined, such as celery (Campuzano et al., 2020), grass (Allan et al., 2021), and green onion (Hickey et al., 2018), and were shown to support muscle cell alignment without the need for biofunctionalization. Decellularized apples were also examined, demonstrating a potential to support myoblast survival (Cheng et al., 2020). Two types of decellularized algae were also shown to allow cell attachment, where cell orientation was affected by their porous versus fibrous architectures (Bar-Shai et al., 2021). Other studies also investigated decellularized plants which were later coated with bio-functional materials to improve cell attachment. For example, decellularized apple constructs coated with poly-L-lysine promoted the cultivation and adipogenesis of pre-adipocytes (Contessi Negrini et al., 2020). In a recent study by Fontana et al., a variety of functionalized decellularized plant leaves or stems were assessed for their potential to support human cell cultivation, suggesting such an approach as a promising course of action to create a diverse array of complex scaffolds in short periods of time and lower costs (Fontana et al., 2017). Full fungal mycelial structures were shown to serve as effective sustainable and low-cost scaffolds for cell cultivation, with fibronectin coating proving useful but not essential (Antinori et al., 2021). Another work found no significant differences between the effects of noncoated spinach leaves and those functionalized with collagen IV or fibronectin on cell behavior (Robbins et al., 2020). Thus, although significant advances have been made in this research field, additional evaluations will be necessary to further profile the behavior of specific CM-relevant cell types on such constructs and to assess the need for construct adaptations to support cell cultivation. Moreover, decellularization techniques should also be developed to enable scale-up production under food-grade conditions. For example, while small-scale decellularization can be conducted either by introducing the reagents through a cannula or by immersion of the tissue, the latter technique will be preferable for large-scale production (Jones et al., 2021). Non-toxic materials and simplified decellularization procedures will also be essential for the large-scale production of edible scaffolding platforms (Antinori et al., 2021). Taken together, further investigation of porous decellularized constructs will likely facilitate the development of low-cost edible scaffolds for CM.

4.3 Fibrous scaffolds

Fibrous constructs fabricated for structured CM purposes can serve as anisotropic scaffolds supportive of cell alignment and enhanced material transfer between seeded cells and the surrounding environment. In contrast to porous scaffolds, fibrous constructs with distinct anisotropy are usually fabricated using spinning techniques, such as electrospinning or rotary jet spinning. Electrospinning exploits electrostatic forces to stretch polymer solution droplets until a jet of the polymer solution is ejected, thinning as the solvent evaporates, and resulting in the formation of solid fibers continuously deposited onto a collector (Levi et al., 2022). The size and structure of the produced filaments can be tailored by adjusting factors such as polymer type, solvent type, temperature, and flow rate (Mendes et al., 2017). Unlike 3D printing, one of the greatest advantages of electrospinning is the ability to create ultrafine fibers with large area-to-volume ratios, allowing the fabrication of nano- to microscale fibers (Reid et al., 2020). Moreover, organized fiber orientation can be achieved by advanced configurations such as using a high-speed rotating collector, which stretches and aligns the produced fibers or by applying external electric or magnetic fields (Reid et al., 2020). Aligned fibers have been shown to induce the unidirectional elongation of seeded cells and promote their myogenesis into myofibers (Riboldi et al., 2008; Langelaan et al., 2010; Reid et al., 2020; Uehara et al., 2020). Recent works have evaluated novel techniques, such as multijet electrospinning or implementations of various electrode geometries for scale-up of fabrication of electrospun fibers (Liu et al., 2016). Additionally, cell electrospinning has been tested to homogenously embed cells in fabricated fibrous structures. A recent work by Yeo et al. presented an *in vitro* muscle model obtained through electrospinning of spheroids embedded in aligned alginate nanofibers (Yeo et al., 2021). High cell viability was measured post-fabrication, and encapsulated spheroids fused to form mature myotubes. Taken together, such fibrous construct fabrication techniques could be especially relevant for the scalable production of muscle tissues for CM.

While electrospinning is a simple, cost-effective, tunable process, most of its researched applications were with materials suitable for TE, which will require adjustment for food-grade systems. Spinning techniques were mostly applied with inedible synthetic or semi-synthetic polymers such as PLGA (Shin et al., 2015), PLA (Badrossamay et al., 2010; Luo et al., 2018), PCL (Tzezana et al., 2008), and GelMA (Ebrahimi et al., 2018), and combinations of such materials with other polymer types (Choi et al., 2008; Gruen et al., 2020; Perez-puyana et al., 2021) (Fig. 3). The electrospinning of several edible materials, such as zein, cellulose, soy protein, starch, pea protein, wheat glutenin, collagen, and gelatin, has been demonstrated as well (Nseir et al., 2013; Xu et al., 2014; Law et al., 2017; Yang et al.,

2017; Ahn et al., 2018; Deng et al., 2018; MacQueen et al., 2019; Senthil Muthu Kumar et al., 2019; Aguilar-Vázquez et al., 2020; Mattice and Marangoni, 2020). For example, cells were shown to adhere to electrospun constructs based on zein-gelatin mixtures (Yang et al., 2017). Another work demonstrated the ability to produce cellulose-soy protein-based fibers through rotary jet spinning, which is supportive of fibroblast adhesion and proliferation (Ahn et al., 2018). A more recent work by MacQueen et al. showed the successful growth of bovine smooth muscle cells and rabbit myoblasts on fibrous gelatin constructs (MacQueen et al., 2019). Despite promising outcomes with electrospinning of edible biopolymers, additional challenges must be resolved before they can be applied for CM. The use of toxic solvents and the potential presence of solvent residues might render it unsafe for ingestion. In contrast, food-grade electrospinning can make use of solutions such as acetic acid, formic acid, or ethanol (Deng et al., 2018; Aguilar-Vázquez et al., 2020; Mattice and Marangoni, 2020), which are considered less toxic. Other components such as carrier polymers, commonly added to improve polymer viscoelasticity, must also be comprised of edible materials, such as polyethylene oxide (PEO) or polyvinyl alcohol (Saquing et al., 2013; Teixeira et al., 2020; Lou et al., 2022). Although such options are edible, more nutritious or cell-compatible alternatives might be preferable. Thus, additional research regarding spinning techniques adjusted for food-grade production is essential before their implementation for large-scale fabrication of fibrous CM tissues.

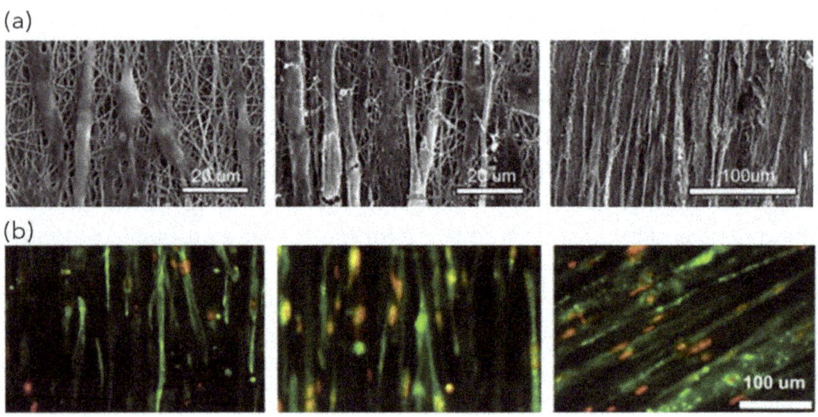

Figure 3 Electrospun constructs for skeletal muscle tissue engineering. Collagen/PCL electrospun aligned nano-fiber mesh scaffolds enable cellular alignment and myogenesis. (a) SEM images of human skeletal muscle cells (hSkMCs) on aligned electrospun PCL/collagen nanofiber mesh 1 (left), 3 (middle), and 7 (right) days after seeding. (b) Immunofluorescent staining images of hSkMCs on electrospun aligned PCL/collagen nanofiber meshes through muscle differentiation. Left to right: desmin, MYH, and sarcomeric actin expressions were observed (green). Source: Adapted with permission from Choi et al. (2008).

4.4 Hydrogels

Hydrogels have long been used for TE research purposes, mainly as flexible, high-water-content, 3D cell cultivation platforms. Such gels are formed upon cross-linking of hydrophilic materials into polymeric matrixes via various physical or chemical mechanisms. Hydrogels can absorb large quantities of water through functional groups present on their polymeric backbones, while maintaining their chemical structure and without dissolving (Chaudhary and Chakraborty, 2022). Thus, as the natural ECM is also a type of complex, multi-component hydrogel with high water content, hydrogels are promising scaffolds that can mimic the desired environment for cell culturing purposes. As scaffolding platforms, hydrogels must be cytocompatible with diffusion kinetics that support sufficient material transfer through the entire construct and with stiffness that promotes cell proliferation and differentiation. Various natural, synthetic, or semi-synthetic materials were successfully used to create hydrogels (Chaudhary and Chakraborty, 2022), including various edible, non-animal-derived materials. Moreover, hydrogel characteristics, such as nutrient diffusion kinetics, stiffness, and degradation rate, can be easily adjusted by changing parameters such as the polymer composition, cross-linking mechanisms, gelation kinetics, and others. For example, while small molecules (e.g. glucose and oxygen) can freely diffuse through ionically cross-linked alginate hydrogels, higher molecular weight molecules (e.g. myoglobin and albumin) show limited diffusion capabilities which decrease with increasing alginate or Ca^{2+} concentrations or as gelation increases (Drury and Mooney, 2003). A study by Chuang et al. demonstrated the ability to design the properties of collagen hydrogels, by altering factors such as the cross-linking mechanism and polymer concentration (Chuang et al., 2018). The microstructure, density, and stiffness of the collagen hydrogels were significantly different when applying physical (thermal) or chemical (covalent) cross-linking. Zhao et al. showed the effect of cross-linking density and resultant stiffness of prepared hydrogels on MSC differentiation. Adipogenic and osteogenic differentiation of MSCs were observed on substrates with stiffnesses similar to those of the respective natural tissues (Zhao et al., 2014). In addition, over time, cells encapsulated within hydrogels degrade the surrounding hydrogel while synthesizing their own ECM – a crucial natural remodeling process that is essential for cell spreading, proliferation, and differentiation. For muscle TE, constructs with mechanical properties characteristic of skeletal muscle were successfully prepared by cross-linking gelatin with genipin, a natural, non-toxic, biocompatible cross-linking agent. The surface of these hydrogels was patterned via soft lithography and resulted in unidirectional alignment and differentiation of myogenic cells (Gattazzo et al., 2018) (Fig. 4). Distler et al. demonstrated the possibility of harnessing alginate, a natural polysaccharide usually nondegradable by

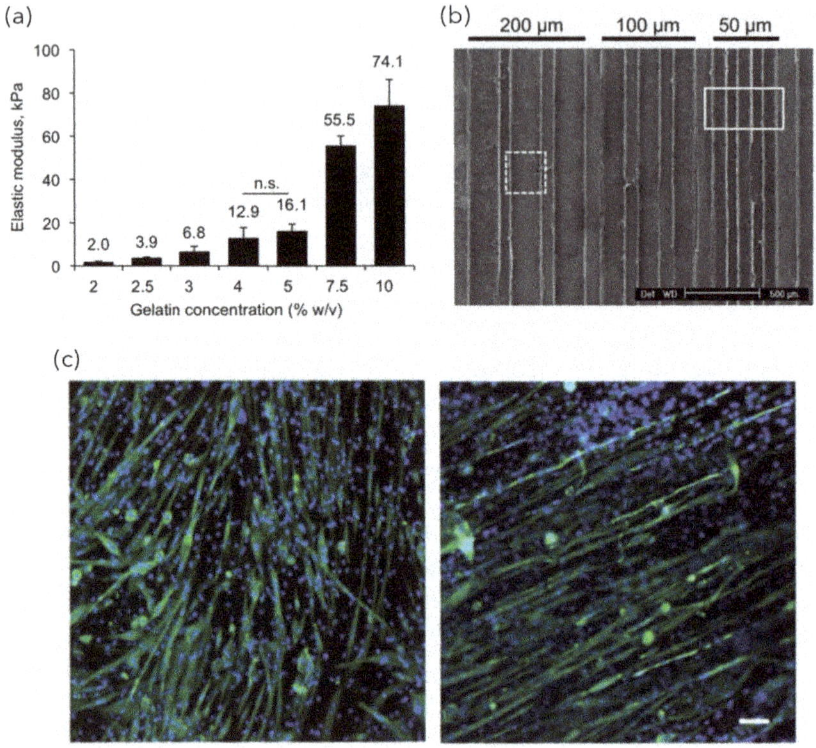

Figure 4 Hydrogels serve as cultivation platforms for muscle tissue development. Gelatin-genipin hydrogels harnessed to promote myogenesis and unidirectional orientation of muscle cells. According to the (a) quantification of compressive elastic modulus, adjusted gelatin concentration allowed the fabrication of constructs with mechanical properties of skeletal muscle (~13 kPa). Such constructs were further patterned to investigate the effect on myotube directionality. (b) Scanning electron microscopy image of dry aligned structures, with varying widths and distances (scale = 500μm). (c) Confocal imaging of immunofluorescent stained C2C12 cells after 1-week maturation on nonpatterned (left) vs. micropatterned (right) constructs with 13 kPa stiffness revealed the ability to direct myotube orientation (Hoechst stained nuclei = blue, MyHC = green. Scale bar =75 mm). Source: Adapted with permission from Gattazzo et al. (2018).

mammalian cells, in an oxidized form to generate alginate-gelatin-based hydrogels (Distler et al., 2020). The hydrogels were shown to support the differentiation of encapsulated C2C12 and were successfully applied as inks for skeletal muscle 3D printing. Thus, the use of hydrogels grants both control over the cultivation platform's characteristics and flexibility in its configuration. Hydrogels can be used to create porous scaffolds, cell-encapsulating microcarriers, cell-laden plugs, matrix fillers of porous constructs, 3D-printable inks, and more. Moreover, they can be advantageous in the fabrication of complex, multi-cell-type constructs. A recent study demonstrated the successful usage of soymilk-gelatin hydrogels as tailored platforms for adipogenesis and myogenesis of 3T3-L1 and C2C12 cells, respectively, which were subsequently stacked to create multi-layered constructs containing muscle and fat tissues (Li et al., 2022). Taken together, as hydrogels can be simply adjusted to create scaffolds for the fabrication of skeletal muscle and adipose tissues, their further development and investigation will be invaluable for the promotion of CM production.

4.5 Three-dimensional printing

In recent years, 3D printing has been explored as an innovative technology to produce complex geometric structures from a variety of materials and is an evolving practice in TE. In general, 3D printing involves a layer-by-layer convergence of materials to eventually fabricate predefined 3D constructs, designed using computer-aided design software. Dedicated printers are suitable for 3D printing of cells, biocompatible materials, and 'bioinks', which are materials naturally found in or that mimic the cell environment, mixed with cells. The well-controlled production process is applied in TE as a reproducible, accurate scaffolding fabrication technique to form complex bio-functional structures. Several bioprinting methods are currently available, such as inkjet bioprinting, stereolithography, laser-assisted deposition, digital light-processing (DLP), and extrusion printing (Bejoy et al., 2021) (Fig. 5a). Although the first four options enable the fabrication of biological structures while preserving relatively high cell viabilities, other restrictions could limit their applicability for CM production (Levi et al., 2022). As inkjet printing includes the individual deposition of small-sized ink droplets onto specific locations on a substrate formed constructs may not provide mechanical integrity sufficient for CM purposes. Other limiting factors include the potential presence of metallic residues when using stereolithography or integration of non-food grade photopolymerizable-resins or light-blocking agents for laser-assisted deposition and DLP and might restrict their use for CM. For example, recent research by Jeong et al. demonstrated DLP printing of thick perfusable constructs with evident myogenesis and adipogenesis of encapsulated immortalized bovine embryonic fibroblast cells transformed with MyoD and peroxisome proliferator-activated receptor γ2 (PPARγ2), respectively. However, the used bioinks were based on GelMa - a semisynthetic photopolymerizable polymer whose edibility is yet to be examined (Jeong et al., 2022).

Extrusion-based bioprinting has gained increasing interest, where the biomaterial is deposited in a continuous manner to produce filaments, layer by layer, eventually fabricating the complete construct. The transition from the liquid phase in which the material leaves the printer nozzle to its solid state afterwards is a crucial part of this bio-printing process. Therefore, commonly

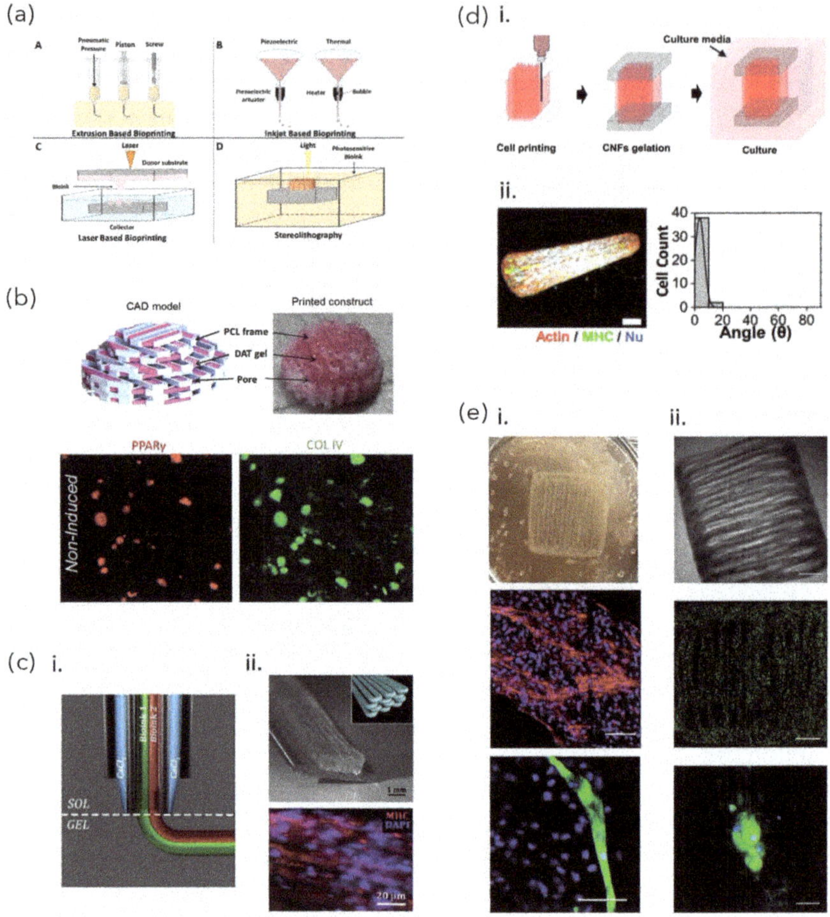

Figure 5 3D Printing for the generation of engineered muscle and adipose tissues. (a) Different techniques for 3D printing: extrusion, inkjet and laser-based printing, and stereolithography. Source: Adapted with permission from Bejoy et al. (2021). (b) Multi-material constructs designed to allow adipose TE. Upper panel: CAD model (left) and printed adipose tissue (right) showing the PCL-based framework, the cell-laden hydrogel (decellularized adipose tissue) and the pores in the construct. Lower panel: Successful differentiation of adipose cells was demonstrated by immunofluorescent staining for PPARg (red), and an essential ECM protein, COL IV (green). Source: Adapted with permission from Pati et al. (2015). (c) Co-axial printing harnessed for skeletal muscle TE. (i) Illustration of a 3D printer equipped with a microfluidic printing head and a coaxial extruder. (ii) Representative image of the printed constructs (top panel) and immunostaining for myogenesis (lower panel: MHC = red, DAPI-stained nuclei = blue) after 21 days of differentiation. The formation of myofibers and sarcomeric organization is evident. Adapted with permission from Costantini et al. (2017). (d) Tendon-gel integrated bioprinting (TIP) for creating cellular fibers. (i) Schematic of tendon-gel integrated bioprinting (TIP), where fabricated fibers are held in place during cultivation via cross-linked collagen nanofiber gels. (ii) Fluorescence image (left) of printed tissue with

used materials for this application are polymers and composites due to their cross-linking abilities through various physical, chemical, or enzymatic mechanisms (Jammalamadaka and Tappa, 2018). Moreover, materials with shear-thinning properties are commonly used, as this non-Newtonian material behavior causes a decrease in viscosity with increased shear rates, allowing them to flow through the nozzle. Upon deposition, the decreased shear rate increases the materials' viscosity, enabling the fabrication of well-defined shapes and geometries (Murphy and Atala, 2014).

The simplest extrusion printing configuration is direct ink deposition, where inks are extruded in the air onto the printer stage or onto a previous layer and are later cross-linked. This technique has already been examined for skeletal muscle and adipose tissue bio-printing, in an attempt to further develop thick, functional, implantable constructs for regenerative medicine purposes (Zhuang et al., 2020; Chen et al., 2021). Apart from the need for adequate material transfer throughout thick constructs to sustain tissue function, scaffold stiffness and geometry must be optimized to best support muscle and adipose tissue maturation (Gilbert et al., 2010; Young et al., 2013; Xie et al., 2018; Bodiou et al., 2020; Jain et al., 2021; Zagury et al., 2022). Although some groups reported on use of single, porous, cell-laden bioinks, others chose to perform multi-material printing and combine several inks in specific construct designs, where one ink encapsulates the cells while the others serve as mechanical supports/sacrificial materials (Pati et al., 2015; Yeo et al., 2016; Choi et al., 2016; Kang et al., 2016; Arab et al., 2018; Kim et al., 2018; Kuss et al., 2018; Contessi Negrini et al., 2019; Tytgat et al., 2020; Zhuang et al., 2020; Negrini et al., 2020; Chen et al., 2021; Li et al., 2021). For example, in a work by Kang et al., PCL was extruded to form support pillars in 3D-printed muscle constructs, apart from its inclusion in a cell-laden bioink for tissue growth. The group also used a pluronic F127 ink to fabricate longitudinal

Figure 5 (*Continued*)

bovine satellite cells (BSCs), stained with actin (red), MHC (green), and nuclei (blue), and cell alignment measurement (right) of the tissue. Adapted with permission from Kang et al., 2021. (e) Plant-protein-enriched bioinks harnessed for bath-based 3D-printing of construct supportive of myogenesis and adipogenesis. (i) Cellular constructs fabricated from bovine satellite cells (BSCs)-containing pea-protein-enriched bioink, using an agar support bath. Fibrous constructs were kept in the bath (top) during cell cultivation. Confocal imaging of stained samples revealed BSCs spreading (middle: phalloidin stained actin = red, DAPI-stained nuclei = blue. Scale = 100 um) and myogenesis (bottom: desmin = green, DAPI stained nuclei = blue. Scale = 50 μm). (ii) Cellular constructs fabricated from bovine mesenchymal stem cells (BMSCs)-containing pea-protein-enriched bioink. Constructs were extracted from the agar bath (top = brightfield image. Scale = 1 mm) and cultivated for further differentiation. Confocal imaging of stained samples revealed the BMSCs had high viability (middle: live cells = green, red cells = dead. Scale = 1mm) and could undergo adipogenesis (bottom: LipidTox stained lipid droplets = green, Draq5 nucleic marker = blue. Scale = 50 um). Source: Adapted with permission from Ianovici et al. (2022).

voids to enhance material transfer (Kang et al., 2016). Pati et al. (Pati et al., 2015) printed a dome-shaped construct by using a supporting framework of PCL bioink, while the main bioink was based on decellularized adipose tissue with adipose-derived MSCs to provide an optimal microenvironment for cell growth and function. Indeed, the authors reported on successful adipogenesis without the need for adipogenic factors (Fig. 5b). Negrini et al. (Contessi Negrini et al., 2019) used alginate as a sacrificial material to fabricate porous gelatin scaffolds with a hollow channel. Dissolution of the alginate left a porous construct with a prevascular channel of specific dimensions, which supported the proliferation and adipogenic differentiation of MSCs. Extrusion-based printing was also used to introduce anisotropy. For example, tissue maintenance and successful myogenesis were evident in 15 mm^3 3D-printed muscle constructs fabricated with fibrinogen-based fibers containing muscle progenitor cells and with voids made by sacrificial gelatin (Kim et al., 2018). The results showed that as cells were supplied with adequate supplies of oxygen and nutrients, cell alignment and myogenesis were significantly promoted compared to non-printed constructs. Other studies on 3D-printed adipose tissue focused on the porosity of printed constructs and demonstrated its centrality in determining the efficacy of adipogenic differentiation (Kuss et al., 2018; Contessi Negrini et al., 2019). The pore size itself can influence cellular infiltration and simultaneous differentiation, where scaffolds with pore sizes >500 μm were shown to enhance such processes (Tytgat et al., 2020). Direct extrusion printing for CM purposes was recently demonstrated in several works. For example, non-animal-derived plant protein-enriched compositions were successfully harnessed as bioinks for printing multi-layered scaffolds supportive of BSCs myogenesis and bovine mesenchymal stem cell (BMSC) adipogenesis (Ianovici et al., 2022). Another study showed the fabrication of multilayered cellular constructs by using cell-laden GelMA – 20% silk fibroin bioink and demonstrated the myogenesis of porcine satellite cells (Li et al., 2021).

In addition to direct printing, advances in material deposition and cross-linking strategies applied in extrusion printing are opening new possibilities for the production of thick and complex constructs. Novel methods such as coaxial printing or support-bath-based printing were developed in recent years to enable quick and precise gelation of soft materials immediately after exiting the printer nozzle, which has resulted in improved printing resolutions. For example, co-axial extrusion, in which several materials are printed simultaneously using an assembly of co-axial needles, can be employed to fabricate hollow channels (Gao et al., 2015; Pi et al., 2018) or cellular strands. For example, coaxial extrusion of alginate as an outer shell stream and calcium cross-linking solution as an inner core stream was conducted, resulting in the fabrication of constructs with built-in microchannels, suitable for media perfusion (Gao et al., 2015). Co-axial printing can also be used to fabricate long, small-diameter cellular strands that allow for sufficient nutrient diffusion to their core. (Onoe et al., 2013; Costantini et al., 2017; Ikeda et al., 2017; Mistry et al., 2017; Choi et al., 2019). A work by Costantini et al. (Costantini et al., 2017) showed the co-axial printing of high-resolution hydrogel fibers laden with muscle precursor cells (C2C12). The core stream consisted of a semi-synthetic PEG-fibrinogen and alginate mixture tailored to allow myogenesis, while the shell stream consisted of a calcium-containing cross-linking solution. After several days of culture, the myoblasts formed multinucleated myotubes within 3D bioprinted fibers, with high degrees of alignment along the direction of the hydrogel fibers (Fig. 5c).

Another currently leading approach to achieve highly complex structures with shape fidelity incorporates granular support baths during printing (Hinton et al., 2015; Choi et al., 2019; Compaan et al., 2019; Lee et al., 2019; Mirdamadi et al., 2019; Kang et al., 2021; Ianovici et al., 2022; Machour et al., 2022). In this configuration, the printer's nozzle is submerged inside the bath while the needle is moving, which provides mechanical support during the cross-linking process, thereby enabling the printing of soft or low-viscosity materials, often with high water contents. The slurry can then be removed, resulting in highly complex printed structures. For example, a complex 3D bioprinted tri-leaflet heart valve was created from collagen using a meltable gelatin-based support bath (Lee et al., 2019). Another work demonstrated the combined configuration of co-axial printing into a gelatin support bath to produce vascularized muscle fibers fabricated with spatially organized muscle and vascular dECM bioinks (Choi et al., 2019). Other reported support baths were based on non-animal-derived materials, such as agarose, gellan-gum, agar, and others, rendering them suitable for CM purposes. A recent study reported on 3D printing of cell-containing fibrinogen-Matrigel bioinks into a modified support bath in a manner termed 'tendon-gel' integrated bioprinting for CM purposes. In this version, a temporary gelatin-based support bath was used for the fabrication of cellular fibers during printing, while collagen nanofiber tendon-like gels were used to withstand cell traction forces during cell differentiation in fabricated fibers (Kang et al., 2021) (Fig. 5d). Recent investigations have assessed the use of support baths not just as a temporary medium during printing but also as a mechanically stabilizing medium for the entire cell cultivation period, to counteract cell-generated traction forces and construct deformation (Ianovici et al., 2022; Machour et al., 2022). For example, the maintenance of 3D-printed cellular constructs within an agar support bath during a 2-week cultivation period allowed for cell spreading and myogenesis, while preserving the printed shapes (Ianovici et al., 2022) (Fig. 5e). Such cultivation technique of printed constructs was also investigated in further depth with a novel carrageenan-based support bath and various cell types, including MSCs (Machour et al., 2022). As the maintenance of structure integrity is crucial during the complex tissue formation process, such cultivation techniques may be of particular interest for whole-cut CM production.

Despite progress in the field, further research will be necessary before 3D bioprinting is applied for CM fabrication. Critical hurdles that remain to be addressed include print time and cell viability post-printing, as well as the complexities of developing suitable bioink compositions and scalable fabrication of multi-cell type constructs. Although extrusion printing is relatively slow at the laboratory scale and may jeopardize cell viabilities due to shear stress applied on encapsulated cells, in-depth optimizations of crucial parameters such as cell concentration, nozzle size and geometry, and printing pressure may reduce cell death and improve the scalability of this fabrication technique (Cidonio et al., 2019; Adhikari et al., 2021). To achieve large-scale printing, simultaneous multi-nozzle extrusion or other solutions may reduce the production time. Despite progress made in 3D printing for TE, literature regarding the application of 3D printing for CM purposes is scarce, especially when as non-animal derived, printable, and potentially edible scaffolding compositions are considered (Ianovici et al., 2022). Thus, further research concerning the development and investigation of such materials is crucial. In terms of multicellular construct fabrication, most studies include MSCs seeding on (Jain et al., 2020) or within (Pati et al., 2015; Contessi Negrini et al., 2019; Negrini et al., 2020; Ianovici et al., 2022) 3D-printed scaffolds, with subsequent induction of adipogenesis to generate engineered adipose tissue. However, to generate multi-tissue constructs that include several types of cells, this approach might be difficult to apply since each cell type has different media requirements for optimal growth and differentiation. As an alternative, it has been suggested that prefabricated matured adipose and muscle tissues be combined after their separate maturation. For example, marble-like constructs of adipose and muscle tissues were successfully created by casting an adipocyte-containing alginate hydrogel on prefabricated 3D-printed scaffolds with matured BSCs (Zagury et al., 2022). In the previously mentioned work involving tendon-gel printing, steak-like tissues were fabricated by assembling 3D-printed cellular fibers containing either BSCs or MSCs, which were first allowed to differentiate separately to form muscle, adipose, and vascular fibers. The structure of the stacked-fibers construct was cross-linked with transglutaminase to ensure stability (Kang et al., 2021). To conclude, further advances in 3D bioprinting of CM-relevant biomaterials, cell types, and printing and cultivation techniques, and their adjustment for industrial-scale production, promise to accelerate the mass-scale production of structured CM.

5 Conclusion and future trends

As CM research holds great promise for a more sustainable meat production industry, meeting key challenges regarding scaffold composition and fabrication techniques used for this purpose is essential. While various scaffolding platforms were already developed for medically oriented applications, this chapter focused on ones relevant for muscle and adipose TE, along with possible routes to address food safety, nutrition, low production costs, and scalability when designing them for CM. Additionally, the suitability of each scaffolding version to serve as a temporary cultivation platform, or alternatively, as an integrated part of the final structured or unstructured CM products, was discussed. For example, as edible compositions may be processed into scaffolds that can be integrated into the final products in some cases, they can contribute their own qualities, such as enhancing fiber or amino acid content, or containing encapsulated vitamins and flavoring agents. The development of recyclable temporary cell expansion platforms which promote process sustainability, or the use of scaffolding biomaterials which can promote circular economy, should be considered as well in future endeavors. In terms of scalability, while a wide array of scaffold fabrication techniques from the TE field is present, significant research is crucial for their potential transformation into industrialized production processes. Additional methods could be harnessed from the food industry sector, such as the use of texturizing technologies for vegetable proteins. However, such techniques are limited in terms of cell encapsulation and are not suited for fabricating all types of desired CM products. Moreover, further investigations are essential to determine the effects of processing and post-processing conditions on properties essential for subsequent cell cultivation on such constructs. Importantly, the diversity of possible scaffolding biomaterials and fabrication methods extends even further than the ones presented so far, as combinations of several techniques to create 3D CM scaffolds should also be considered. Together with the fact that the desired properties vary between different CM products, it is likely that a variety of solutions will be developed, where each production methodology is tailored to a specific product. In conclusion, while efforts are already made in the field of CM development, further investigations are essential to address unique challenges in scaffold-based CM fabrication. Thus, interdisciplinary collaborations would be invaluable for enabling research conducted both in industry and academia to reach significant advances.

6 Where to look for further information

As efforts in the development of cultured meat are currently expanding within the academic community, the most up-to-date research is presented in scientific conferences. Therefore, the participation in conferences which are specifically focused on this topic or other leading conferences which overlap with this field is highly recommended, as this is a scientific endeavor. For example:

- The International Scientific Conference on Cultured Meat (ISCCM).
- Tissue Engineering and Regenerative Medicine International Society (TERMIS) International Conference.
- The Society for Biomaterials (SFB) Annual Meeting and Exposition.
- The European Federation of Food Science and Technology (EFFoST) International Conference.

Additional information, further funding, and research collaboration opportunities are also available at international non-profit organizations and institutes, such as the Good Food institute (GFI) or New Harvest, who aim to accelerate alternative protein and cellular agriculture innovations.

7 References

Acarregui, A., Murua, A., Pedraz, J. L., Orive, G. and Hernández, R. M. (2012). A perspective on bioactive cell microencapsulation, *BioDrugs* 26(5), 283–301. doi: 10.1007/BF03261887.

Adhikari, J., Roy, A., Das, A., Ghosh, M., Thomas, S., Sinha, A., Kim, J. and Saha, P. (2021). Effects of processing parameters of 3D bioprinting on the cellular activity of bioinks, *Macromolecular Bioscience* 21(1), e2000179. doi: 10.1002/mabi.202000179.

Aguilar-Vázquez, G., Ortiz-Frade, L., Figueroa-Cárdenas, J. D., López-Rubio, A. and Mendoza, S. (2020). Electrospinnability study of pea (Pisum sativum) and common bean (Phaseolus vulgaris L.) using the conformational and rheological behavior of their protein isolates, *Polymer Testing* 81. doi: 10.1016/j.polymertesting.2019.106217.

Ahmad, K., Lim, J. H., Lee, E. J., Chun, H. J., Ali, S., Ahmad, S. S., Shaikh, S. and Choi, I. (2021). Extracellular matrix and the production of cultured meat, *Foods* 10(12), 1–14. doi: 10.3390/foods10123116.

Ahn, S., Chantre, C. O., Gannon, A. R., Lind, J. U., Campbell, P. H., Grevesse, T., O'Connor, B. B. and Parker, K. K. (2018). Soy protein/cellulose nanofiber scaffolds mimicking skin extracellular matrix for enhanced wound healing, *Advanced Healthcare Materials* 7(9), e1701175. doi: 10.1002/adhm.201701175.

Albrecht, F. B., Schmidt, F. F., Volz, A. C. and Kluger, P. J. (2022) Bioprinting of 3D adipose tissue models using a GelMA-Bioink with human mature adipocytes or human adipose-derived stem cells, *Gels* 8(10), 611. doi: 10.3390/gels8100611.

Ali, A. and Ahmed, S. (2018). Recent advances in edible polymer based hydrogels as a sustainable alternative to conventional polymers, *Journal of Agricultural and Food Chemistry* 66(27), 6940–6967. doi: 10.1021/acs.jafc.8b01052.

Allan, S. J., Ellis, M. J. and De Bank, P. A. (2021). Decellularized grass as a sustainable scaffold for skeletal muscle tissue engineering, *Journal of Biomedical Materials Research. Part A* 109(12), 2471–2482. doi: 10.1002/jbm.a.37241.

Andreassen, R. C., Rønning, S. B., Solberg, N. T., Grønlien, K. G., Kristoffersen, K. A., Høst, V., Kolset, S. O. and Pedersen, M. E. (2022). Production of food-grade microcarriers based on by-products from the food industry to facilitate the expansion of bovine skeletal muscle satellite cells for cultured meat production, *Biomaterials*. 286, 121602. doi: 10.1016/j.biomaterials.2022.121602.

Ansari, S., Chen, C., Xu, X., Annabi, N., Zadeh, H. H., Wu, B. M., Khademhosseini, A., Shi, S. and Moshaverinia, A. (2016). Muscle tissue engineering using gingival mesenchymal stem cells encapsulated in alginate hydrogels containing multiple growth factors, *Annals of Biomedical Engineering* 44(6), 1908–1920. doi: 10.1007/s10439-016-1594-6.

Antinori, M. E., Contardi, M., Suarato, G., Armirotti, A., Bertorelli, R., Mancini, G., Debellis, D. and Athanassiou, A. (2021) Advanced mycelium materials as potential self-growing biomedical scaffolds, *Scientific Reports*. 11(1), pp. 1–14. doi: 10.1038/s41598-021-91572-x.

Arab, W., Rauf, S., Al-Harbi, O. and Hauser, C. A. E. (2018). Novel ultrashort self-assembling peptide bioinks for 3D culture of muscle myoblast cells, *International Journal of Bioprinting* 4(2), 129. doi: 10.18063/IJB.v4i2.129.

Aubin, H., Nichol, J. W., Hutson, C. B., Bae, H., Sieminski, A. L., Cropek, D. M., Akhyari, P. and Khademhosseini, A. (2010). Directed 3D cell alignment and elongation in microengineered hydrogels, *Biomaterials*. 31(27), 6941–6951. doi: 10.1016/j.biomaterials.2010.05.056.

Badrossamay, M. R., McIlwee, H. A., Goss, J. A. and Parker, K. K. (2010). Nanofiber assembly by rotary jet-spinning, *Nano Letters* 10(6), 2257–2261. doi: 10.1021/nl101355x.

Bar-Shai, N., Sharabani-Yosef, O., Zollmann, M., Lesman, A. and Golberg, A. (2021). Seaweed cellulose scaffolds derived from green macroalgae for tissue engineering, *Scientific Reports*. 11(1), 11843. doi: 10.1038/s41598-021-90903-2.

Ben-Arye, T., Shandalov, Y., Ben-Shaul, S., Landau, S., Zagury, Y., Ianovici, I., Lavon, N. and Levenberg, S. (2020). Textured soy protein scaffolds enable the generation of three-dimensional bovine skeletal muscle tissue for cell-based meat, *Nature Food*. 1(4), 1. doi: 10.1038/s43016-020-0046-5.

Ben-Arye, T. and Levenberg, S. (2019). Tissue engineering for clean meat production, *Frontiers in Sustainable Food Systems* 3(June), 1–19. doi: 10.3389/fsufs.2019.00046.

Bodiou, V., Moutsatsou, P. and Post, M. J. (2020). Microcarriers for upscaling cultured meat production, *Frontiers in Nutrition* 7, 10. doi: 10.3389/fnut.2020.00010.

Bolívar-Monsalve, E. J., Ceballos-González, C. F., Borrayo-Montaño, K. I., Quevedo-Moreno, D. A., Yee-de León, J. F., Khademhosseini, A., Weiss, P. S., Alvarez, M. M. and Trujillo-de Santiago, G. (2021). Continuous chaotic bioprinting of skeletal muscle-like constructs, *Bioprinting* 21. doi: 10.1016/j.bprint.2020.e00125.

Bomkamp, C., Skaalure, S. C., Fernando, G. F., Ben-Arye, T., Swartz, E. W. and Specht, E. A. (2022). Scaffolding biomaterials for 3D cultivated meat: prospects and challenges, *Advanced Science* 9(3), e2102908. doi: 10.1002/advs.202102908.

Cahn, F. (1990). Biomaterials aspects of porous microcarriers for animal cell culture, *Trends in Biotechnology* 8(5), 131–136. doi: 10.1016/0167-7799(90)90154-P.

Campodoni, E., Montanari, M., Dozio, S. M., Heggset, E. B., Panseri, S., Montesi, M., Tampieri, A., Syverud, K. and Sandri, M. (2020). Blending gelatin and cellulose nanofibrils: biocomposites with tunable degradability and mechanical behavior, *Nanomaterials* 10(6), 1–18. doi: 10.3390/nano10061219.

Campuzano, S., Mogilever, N. B. and Pelling, A. E. (2020). Decellularized plant-based scaffolds for guided alignment of myoblast cells, *bioRxiv*. doi: 10.1101/2020.02.23.958686.

Campuzano, S. and Pelling, A. E. (2019). Scaffolds for 3D cell culture and cellular agriculture applications derived from non-animal sources, *Frontiers in Sustainable Food Systems* 3(May), 1–9. doi: 10.3389/fsufs.2019.00038.

Chandler, E. M., Berglund, C. M., Lee, J. S., Polacheck, W. J., Gleghorn, J. P., Kirby, B. J. and Fischbach, C. (2011). Stiffness of photocrosslinked RGD-alginate gels regulates adipose progenitor cell behavior, *Biotechnology and Bioengineering* 108(7), 1683–1692. doi: 10.1002/bit.23079.

Chatterjee, C., Gleddie, S. and Xiao, C. W. (2018). Soybean bioactive peptides and their functional properties, *Nutrients* 10(9), 8–11. doi: 10.3390/nu10091211.

Chaudhary, S. and Chakraborty, E. (2022). Hydrogel based tissue engineering and its future applications in personalized disease modeling and regenerative therapy, *Beni-Suef University Journal of Basic and Applied Sciences*. 11(1), 3. doi: 10.1186/s43088-021-00172-1.

Chen, L., Guttieres, D., Koenigsberg, A., Barone, P. W., Sinskey, A. J. and Springs, S. L. (2022). Large-scale cultured meat production: Trends, challenges and promising biomanufacturing technologies, *Biomaterials*. doi: 10.1016/j.biomaterials.2021.121274.

Chen, A. K. L., Chen, X., Choo, A. B., Reuveny, S. and Oh, S. K. (2011). Critical microcarrier properties affecting the expansion of undifferentiated human embryonic stem cells, *Stem Cell Research*. 7(2), 97–111. doi: 10.1016/j.scr.2011.04.007.

Chen, H., Zhong, J., Wang, J., Huang, R., Qiao, X., Wang, H. and Tan, Z. (2019). Enhanced growth and differentiation of myoblast cells grown on E-jet 3D printed platforms, *International Journal of Nanomedicine* 14, 937–950.

Chen, L., Guttieres, D., Koenigsberg, A., Barone, P. W., Sinskey, A. J. and Springs, S. L. (2022). Large-scale cultured meat production: trends, challenges and promising biomanufacturing technologies, *Biomaterials* 280, 121274. doi: 10.1016/j.biomaterials.2021.121274.

Chen, S., Nakamoto, T., Kawazoe, N. and Chen, G. (2015). Engineering multi-layered skeletal muscle tissue by using 3D microgrooved collagen scaffolds, *Biomaterials*. 73, 23–31. doi: 10.1016/j.biomaterials.2015.09.010.

Chen, Y., Liu, Y., Zhang, J., Liu, H., Wang, J., Liu, Q. and Zhang, Y. (2021). Three-dimensional bioprinting adipose tissue and mammary organoids feasible for artificial breast structure regeneration, *Materials and Design*. 200, 109467. doi: 10.1016/J.MATDES.2021.109467.

Chen, Y. M., Xi, T., Zheng, Y., Guo, T., Hou, J., Wan, Y. and Gao, C. (2009). In vitro cytotoxicity of bacterial cellulose scaffolds used for tissue-engineered bone, *Journal of Bioactive and Compatible Polymers* 24 (Suppl.1), 137–145. doi: 10.1177/0883911509102710.

Cheng, Y. W., Shiwarski, D. J., Ball, R. L., Whitehead, K. A. and Feinberg, A. W. (2020). Engineering aligned skeletal muscle tissue using decellularized plant-derived scaffolds, *ACS Biomaterials Science and Engineering* 6(5), 3046–3054. doi: 10.1021/acsbiomaterials.0c00058.

Choi, J. S., Lee, S. J., Christ, G. J., Atala, A. and Yoo, J. J. (2008). The influence of electrospun aligned poly(ε-caprolactone)/collagen nanofiber meshes on the formation of self-aligned skeletal muscle myotubes, *Biomaterials* 29(19), 2899–2906. doi: 10.1016/j.biomaterials.2008.03.031.

Choi, S. H. and Park, T. G. (2002). Synthesis and characterization of elastic PLGA/PCL/PLGA tri-block copolymers, *Journal of Biomaterials Science. Polymer Edition* 13(10), 1163–1173. doi: 10.1163/156856202320813864.

Choi, S. W., Zhang, Y., Yeh, Y., Lake Wooten, A. and Xia, Y. (2012). Biodegradable porous beads and their potential applications in regenerative medicine, *Journal of Materials Chemistry* 22(23), 11442–11451. doi: 10.1039/c2jm16019f.

Choi, Y. J., Kim, T. G., Jeong, J., Yi, H. G., Park, J. W., Hwang, W. and Cho, D. W. (2016). 3D cell printing of functional skeletal muscle constructs using skeletal muscle-derived bioink, *Advanced Healthcare Materials* 5(20), 2636–2645. doi: 10.1002/adhm.201600483.

Choi, Y. J., Jun, Y. J., Kim, D. Y., Yi, H. G., Chae, S. H., Kang, J., Lee, J., Gao, G., Kong, J. S., Jang, J., Chung, W. K., Rhie, J. W. and Cho, D. W. (2019). A 3D cell printed muscle construct with tissue-derived bioink for the treatment of volumetric muscle loss, *Biomaterials*. Elsevier Ltd 206, 160–169. doi: 10.1016/j.biomaterials.2019.03.036.

Chuang, C. H., Lin, R. Z., Melero-Martin, J. M. and Chen, Y. C. (2018). Comparison of covalently and physically cross-linked collagen hydrogels on mediating vascular network formation for engineering adipose tissue, *Artificial Cells, Nanomedicine, and Biotechnology*. 46 (sup3), S434–S447. doi: 10.1080/21691401.2018.1499660.

Chui, C. Y., Odeleye, A., Nguyen, L., Kasoju, N., Soliman, E. and Ye, H. (2019). Electrosprayed genipin cross-linked alginate–chitosan microcarriers for ex vivo expansion of mesenchymal stem cells, *Journal of Biomedical Materials Research. Part A* 107(1), 122–133. doi: 10.1002/jbm.a.36539.

Chung, H. J. and Park, T. G. (2009). Injectable cellular aggregates prepared from biodegradable porous microspheres for adipose tissue engineering. Tissue Engineering Part A 15(6), 1391–1400.

Cidonio, G., Glinka, M., Dawson, J. I. and Oreffo, R. O. C. (2019). The cell in the ink: improving biofabrication by printing stem cells for skeletal regenerative medicine, *Biomaterials*. 209(March), 10–24. doi: 10.1016/j.biomaterials.2019.04.009.

Clevenger, T. N., Luna, G., Boctor, D., Fisher, S. K. and Clegg, D. O. (2016). Cell-mediated remodeling of biomimetic encapsulating hydrogels triggered by adipogenic differentiation of adipose stem cells, *Journal of Tissue Engineering* 7, 2041731416670482. doi: 10.1177/2041731416670482.

Clevenger, T. N., Hinman, C. R., Ashley Rubin, R. K., Smither, K., Burke, D. J., Hawker, C. J., Messina, D., Van Epps, D. and Clegg, D. O. (2016). Vitronectin-based, biomimetic encapsulating hydrogel scaffolds support adipogenesis of adipose stem cells, *Tissue Engineering - Part A* 22(7–8), 597–609. doi: 10.1089/ten.TEA.2015.0550.

Colle, J., Blondeel, P., De Bruyne, A., Bochar, S., Tytgat, L., Vercruysse, C., Van Vlierberghe, S., Dubruel, P. and Declercq, H. (2020). Bioprinting predifferentiated adipose-derived mesenchymal stem cell spheroids with methacrylated gelatin ink for adipose tissue engineering, *Journal of Materials Science: Materials in Medicine*. 31(4), 1–15. doi: 10.1007/S10856-020-06374-W/FIGURES/10.

Compaan, A. M., Song, K. and Huang, Y. (2019). Gellan fluid gel as a versatile support bath material for fluid extrusion bioprinting, *ACS Applied Materials and Interfaces* 11(6), 5714–5726. doi: 10.1021/acsami.8b13792.

Contessi Negrini, N., Bonnetier, M., Giatsidis, G., Orgill, D. P., Farè, S. and Marelli, B. (2019). Tissue-mimicking gelatin scaffolds by alginate sacrificial templates for adipose tissue engineering, *Acta Biomaterialia*. 87, 61–75. doi: 10.1016/J.ACTBIO.2019.01.018.

Contessi Negrini, N., Toffoletto, N., Farè, S. and Altomare, L. (2020). Plant tissues as 3D natural scaffolds for adipose, bone and tendon tissue regeneration, *Frontiers in Bioengineering and Biotechnology* 8, 723. doi: 10.3389/fbioe.2020.00723.

Costantini, M., Testa, S., Mozetic, P., Barbetta, A., Fuoco, C., Fornetti, E., Tamiro, F., Bernardini, S., Jaroszewicz, J., Święszkowski, W., Trombetta, M., Castagnoli, L., Seliktar, D., Garstecki, P., Cesareni, G., Cannata, S., Rainer, A. and Gargioli, C. (2017). Microfluidic-enhanced 3D bioprinting of aligned myoblast-laden hydrogels leads to functionally organized myofibers in vitro and in vivo, *Biomaterials* 131, 98–110. doi: 10.1016/j.biomaterials.2017.03.026.

Custódio, C. A., Cerqueira, M. T., Marques, A. P., Reis, R. L. and Mano, J. F. (2015). Cell selective chitosan microparticles as injectable cell carriers for tissue regeneration, *Biomaterials* 43(1), 23–31. doi: 10.1016/j.biomaterials.2014.11.047.

D'Andrea, P., Civita, D., Cok, M., Ulloa Severino, L., Vita, F., Scaini, D., Casalis, L., Lorenzon, P., Donati, I. and Bandiera, A. (2017). Myoblast adhesion, proliferation and differentiation on human elastin-like polypeptide (HELP) hydrogels, *Journal of Applied Biomaterials and Functional Materials* 15(1), e43–e53. doi: 10.5301/jabfm.5000331.

de Lucena-Thomas, J. P., Boonprasirt, P., Luetchford, K., De Bank, P. and Ellis, M. (2020). Bed expansion properties of tissue engineering particles in a fluidised bed bioreactor, *Biochemical Engineering Journal*. 160, 107632. doi: 10.1016/j.bej.2020.107632.

Dekkers, B. L., Boom, R. M. and van der Goot, A. J. (2018). Structuring processes for meat analogues, *Trends in Food Science and Technology*. 81(May), 25–36. doi: 10.1016/j.tifs.2018.08.011.

Deng, L., Kang, X., Liu, Y., Feng, F. and Zhang, H. (2018). Characterization of gelatin/zein films fabricated by electrospinning vs solvent casting, *Food Hydrocolloids*. 74, 324–332. doi: 10.1016/j.foodhyd.2017.08.023.

Derakhti, S., Safiabadi-Tali, S. H., Amoabediny, G. and Sheikhpour, M. (2019). Attachment and detachment strategies in microcarrier-based cell culture technology: A comprehensive review, *Materials Science and Engineering. Part C* 103(December 2018), 109782. doi: 10.1016/j.msec.2019.109782.

Devaere, J., De Winne, A., Dewulf, L., Fraeye, I., Šoljić, I., Lauwers, E., de Jong, A. and Sanctorum, H. (2022). Improving the aromatic profile of plant-based meat alternatives: effect of myoglobin addition on volatiles, *Foods* 11(13), 1–12. doi: 10.3390/foods11131985.

Dhingra, D., Michael, M., Rajput, H. and Patil, R. T. (2012). Dietary fibre in foods: a review, *Journal of Food Science and Technology* 49(3), 255–266. doi: 10.1007/s13197-011-0365-5.

Dimida, S., Barca, A., Cancelli, N., De Benedictis, V., Raucci, M. G. and Demitri, C. (2017). Effects of genipin concentration on cross-linked chitosan scaffolds for bone tissue engineering: structural characterization and evidence of biocompatibility features, *International Journal of Polymer Science*. 2017, 1–8. doi: 10.1155/2017/8410750.

Distler, T., Solisito, A. A., Schneidereit, D., Friedrich, O., Detsch, R. and Boccaccini, A. R. (2020). 3D printed oxidized alginate-gelatin bioink provides guidance for C2C12 muscle precursor cell orientation and differentiation via shear stress during bioprinting, *Biofabrication* 12(4), 045005. doi: 10.1088/1758-5090/ab98e4.

Doi, M., Thyboll, J., Kortesmaa, J., Jansson, K., Iivanainen, A., Parvardeh, M., Timpl, R., Hedin, U., Swedenborg, J. and Tryggvason, K. (2002). Recombinant human laminin-10 ($\alpha5\beta1\gamma1$). Production, purification, and migration-promoting activity on vascular endothelial cells, *Journal of Biological Chemistry*. 277(15), 12741–12748. doi: 10.1074/jbc.M111228200.

Drury, J. L. and Mooney, D. J. (2003). Hydrogels for tissue engineering: scaffold design variables and applications, *Biomaterials* 24(24), 4337–4351. doi: 10.1016/S0142-9612(03)00340-5.

Duffy, R. M., Sun, Y. and Feinberg, A. W. (2016). Understanding the role of ECM protein composition and geometric micropatterning for engineering human skeletal muscle, *Annals of Biomedical Engineering* 44(6), 2076–2089. doi: 10.1007/s10439-016-1592-8.

Dutta, S. D., Ganguly, K., Jeong, M. S., Patel, D. K., Patil, T. V., Cho, S. J. and Lim, K. T. (2022). Bioengineered lab-grown meat-like constructs through 3D bioprinting of antioxidative protein hydrolysates, *ACS Applied Materials and Interfaces* 14(30), 34513–34526. doi: 10.1021/acsami.2c10620.

Ebrahimi, M., Ostrovidov, S., Salehi, S., Kim, S. B., Bae, H. and Khademhosseini, A. (2018). Enhanced skeletal muscle formation on microfluidic spun gelatin methacryloyl (GelMA) fibres using surface patterning and agrin treatment, *Journal of Tissue Engineering and Regenerative Medicine* 12(11), 2151–2163. doi: 10.1002/term.2738.

Elmowafy, E. M., Tiboni, M. and Soliman, M. E. (2019). *Biocompatibility, Biodegradation and Biomedical Applications of Poly (Lactic Acid Poly (Lactic co Glycolic Acid) Micro and Nanoparticles, Journal of Pharmaceutical Investigation*. Springer, Singapore.

Enrione, J., Blaker, J. J., Brown, D. I., Weinstein-Oppenheimer, C. R., Pepczynska, M., Olguín, Y., Sánchez, E. and Acevedo, C. A. (2017). Edible scaffolds based on non-mammalian biopolymers for myoblast growth, *Materials* 10(12), 1–15. doi: 10.3390/ma10121404.

Fan, Y., Zhang, F. and Tzanakakis, E. S. (2017). Engineering xeno-free microcarriers with recombinant vitronectin, albumin and UV irradiation for human pluripotent stem cell bioprocessing, *ACS Biomaterials Science and Engineering* 3(8), 1510–1518. doi: 10.1021/acsbiomaterials.6b00253.

Fontana, G., Gershlak, J., Adamski, M., Lee, J. S., Matsumoto, S., Le, H. D., Binder, B., Wirth, J., Gaudette, G. and Murphy, W. L. (2017). Biofunctionalized plants as diverse biomaterials for human cell culture, *Advanced Healthcare Materials* 6(8). doi: 10.1002/adhm.201601225.

Fraeye, I., Kratka, M., Vandenburgh, H. and Thorrez, L. (2020). Sensorial and nutritional aspects of cultured meat in comparison to traditional meat: much to be inferred, *Frontiers in Nutrition* 7(March), 35. doi: 10.3389/fnut.2020.00035.

Fraser, R. Z., Shitut, M., Agrawal, P., Mendes, O. and Klapholz, S. (2018). Safety evaluation of soy leghemoglobin protein preparation derived from pichia pastoris, intended for use as a flavor catalyst in plant-based meat, *International Journal of Toxicology* 37(3), 241–262. doi: 10.1177/1091581818766318.

Freeman, F. E. and Kelly, D. J. (2017). Tuning alginate bioink stiffness and composition for controlled growth factor delivery and to spatially direct MSC Fate within bioprinted tissues, *Scientific Reports*. 7(1), 17042. doi: 10.1038/s41598-017-17286-1.

Fuoco, C., Sangalli, E., Vono, R., Testa, S., Sacchetti, B., Latronico, M. V., Bernardini, S., Madeddu, P., Cesareni, G., Seliktar, D., Rizzi, R., Bearzi, C., Cannata, S. M., Spinetti, G. and Gargioli, C. (2014). 3D hydrogel environment rejuvenates aged pericytes for skeletal muscle tissue engineering, *Frontiers in Physiology* 5(May), 203. doi: 10.3389/fphys.2014.00203.

Gao, Q., He, Y., Fu, J. Z., Liu, A. and Ma, L. (2015). Coaxial nozzle-assisted 3D bioprinting with built-in microchannels for nutrients delivery, *Biomaterials*. 61, 203–215. doi: 10.1016/j.biomaterials.2015.05.031.

Garg, K. (2017). Laminin enriched scaffolds for tissue engineering applications, *Advances in Tissue Engineering and Regenerative Medicine: Open Access* 2(3), 194–200. doi: 10.15406/atroa.2017.02.00033.

Gattazzo, F., De Maria, C., Rimessi, A., Donà, S., Braghetta, P., Pinton, P., Vozzi, G. and Bonaldo, P. (2018). Gelatin-genipin-based biomaterials for skeletal muscle tissue engineering, *Journal of Biomedical Materials Research. Part B, Applied Biomaterials* 106(8), 2763–2777. doi: 10.1002/jbm.b.34057.

Gaudet, I. D. and Shreiber, D. I. (2012). Characterization of methacrylated type-I collagen as a dynamic, photoactive hydrogel, *Biointerphases* 7(1–4), 25. doi: 10.1007/s13758-012-0025-y.

Gershlak, J. R., Hernandez, S., Fontana, G., Perreault, L. R., Hansen, K. J., Larson, S. A., Binder, B. Y., Dolivo, D. M., Yang, T., Dominko, T., Rolle, M. W., Weathers, P. J., Medina-Bolivar, F., Cramer, C. L., Murphy, W. L. and Gaudette, G. R. (2017). Crossing kingdoms: using decellularized plants as perfusable tissue engineering scaffolds, *Biomaterials*. 125, 13–22. doi: 10.1016/j.biomaterials.2017.02.011.

Gilbert, P. M., Havenstrite, K. L., Magnusson, K. E., Sacco, A., Leonardi, N. A., Kraft, P., Nguyen, N. K., Thrun, S., Lutolf, M. P. and Blau, H. M. (2010). Substrate elasticity regulates skeletal muscle stem cell self-renewal in culture, *Science* 329(5995), 1078–1081. doi: 10.1126/science.1191035.

Gong, S., Wang, H., Sun, Q., Xue, S. T. and Wang, J. Y. (2006). Mechanical properties and in vitro biocompatibility of porous zein scaffolds, *Biomaterials* 27(20), 3793–3799. doi: 10.1016/j.biomaterials.2006.02.019.

Gröhn, P., Klöck, G. and Zimmermann, U. (1997). Collagen-coated Ba2+-alginate microcarriers for the culture of anchorage-dependent mammalian cells, *BioTechniques* 22(5), 970–975. doi: 10.2144/97225rr06.

Gruen, V., Apsite, I., Uribe, J. M., Posada, A. F., Rosenfeldt, S., Salehi, S. and Ionov, L. (2020). *4D Biofabrication of Skeletal Muscle Microtissues*. IOP Publishing.

Guarino, V., Altobelli, R. and Ambrosio, L. (2016). Chitosan microgels and nanoparticles via electrofluidodynamic techniques for biomedical applications, *Gels* 2(1). doi: 10.3390/gels2010002.

Halford, M. M., He, Y. C. and Stacker, S. A. (2014). Expression and purification of bioactive, low-endotoxin recombinant human vitronectin, *BioTechniques* 56(6), 331–333. doi: 10.2144/000114181.

Han, D., McMillin, K. W. and Godber, J. S. (1994). Hemoglobin, myoglobin, and total pigments in beef and chicken muscles: chromatographic determination, *Journal of Food Science* 59(6), 1279–1282. doi: 10.1111/j.1365-2621.1994.tb14695.x.

Hickey, R. J., Modulevsky, D. J., Cuerrier, C. M. and Pelling, A. E. (2018). Customizing the shape and microenvironment biochemistry of biocompatible macroscopic plant-derived cellulose scaffolds, *ACS Biomaterials Science and Engineering* 4(11), 3726–3736. doi: 10.1021/acsbiomaterials.8b00178.

Hinton, T. J., Jallerat, Q., Palchesko, R. N., Park, J. H., Grodzicki, M. S., Shue, H. J., Ramadan, M. H., Hudson, A. R. and Feinberg, A. W. (2015). Three-dimensional printing of complex biological structures by freeform reversible embedding of suspended hydrogels, *Science Advances* 1(9), e1500758. doi: 10.1126/sciadv.1500758.

Hong, Y., Gao, C., Xie, Y., Gong, Y. and Shen, J. (2005). Collagen-coated polylactide microspheres as chondrocyte microcarriers, *Biomaterials* 26(32), 6305–6313. doi: 10.1016/j.biomaterials.2005.03.038.

Huang, L., Abdalla, A. M. E., Xiao, L. and Yang, G. (2020). Biopolymer-based microcarriers for three-dimensional cell culture and engineered tissue formation, *International Journal of Molecular Sciences* 21(5).

Hwang, C. M., Sant, S., Masaeli, M., Kachouie, N. N., Zamanian, B., Lee, S. H. and Khademhosseini, A. (2010). Fabrication of three-dimensional porous cell-laden hydrogel for tissue engineering, *Biofabrication* 2(3), 035003. doi: 10.1088/1758-5082/2/3/035003.

Hwangbo, H., Lee, H., Jin, E. J., Lee, J., Jo, Y., Ryu, D. and Kim, G. (2022). Bio-printing of aligned GelMa-based cell-laden structure for muscle tissue regeneration. KeAi Communications Co., Ltd, *Bioactive Materials* 8(June 2021), 57–70. doi: 10.1016/j.bioactmat.2021.06.031.

Ianovici, I., Zagury, Y., Redenski, I., Lavon, N. and Levenberg, S. (2022). 3D-printable plant protein-enriched scaffolds for cultivated meat development, *Biomaterials*. 284(March), 121487. doi: 10.1016/j.biomaterials.2022.121487.

Ikeda, K., Nagata, S., Okitsu, T. and Takeuchi, S. (2017). Cell fiber-based three-dimensional culture system for highly efficient expansion of human induced pluripotent stem cells, *Scientific Reports*. 7(1), 2850. doi: 10.1038/s41598-017-03246-2.

Jahangirian, H., Azizi, S., Rafiee-Moghaddam, R., Baratvand, B. and Webster, T. J. (2019). Status of plant protein-based green scaffolds for regenerative medicine applications, *Biomolecules* 9(10). doi: 10.3390/biom9100619.

Jain, S., Yassin, M. A., Fuoco, T., Liu, H., Mohamed-Ahmed, S., Mustafa, K. and Finne-Wistrand, A. (2020). Engineering 3D degradable, pliable scaffolds toward adipose tissue regeneration; optimized printability, simulations and surface modification, *Journal of Tissue Engineering*. 11, 2041731420954316. doi: 10.1177/2041731420954316.

Jain, S., Yassin, M. A., Fuoco, T., Mohamed-Ahmed, S., Vindenes, H., Mustafa, K. and Finne-Wistrand, A. (2021). Understanding of how the properties of medical grade lactide based copolymer scaffolds influence adipose tissue regeneration: sterilization and a systematic in vitro assessment, *Materials Science and Engineering. Part C*. 124, 112020. doi: 10.1016/j.msec.2021.112020.

Jammalamadaka, U. and Tappa, K. (2018). Recent advances in biomaterials for 3D printing and tissue engineering, *Journal of Functional Biomaterials* 9(1). doi: 10.3390/jfb9010022.

Jana, S., Cooper, A. and Zhang, M. (2013). Chitosan scaffolds with unidirectional microtubular pores for large skeletal myotube generation, *Advanced Healthcare Materials* 2(4), 557–561. doi: 10.1002/adhm.201200177.

Jeong, D., Seo, J. W., Lee, H. G., Jung, W. K., Park, Y. H. and Bae, H. (2022). Efficient myogenic/adipogenic transdifferentiation of bovine fibroblasts in a 3D bioprinting system for steak-type cultured meat production, *Advanced Science* 9(31), e2202877. doi: 10.1002/advs.202202877.

Jin, M., Shi, J., Zhu, W., Yao, H. and Wang, D. A. (2021). Polysaccharide-based biomaterials in tissue engineering: a review, *Tissue Engineering. Part B, Reviews* 27(6), 604–626. doi: 10.1089/ten.TEB.2020.0208.

Jones, J. D., Rebello, A. S. and Gaudette, G. R. (2021). Decellularized spinach: an edible scaffold for laboratory-grown meat, *Food Bioscience*. 41, 100986. doi: 10.1016/j.fbio.2021.100986.

Kaczmarska, K., Taylor, M., Piyasiri, U. and Frank, D. (2021). Flavor and metabolite profiles of meat, meat substitutes, and traditional plant-based high-protein food products available in Australia, *Foods* 10(4). doi: 10.3390/foods10040801.

Kanematsu, A., Marui, A., Yamamoto, S., Ozeki, M., Hirano, Y., Yamamoto, M., Ogawa, O., Komeda, M. and Tabata, Y. (2004). Type I collagen can function as a reservoir of basic fibroblast growth factor, *Journal of Controlled Release* 99(2), 281–292. doi: 10.1016/j.jconrel.2004.07.008.

Kang, D. H., Louis, F., Liu, H., Shimoda, H., Nishiyama, Y., Nozawa, H., Kakitani, M., Takagi, D., Kasa, D., Nagamori, E., Irie, S., Kitano, S. and Matsusaki, M. (2021). Engineered whole cut meat-like tissue by the assembly of cell fibers using tendon-gel integrated bioprinting, *Nature Communications*. 12(1), 5059. doi: 10.1038/s41467-021-25236-9.

Kang, H. W., Lee, S. J., Ko, I. K., Kengla, C., Yoo, J. J. and Atala, A. (2016). A 3D bioprinting system to produce human-scale tissue constructs with structural integrity, *Nature Biotechnology*. 34(3), 312–319. doi: 10.1038/nbt.3413.

Kankala, R. K., Zhao, J., Liu, C. G., Song, X. J., Yang, D. Y., Zhu, K., Wang, S. B., Zhang, Y. S. and Chen, A. Z. (2019). Highly porous microcarriers for minimally invasive in situ skeletal muscle cell delivery, *Small* 15(25), e1901397. doi: 10.1002/smll.201901397.

Khan, S., Ul-Islam, M., Ikram, M., Ullah, M. W., Israr, M., Subhan, F., Kim, Y., Jang, J. H., Yoon, S. and Park, J. K. (2016). Three-dimensionally microporous and highly biocompatible bacterial cellulose-gelatin composite scaffolds for tissue engineering applications, *RSC Advances*. 6(112), 110840–110849. doi: 10.1039/C6RA18847H.

Kim, J. H., Seol, Y. J., Ko, I. K., Kang, H. W., Lee, Y. K., Yoo, J. J., Atala, A. and Lee, S. J. (2018). 3D bioprinted human skeletal muscle constructs for muscle function restoration, *Scientific Reports*. 8(1), 12307. doi: 10.1038/s41598-018-29968-5.

Kim, M. J., Choi, Y. S., Yang, S. H., Hong, H. N., Cho, S. W., Cha, S. M., Pak, J. H., Kim, C. W., Kwon, S. W. and Park, C. J. (2006). Muscle regeneration by adipose tissue-derived adult stem cells attached to injectable PLGA spheres, *Biochemical and Biophysical Research Communications*. 348(2), 386–392. doi: 10.1016/j.bbrc.2006.07.063.

Kim, Y. and Kim, G. (2013). Collagen/alginate scaffolds comprising core (PCL)–shell (collagen/alginate) struts for hard tissue regeneration: Fabrication, characterisation, and cellular activities, *Journal of Materials Chemistry. B* 1(25), 3185–3194. doi: 10.1039/c3tb20485e.

Kinsella, M. and Monk, C. (2012). Laminin121 – recombinant expression and interactions with integrins 23(1), 1–7. doi: 10.1016/j.matbio.2010.05.004.Laminin121.

Kolasa, M., Czerczak, K., Fraczyk, J., Szymanski, L., Lewicki, S., Bednarowicz, A., Tarzynska, N., Sikorski, D., Szparaga, G., Draczynski, Z., Cierniak, S., Brzoskowska, U., Galita, G., Majsterek, I., Bociaga, D., Krol, P. and Kolesinska, B. (2022). Evaluation of polysaccharide-peptide conjugates containing the RGD motif for potential use in muscle tissue regeneration, *Materials* 15(18), 1–36. doi: 10.3390/ma15186432.

Kong, Y., Ong, S., Liu, M. H., Yu, H. and Huang, D. (2022). Functional composite microbeads for cell-based meat culture: Effect of animal gelatin coating on cell proliferation and differentiation, *Journal of Physics. Part D*. 55(34). doi: 10.1088/1361-6463/ac7011.

Kroehne, V., Heschel, I., Schügner, F., Lasrich, D., Bartsch, J. W. and Jockusch, H. (2008). Use of a novel collagen matrix with oriented pore structure for muscle cell differentiation in cell culture and in grafts, *Journal of Cellular and Molecular Medicine* 12(5A), 1640–1648. doi: 10.1111/j.1582-4934.2008.00238.x.

Krontiras, P., Gatenholm, P. and Hagg, D. A. (2015). Adipogenic differentiation of stem cells in three-dimensional porous bacterial nanocellulose scaffolds, *Journal of Biomedical Materials Research. Part B, Applied Biomaterials* 103(1), 195–203. doi: 10.1002/jbm.b.33198.

Kuss, M., Kim, J., Qi, D., Wu, S., Lei, Y., Chung, S. and Duan, B. (2018). Effects of tunable, 3D-bioprinted hydrogels on human brown adipocyte behavior and metabolic function, *Acta Biomaterialia*. 71, 486–495. doi: 10.1016/J.ACTBIO.2018.03.021.

Kyriakopoulou, K., Dekkers, B. and van der Goot, A. J. (2018). *Plant-based Meat Analogues, Sustainable Meat Production and Processing*, Elsevier Inc.

Langelaan, M. L. P., Boonen, K. J. M., Polak, R. B., Baaijens, F. P. T., Post, M. J. and van der Schaft, D. W. J. (2010). Meet the new meat: Tissue engineered skeletal muscle, *Trends in Food Science and Technology*. 21(2), 59–66. doi: 10.1016/j.tifs.2009.11.001.

Law, J. X., Liau, L. L., Saim, A., Yang, Y. and Idrus, R. (2017). Electrospun collagen nanofibers and their applications in skin tissue engineering, *Tissue Engineering and Regenerative Medicine* 14(6), 699–718. doi: 10.1007/s13770-017-0075-9.

Lee, A., Hudson, A. R., Shiwarski, D. J., Tashman, J. W., Hinton, T. J., Yerneni, S., Bliley, J. M., Campbell, P. G. and Feinberg, A. W. (2019). 3D bioprinting of collagen to rebuild components of the human heart, *Science* 365(6452), 482–487. doi: 10.1126/science.aav9051.

Lee, H. J., Yong, H. I., Kim, M., Choi, Y. S. and Jo, C. (2020). Status of meat alternatives and their potential role in the future meat market - a review, *Asian-Australasian Journal of Animal Sciences* 33(10), 1533–1543. doi: 10.5713/ajas.20.0419.

Lee, P. Y., Cobain, E., Huard, J. and Huang, L. (2007). Thermosensitive hydrogel PEG – PLGA – PEG enhances engraftment of muscle-derived stem cells and promotes healing in diabetic wound, *Molecular Therapy*. 15(6), 1189–1194. doi: 10.1038/sj.mt.6300156.

Lev, R. and Seliktar, D. (2018). Hydrogel biomaterials and their therapeutic potential for muscle injuries and muscular dystrophies, *Journal of the Royal Society, Interface* 15(138). doi: 10.1098/rsif.2017.0380.

Levi, S., Yen, F., Baruch, L. and Machluf, M. (2022). Scaffolding technologies for the engineering of cultured meat: towards a safe, sustainable, and scalable production, *Trends in Food Science and Technology*. 126(March), 13–25. doi: 10.1016/j.tifs.2022.05.011.

Levy-Mishali, M., Zoldan, J. and Levenberg, S. (2009). Effect of scaffold stiffness on myoblast differentiation, *Tissue Engineering. Part A* 15(4), 23–27.

Li, C. H., Yang, I. H., Ke, C. J., Chi, C. Y., Matahum, J., Kuan, C. Y., Celikkin, N., Swieszkowski, W. and Lin, F. H. (2022). The production of fat-containing cultured meat by stacking aligned muscle layers and adipose layers formed from gelatin-soymilk scaffold, *Frontiers in Bioengineering and Biotechnology* 10(April), 875069. doi: 10.3389/fbioe.2022.875069.

Li, W., Han, Y., Yang, H., Wang, G., Lan, R. and Wang, J. Y. (2016). Preparation of microcarriers based on zein and their application in cell culture, *Materials Science and Engineering. C, Materials for Biological Applications*. 58, 863–869. doi: 10.1016/j.msec.2015.09.045.

Li, Y., Liu, W., Li, S., Zhang, M., Yang, F. and Wang, S. (2021). Porcine skeletal muscle tissue fabrication for cultured meat production using three-dimensional bioprinting technology, *Journal of Future Foods*. 1(1), 88–97. doi: 10.1016/j.jfutfo.2021.09.005.

Liu, S., Wang, T., Li, S. and Wang, X. (2022). Application status of sacrificial biomaterials in 3D bioprinting, *Polymers* 14(11). doi: 10.3390/polym14112182.

Liu, Y., Wang, R., Ding, S., Deng, L., Zhang, Y., Li, J., Shi, Z., Wu, Z., Liang, K., Yan, X., Liu, W. and Du, Y. (2022). Engineered meatballs via scalable skeletal muscle cell expansion and modular micro-tissue assembly using porous gelatin micro-carriers, *Biomaterials*. 287, 121615. doi: 10.1016/j.biomaterials.2022.121615.

Liu, Z., Chen, R. and He, J. (2016). Active generation of multiple jets for producing nanofibres with high quality and high throughput, *Materials and Design*. 94, 496–501. doi: 10.1016/j.matdes.2016.01.075.

Lou, C. W., Lin, M. C., Huang, C. H., Lai, M. F., Shiu, B. C. and Lin, J. H. (2022). Preparation of needleless electrospinning polyvinyl alcohol/water-soluble chitosan nanofibrous membranes: antibacterial property and filter efficiency, *Polymers* 14(5). doi: 10.3390/polym14051054.

Luo, B., Tian, L., Chen, N., Ramakrishna, S., Thakor, N. and Yang, I. H. (2018). Electrospun nanofibers facilitate better alignment, differentiation, and long-term culture in an in vitro model of the neuromuscular junction (NMJ), *Biomaterials Science*. 6(12), 3262-3272. doi: 10.1039/c8bm00720a.

Lutzweiler, G., Ndreu Halili, A. and Engin Vrana, N. (2020). The overview of porous, bioactive scaffolds as instructive biomaterials for tissue regeneration and their clinical translation, *Pharmaceutics* 12(7), 1–29. doi: 10.3390/pharmaceutics12070602.

Ma, W., Qi, B., Sami, R., Jiang, L., Li, Y. and Wang, H. (2018). Conformational and functional properties of soybean proteins produced by extrusion-hydrolysis approach, *International Journal of Analytical Chemistry*. 2018, 9182508. doi: 10.1155/2018/9182508.

Machour, M., Hen, N., Goldfracht, I., Safina, D., Davidovich-Pinhas, M., Bianco-Peled, H. and Levenberg, S. (2022). Print-and-grow within a novel support material for 3D bioprinting and post-printing tissue growth, *Advanced Science* 9(34), e2200882. doi: 10.1002/advs.202200882.

MacQueen, L. A., Alver, C. G., Chantre, C. O., Ahn, S., Cera, L., Gonzalez, G. M., O'Connor, B. B., Drennan, D. J., Peters, M. M., Motta, S. E., Zimmerman, J. F. and Parker, K. K. (2019). Muscle tissue engineering in fibrous gelatin: Implications for meat analogs, *npj Science of Food*. 3, 20. doi: 10.1038/s41538-019-0054-8.

Marga, F. S., Purcell, B. P., Forgacs, G. and Forgacs, A. (2017). Inventors; Modern Meadow Inc, assignee, *Edible and Animal-Product-Free Microcarriers for Engineered Meat*. https://patents.google.com/patent/WO2015038988A1/en.

Mattice, K. D. and Marangoni, A. G. (2020). Comparing methods to produce fibrous material from zein, *Food Research International*. 128, 108804. doi: 10.1016/j.foodres.2019.108804.

Mehta, F., Theunissen, R. and Post, M. J. (n.d.). *Chapter 8 Adipogenesis from Bovine Precursors* (vol. 1889), pp. 111–125.

Mendes, A. C., Stephansen, K. and Chronakis, I. S. (2017). Electrospinning of food proteins and polysaccharides, *Food Hydrocolloids*. 68, 53–68. doi: 10.1016/j.foodhyd.2016.10.022.

Miranda-Nieves, D. and Chaikof, E. L. (2017). Collagen and elastin biomaterials for the fabrication of engineered living tissues, *ACS Biomaterials Science and Engineering* 3(5), 694–711. doi: 10.1021/acsbiomaterials.6b00250.

Mirdamadi, E., Muselimyan, N., Koti, P., Asfour, H. and Sarvazyan, N. (2019). Agarose slurry as a support medium for bioprinting and culturing freestanding cell-laden hydrogel constructs, *3D Printing and Additive Manufacturing* 6(3), 158–164. doi: 10.1089/3dp.2018.0175.

Mistry, P., Aied, A., Alexander, M., Shakesheff, K., Bennett, A. and Yang, J. (2017). Bioprinting using mechanically robust core-shell cell-laden hydrogel strands, *Macromolecular Bioscience* 17(6), 1–8. doi: 10.1002/mabi.201600472.

Mohammadinejad, R., Kumar, A., Ranjbar-Mohammadi, M., Ashrafizadeh, M., Han, S. S., Khang, G. and Roveimiab, Z. (2020). Recent advances in natural gum-based biomaterials for tissue engineering and regenerative medicine: a review, *Polymers* 12(1). doi: 10.3390/polym12010176.

Molnar, P., Wang, W., Natarajan, A., Rumsey, J. W. and Hickman, J. J. (2007). Photolithographic patterning of C2C12 myotubes using vitronectin as growth substrate in serum-free medium, *Biotechnology Progress* 23(1), 265–268. doi: 10.1021/bp060302q.

Muoio, F., Panella, S., Lindner, M., Jossen, V., Harder, Y., Moccetti, T., Eibl, R., Müller, M. and Tallone, T. (2021). Development of a biodegradable microcarrier for the cultivation of human adipose stem cells (hASCs) with a defined Xeno- and serum-free medium, *Applied Sciences* 11(3), 925. https://doi.org/10.3390/app11030925.

Murphy, S. V. and Atala, A. (2014). 3D bioprinting of tissues and organs, *Nature Biotechnology*. 32(8), 773–785. doi: 10.1038/nbt.2958.

Negrini, N. C., Celikkin, N., Tarsini, P., Farè, S. and Święszkowski, W. (2020). Three-dimensional printing of chemically crosslinked gelatin hydrogels for adipose tissue engineering, *Biofabrication*. 2, 12, 025001. doi: 10.1088/1758-5090/AB56F9.

Ng, S. and Kurisawa, M. (2021). Integrating biomaterials and food biopolymers for cultured meat production, *Acta Biomaterialia* 124, 108–129. doi: 10.1016/j.actbio.2021.01.017.

Nienow, A. W., Rafiq, Q. A., Coopman, K. and Hewitt, C. J. (2014). A potentially scalable method for the harvesting of hMSCs from microcarriers, *Biochemical Engineering Journal* 85, 79–88. doi: 10.1016/j.bej.2014.02.005.

Norris, S. C. P., Kawecki, N. S., Davis, A. R., Chen, K. K. and Rowat, A. C. (2022). Emulsion-templated microparticles with tunable stiffness and topology: applications as edible microcarriers for cultured meat, *Biomaterials*. 287(June), 121669. doi: 10.1016/j.biomaterials.2022.121669.

Nseir, N., Regev, O., Kaully, T., Blumenthal, J., Levenberg, S. and Zussman, E. (2013). Biodegradable scaffold fabricated of electrospun albumin fibers: mechanical and biological characterization, *Tissue Engineering. Part C, Methods* 19(4), 257–264. doi: 10.1089/ten.TEC.2012.0118.

Olsen, D., Yang, C., Bodo, M., Chang, R., Leigh, S., Baez, J., Carmichael, D., Perälä, M., Hämäläinen, E. R., Jarvinen, M. and Polarek, J. (2003). Recombinant collagen and gelatin for drug delivery, *Advanced Drug Delivery Reviews* 55(12), 1547–1567. doi: 10.1016/j.addr.2003.08.008.

Onoe, H., Okitsu, T., Itou, A., Kato-Negishi, M., Gojo, R., Kiriya, D., Sato, K., Miura, S., Iwanaga, S., Kuribayashi-Shigetomi, K., Matsunaga, Y. T., Shimoyama, Y. and Takeuchi, S. (2013). Metre-long cell-laden microfibres exhibit tissue morphologies and functions, *Nature Materials*. 12(6), 584–590. doi: 10.1038/nmat3606.

Parmar, I. A., Shedge, A. S., Badiger, M. V., Wadgaonkar, P. P. and Lele, A. K. (2017). Thermo-reversible sol-gel transition of aqueous solutions of patchy polymers, *RSC Advances*. 7(9), 5101–5110. doi: 10.1039/C6RA27030A.

Pati, F., Ha, D. H., Jang, J., Han, H. H., Rhie, J. W. and Cho, D. W. (2015). Biomimetic 3D tissue printing for soft tissue regeneration, *Biomaterials*. 62, 164–175. doi: 10.1016/J.BIOMATERIALS.2015.05.043.

Peng, C. A., Kozubowski, L. and Marcotte, W. R. (2020). Advances in plant-derived scaffold proteins, *Frontiers in Plant Science* 11(February), 122. doi: 10.3389/fpls.2020.00122.

Penton, C. M., Badarinarayana, V., Prisco, J., Powers, E., Pincus, M., Allen, R. E. and August, P. R. (2016). Laminin 521 maintains differentiation potential of mouse and human satellite cell-derived myoblasts during long-term culture expansion, *Skeletal Muscle* 6(1), 44. doi: 10.1186/s13395-016-0116-4.

Pérez-Guzmán, C. J. and Castro-Muñoz, R. (2020). A review of zein as a potential biopolymer for tissue engineering and nanotechnological applications, *Processes* 8(11), 1–21. doi: 10.3390/pr8111376.

Perez-puyana, V., Wieringa, P., Yuste, Y., de la Portilla, F., Guererro, A., Romero, A. and Moroni, L. (2021). Fabrication of hybrid scaffolds obtained from combinations of PCL with gelatin or collagen via electrospinning for skeletal muscle tissue engineering, *Journal of Biomedical Materials Research. Part A* 109(9), 1600–1612. doi: 10.1002/jbm.a.37156.

Perez-Puyana, V., Rubio-Valle, J. F., Jiménez-Rosado, M., Guerrero, A. and Romero, A. (2020). Alternative processing methods of hybrid porous scaffolds based on gelatin and chitosan, *Journal of the Mechanical Behavior of Biomedical Materials* 102(August 2019), 103472. doi: 10.1016/j.jmbbm.2019.103472.

Perez-Puyana, V., Villanueva, P., Jiménez-Rosado, M., de la Portilla, F. and Romero, A. (2021). Incorporation of elastin to improve polycaprolactone-based scaffolds for skeletal muscle via electrospinning, *Polymers* 13(9), 1–10. doi: 10.3390/polym13091501.

Perteghella, S., Martella, E., de Girolamo, L., Perucca Orfei, C., Pierini, M., Fumagalli, V., Pintacuda, D. V., Chlapanidas, T., Viganò, M., Faragò, S., Torre, M. L. and Lucarelli, E. (2017). Fabrication of innovative silk/alginate microcarriers for mesenchymal stem cell delivery and tissue regeneration, *International Journal of Molecular Sciences* 18(9), 1–18. doi: 10.3390/ijms18091829.

Phan, H. P., Sugino, M. and Niimi, T. (2009). The production of recombinant human laminin-332 in a Leishmania tarentolae expression system, *Protein Expression and Purification*. 68(1), 79–84. doi: 10.1016/j.pep.2009.07.005.

Pi, Q., Maharjan, S., Yan, X., Liu, X., Singh, B., van Genderen, A. M., Robledo-Padilla, F., Parra-Saldivar, R., Hu, N., Jia, W., Xu, C., Kang, J., Hassan, S., Cheng, H., Hou, X., Khademhosseini, A. and Zhang, Y. S. (2018). Digitally tunable microfluidic bioprinting of multilayered cannular tissues, *Advanced Materials* 30(43), e1706913. doi: 10.1002/adma.201706913.

Pot, M. W., Faraj, K. A., Adawy, A., van Enckevort, W. J., van Moerkerk, H. T., Vlieg, E., Daamen, W. F. and van Kuppevelt, T. H. (2015). Versatile wedge-based system for the construction of unidirectional collagen scaffolds by directional freezing: practical and theoretical considerations, *ACS Applied Materials and Interfaces* 7(16), 8495–8505. doi: 10.1021/acsami.5b00169.

Qi, X., Su, T., Zhang, M., Tong, X., Pan, W., Zeng, Q., Zhou, Z., Shen, L., He, X. and Shen, J. (2020). Macroporous hydrogel scaffolds with tunable physicochemical properties for tissue engineering constructed using renewable polysaccharides, *ACS Applied Materials and Interfaces* 12(11), 13256–13264. doi: 10.1021/acsami.9b20794.

Radaei, P., Mashayekhan, S. and Vakilian, S. (2017). Modeling and optimization of gelatin-chitosan micro-carriers preparation for soft tissue engineering: using Response Surface Methodology, *Materials Science and Engineering. C, Materials for Biological Applications*. 75, 545–553. doi: 10.1016/j.msec.2017.02.108.

Reddy, M. S. B., Ponnamma, D., Choudhary, R. and Sadasivuni, K. K. (2021). A comparative review of natural and synthetic biopolymer composite scaffolds, *Polymers* 13(7). doi: 10.3390/polym13071105.

Reid, G., Magarotto, F., Marsano, A. and Pozzobon, M. (2020). Next stage approach to tissue engineering skeletal muscle, *Bioengineering* 7(4). doi: 10.3390/bioengineering7040118.

Reiss, J., Robertson, S. and Suzuki, M. (2021). Cell sources for cultivated meat: applications and considerations throughout the production workflow, *International Journal of Molecular Sciences* 22(14). doi: 10.3390/ijms22147513.

Riboldi, S. A., Sadr, N., Pigini, L., Neuenschwander, P., Simonet, M., Mognol, P., Sampaolesi, M., Cossu, G. and Mantero, S. (2008). Skeletal myogenesis on highly orientated microfibrous polyesterurethane scaffolds, *Journal of Biomedical Materials Research. Part A* 84(4), 1094–1101. doi: 10.1002/jbm.a.31534.

Risner, D., Li. F., Fell, J. S., Pace, S. A., Siegel, J. B., Tagkopoulos, I. and Spang, E. S. (2020). Preliminary techno-economic assessment of animal cell-based meat, *Foods*. 10(1), 3. doi: 10.3390/foods10010003.

Robbins, E. R., Pins, G. D., Laflamme, M. A. and Gaudette, G. R. (2020). Creation of a contractile biomaterial from a decellularized spinach leaf without ECM protein coating: an in vitro study, *Journal of Biomedical Materials Research. Part A* 108(10), 2123–2132. doi: 10.1002/jbm.a.36971.

Ronzoni, F. L., Aliberti, F., Scocozza, F., Benedetti, L., Auricchio, F., Sampaolesi, M., Cusella, G., Redwan, I. N., Ceccarelli, G. and Conti, M. (2022). Myoblast 3D bioprinting to burst in vitro skeletal muscle differentiation, *Journal of Tissue Engineering and Regenerative Medicine* 16(5), 484–495. doi: 10.1002/term.3293.

Roy, D. C. and Hocking, D. C. (2013). Recombinant fibronectin matrix mimetics specify integrin adhesion and extracellular matrix assembly, *Tissue Engineering - Part A* 19(3–4), 558–570. doi: 10.1089/ten.TEA.2012.0257.

Rubio, N. R., Fish, K. D., Trimmer, B. A. and Kaplan, D. L. (2019). In vitro insect muscle for tissue engineering applications, *ACS Biomaterials Science and Engineering* 5(2), 1071–1082. doi: 10.1021/acsbiomaterials.8b01261.

Ruoslahti, E. and Pierschbacher, M. D. (1986). Argining-glycine-aspartic acid: A versatile cell recognition signal minireview, *Cell* 44(4), 517–518.

Saquing, C. D., Tang, C., Monian, B., Bonino, C. A., Manasco, J. L., Alsberg, E. and Khan, S. A. (2013). Alginate-polyethylene oxide blend nanofibers and the role of the carrier polymer in electrospinning, *Industrial and Engineering Chemistry Research* 52(26), 8692–8704. doi: 10.1021/ie302385b.

Schmid, E. M., Farahnaky, A., Adhikari, B. and Torley, P. J. (2022). High moisture extrusion cooking of meat analogs: a review of mechanisms of protein texturization, *Comprehensive Reviews in Food Science and Food Safety*, 21(6), 1–37. doi: 10.1111/1541-4337.13030.

Seah, J. S. H., Singh, S., Tan, L. P. and Choudhury, D. (2022). Scaffolds for the manufacture of cultured meat, *Critical Reviews in Biotechnology*. 42(2), 311–323. doi: 10.1080/07388551.2021.1931803.

Senthil Muthu Kumar, T., Senthil Kumar, K., Rajini, N., Siengchin, S., Ayrilmis, N. and Varada Rajulu, A. (2019). A comprehensive review of electrospun nanofibers: food and packaging perspective, *Composites Part B: Engineering*. 175(June), 107074. doi: 10.1016/j.compositesb.2019.107074.

Shi, X., Cui, L., Sun, H., Jiang, N., Heng, L., Zhuang, X., Gan, Z. and Chen, X. (2019). Promoting cell growth on porous PLA microspheres through simple degradation methods, *Polymer Degradation and Stability*. 161, 319–325. doi: 10.1016/j.polymdegradstab.2019.01.003.

Shim, T. S., Kim, S. H. and Yang, S. M. (2013). Elaborate design strategies toward novel microcarriers for controlled encapsulation and release, *Particle and Particle Systems Characterization* 30(1), 9–45. doi: 10.1002/ppsc.201200044.

Shin, Y. C., Lee, J. H., Kim, M. J., Hong, S. W., Kim, B., Hyun, J. K., Choi, Y. S., Park, J. C. and Han, D. W. (2015). Stimulating effect of graphene oxide on myogenesis of C2C12 myoblasts on RGD peptide-decorated PLGA nanofiber matrices, *Journal of Biological Engineering* 9, 22. doi: 10.1186/s13036-015-0020-1.

Shit, S. C. and Shah, P. M. (2014). Edible polymers: challenges and opportunities, *Journal of Polymers* 2014, 1–13. doi: 10.1155/2014/427259.

Simsa, R., Yuen, J., Stout, A., Rubio, N., Fogelstrand, P. and Kaplan, D. L. (2019). Extracellular heme proteins influence bovine myosatellite cell proliferation and the color of cell-based meat, *Foods* 8(10). doi: 10.3390/foods8100521.

Song, X., Zhu, C., Fan, D., Mi, Y., Li, X., Fu, R. Z., Duan, Z., Wang, Y. and Feng, R. R. (2017). A novel human-like collagen hydrogel scaffold with porous structure and sponge-like properties, *Polymers* 9(12), 1–16. doi: 10.3390/polym9120638.

Srutee, R., Sowmya, R. S., Uday, S. and A. (2021). Clean meat: Techniques for meat production and its upcoming challenges, *Animal Biotechnology*. 0(0), 1–9. doi: 10.1080/10495398.2021.1911810.

Staunton, D., Millard, C. J., Aricescu, A. R. and Campbell, I. D. (2009). Preparation of recombinant fibronectin fragments for functional and structural studies, *Methods in Molecular Biology (Clifton, N.J.)* 522, 73–99. doi: 10.1007/978-1-59745-413-1_5.

Sze, J. H., Brownlie, J. C. and Love, C. A. (2016). Biotechnological production of hyaluronic acid: A mini review, *3 Biotech* 6(1), 67. doi: 10.1007/s13205-016-0379-9.

Szklanny, A. A., Machour, M., Redenski, I., Chochola, V., Goldfracht, I., Kaplan, B., Epshtein, M., Simaan Yameen, H., Merdler, U., Feinberg, A., Seliktar, D., Korin, N., Jaroš, J. and Levenberg, S. (2021). 3D bioprinting of engineered tissue flaps with hierarchical vessel networks (VesselNet) for direct host-to-implant perfusion, *Advanced Materials* 33(42), e2102661. doi: 10.1002/adma.202102661.

Tavassoli, H., Alhosseini, S. N., Tay, A., Chan, P. P. Y., Weng Oh, S. K. and Warkiani, M. E. (2018). Large-scale production of stem cells utilizing microcarriers: A biomaterials engineering perspective from academic research to commercialized products, *Biomaterials*. 181, 333–346. doi: 10.1016/j.biomaterials.2018.07.016.

Teixeira, M. A., Amorim, M. T. P. and Felgueiras, H. P. (2020). Poly (Vinyl Alcohol)- Based Nanofibrous Electrospun Scaffolds for Tissue Engineering Application, *Polymers*. 12, 7. https://doi.org/10.3390/polym12010007.

Teßmar, K. V., Seitz, A.K., Teßmar, J.K., Blunk, T. and Göpferich, A.M. (2010). Enzymatically degradable poly (ethylene glycol) based hydrogels for adipose tissue engineering, *Biomaterials*. 31, 3957–3966. doi: 10.1016/j.biomaterials.2010.01.128.

Tytgat, L., Kollert, M. R., Van Damme, L., Thienpont, H., Ottevaere, H., Duda, G. N., Geissler, S., Dubruel, P., Van Vlierberghe, S. and Qazi, T. H. (2020). Evaluation of 3D printed gelatin-based scaffolds with varying pore size for MSC-based adipose tissue engineering, *Macromolecular Bioscience*. 20(4), e1900364. doi: 10.1002/MABI.201900364.

Tzezana, R., Reznik, S., Blumenthal, J., Zussman, E. and Levenberg, S. (2012). Regulation of stem cell differentiation by control of retinoic acid gradients in hydrospun 3D scaffold, *Macromolecular Bioscience* 12(5), 598–607. doi: 10.1002/mabi.201100312.

Tzezana, R., Zussman, E. and Levenberg, S. (2008). A layered ultra-porous scaffold for tissue engineering, created via a hydrospinning method, *Tissue Engineering. Part C, Methods* 14(4), 281–288. doi: 10.1089/ten.tec.2008.0201.

Uehara, T. M., Paino, I. M. M., Santos, F. A., Scagion, V. P., Correa, D. S. and Zucolotto, V. (2020). Fabrication of random and aligned electrospun nanofibers containing graphene oxide for skeletal muscle cells scaffold, *Polymers for Advanced Technologies* 31(6), 1437–1443. doi: 10.1002/pat.4874.

Verbruggen, S., Luining, D., van Essen, A. and Post, M. J. (2018). Bovine myoblast cell production in a microcarriers-based system, *Cytotechnology*. 70(2), 503–512. doi: 10.1007/s10616-017-0101-8.

Wang, C., Gong, Y., Lin, Y., Shen, J. and Wang, D. A. (2008). A novel gellan gel-based microcarrier for anchorage-dependent cell delivery, *Acta Biomaterialia* 4(5), 1226-1234. doi: 10.1016/j.actbio.2008.03.008.

Wang, Z., Zhang, X., Ding, Y., Ren, Z. and Wei, D. (2023). Natural biopolyester microspheres with diverse structures and surface topologies as micro-devices for biomedical applications. *Smart Materials in Medicine* 4, 15–36. doi: 10.1016/j.smaim.2022.07.004.

Wezel, A. L. Van (1967). Growth of cell-strains and primary cells on micro-carriers in homogeneous culture, *Nature* 216(5110), 64–65.

Wilschut, K. J., Haagsman, H. P. and Roelen, B. A. J. (2010). Extracellular matrix components direct porcine muscle stem cell behavior, *Experimental Cell Research*. 316(3), 341-352. doi: 10.1016/j.yexcr.2009.10.014.

Wollschlaeger, J. O., Maatz, R., Albrecht, F. B., Klatt, A., Heine, S., Blaeser, A. and Kluger, P. J. (2022). Scaffolds for cultured meat on the basis of polysaccharide hydrogels enriched with plant-based proteins, *Gels* 8(2), 1–21. doi: 10.3390/gels8020094.

Wongkanya, R., Chuysinuan, P., Pengsuk, C., Techasakul, S., Lirdprapamongkol, K., Svasti, J. and Nooeaid, P. (2017). Electrospinning of alginate/soy protein isolated nanofibers and their release characteristics for biomedical applications, *Journal of Science: Advanced Materials and Devices*. 2(3), 309–316. doi: 10.1016/j.jsamd.2017.05.010.

Xiang, N., Yuen, J. S. K., Stout, A. J., Rubio, N. R., Chen, Y. and Kaplan, D. L. (2022). 3D porous scaffolds from wheat glutenin for cultured meat applications, *Biomaterials*. 285(March), 121543. doi: 10.1016/j.biomaterials.2022.121543.

Xiang, N., Yao, Y., Yuen, J. S. K., Stout, A. J., Fennelly, C., Sylvia, R., Schnitzler, A., Wong, S. and Kaplan, D. L. (2022). Edible films for cultivated meat production, *Biomaterials*. 287(May), 121659. doi: 10.1016/j.biomaterials.2022.121659.

Xie, J., Zhang, D., Zhou, C., Yuan, Q., Ye, L. and Zhou, X. (2018). Substrate elasticity regulates adipose-derived stromal cell differentiation towards osteogenesis and adipogenesis through β-catenin transduction, *Acta Biomaterialia* 79, 83–95. doi: 10.1016/j.actbio.2018.08.018.

Xu, H., Cai, S., Sellers, A. and Yang, Y. (2014). Electrospun ultrafine fibrous wheat glutenin scaffolds with three-dimensionally random organization and water stability for soft tissue engineering, *Journal of Biotechnology*. 184, 179–186. doi: 10.1016/j.jbiotec.2014.05.011.

Yang, F., Miao, Y., Wang, Y., Zhang, L. M. and Lin, X. (2017). Electrospun zein/gelatin scaffold-enhanced cell attachment and growth of human periodontal ligament stem cells, *Materials* 10(10), 1–14. doi: 10.3390/ma10101168.

Yang, G. H., Kim, W., Kim, J. and Kim, G. (2021). A skeleton muscle model using GelMA-based cell-aligned bioink processed with an electric-field assisted 3D/4D bioprinting, *Theranostics* 11(1), 48–63. doi: 10.7150/THNO.50794.

Yang, Z., Yuan, S., Liang, B., Liu, Y., Choong, C. and Pehkonen, S. O. (2014). Chitosan microsphere scaffold tethered with RGD-conjugated poly(methacrylic acid) brushes as effective carriers for the endothelial cells, *Macromolecular Bioscience* 14(9), 1299–1311. doi: 10.1002/mabi.201400136.

Yao, R., Zhang, R., Luan, J. and Lin, F. (2012). Alginate and alginate/gelatin microspheres for human adipose-derived stem cell encapsulation and differentiation, *Biofabrication* 4(2), 025007. doi: 10.1088/1758-5082/4/2/025007.

Yeo, M., Chae, S. J. and Kim, G. H. (2021). An in vitro model using spheroids-laden nanofibrous structures for attaining high degree of myoblast alignment and differentiation, *Theranostics* 11(7), 3331–3347. doi: 10.7150/THNO.53928.

Yeo, M., Lee, H. and Kim, G. H. (2016). Combining a micro/nano-hierarchical scaffold with cell-printing of myoblasts induces cell alignment and differentiation favorable to skeletal muscle tissue regeneration, *Biofabrication*. 3, 8. doi: 10.1088/1758-5090/8/3/035021.

Young, D. A., Choi, Y. S., Engler, A. J. and Christman, K. L. (2013). Stimulation of adipogenesis of adult adipose-derived stem cells using substrates that mimic the stiffness of adipose tissue, *Biomaterials*. 34(34), 8581–8588. doi: 10.1016/J.BIOMATERIALS.2013.07.103.

Yuan, Y., He, N., Xue, Q., Guo, Q., Dong, L., Haruna, M. H., Zhang, X., Li, B. and Li, L. (2021). Shellac: A promising natural polymer in the food industry, *Trends in Food Science and Technology* 109, 139–153. doi: 10.1016/j.tifs.2021.01.031.

Zagury, Y., Ianovici, I., Landau, S., Lavon, N. and Levenberg, S. (2022). Engineered marble-like bovine fat tissue for cultured meat, *Communications Biology*. 1, 5. doi: 10.1038/s42003-022-03852-5.

Zamri, M. F. M. A., Bahru, R., Amin, R., Aslam Khan, M. U., Razak, S. I. A., Hassan, S. A., Kadir, M. R. A. and Nayan, N. H. M. (2021). Waste to health: a review of waste derived materials for tissue engineering, *Journal of Cleaner Production* 290. doi: 10.1016/j.jclepro.2021.125792.

Zernov, A., Baruch, L. and Machluf, M. (2022). Chitosan-collagen hydrogel microparticles as edible cell microcarriers for cultured meat, *Food Hydrocolloids*. 129(February), 107632. doi: 10.1016/j.foodhyd.2022.107632.

Zhang, Z., Eyster, T. W. and Ma, P. X. (2016). Nanostructured injectable cell microcarriers for tissue regeneration, *Nanomedicine* 11(12), 1611–1628. doi: 10.2217/nnm-2016-0083.

Zhao, W., Li, X., Liu, X., Zhang, N. and Wen, X. (2014). Effects of substrate stiffness on adipogenic and osteogenic differentiation of human mesenchymal stem cells, *Materials Science and Engineering. C, Materials for Biological Applications*. 40, 316–323. doi: 10.1016/j.msec.2014.03.048.

Zheng, Y. Y., Chen, Y., Zhu, H. Z., Li, C. B., Song, W. J., Ding, S. J. and Zhou, G. H. (2022). Production of cultured meat by culturing porcine smooth muscle cells in vitro with food grade peanut wire-drawing protein scaffold, *Food Research International*. 159(June), 111561. doi: 10.1016/j.foodres.2022.111561.

Zhou, A., Ye, Z., Zhou, Y. and Tan, W. S. (2019). Bioactive poly (e-caprolactone) microspheres with tunable open pores as microcarriers for tissue regeneration, *Journal of Biomaterials Applications* 33(9), 1242–1251. doi: 10.1177/0885328218825371.

Zhuang, P., An, J., Chua, C. K. and Tan, L. P. (2020). Bioprinting of 3D in vitro skeletal muscle models: a review, *Materials and Design*. 193, 108794. doi: 10.1016/j.matdes.2020.108794.

Chapter 9

Scaling cell production sustainably in cultured meat product development

Panagiota Moutsatsou, Mosa Meat, The Netherlands; Paul Cameron, University College London, UK; Ben Dages, Aston University, UK; Alvin W. Nienow, University College London and University of Birmingham, UK; Gary Lye, University College London, UK; Eirini Theodosiou, Aston University, UK; and Mariana P. Hanga, University College London and Quest Meat Ltd, UK*

1 Introduction

Cultured meat is an emerging novel food with potential to address many of the shortcomings of meat produced through intensified animal farming (Post, 2014; Stephens et al., 2018; Stephens and Ellis, 2020). To date, only one cultured meat product has received regulatory approval for commercialisation in Singapore and that was the cultured chicken nuggets from Eat Just (BBC, 2020). However, these products are still available on an order-by-order basis and at a relatively high price of US$50 per nugget (BBC, 2020). The vision is that the cultured meat products will become affordable and readily available to the population. However, in order to achieve that vision and to reap all the benefits of cultured meat, several challenges need to be addressed, and these are: (1) scalable production to produce large quantities of cultured meat; (2)

* Department of Biochemical Engineering, University College London, London, UK and Department of Chemical Engineering, University of Birmingham, Birmingham, UK.

http://dx.doi.org/10.19103/AS.2023.0130.11

cost to achieve affordability of these products; (3) nutritional and sensorial characteristics to achieve consumer acceptance; and (4) sustainability of the resources needed for production of these products.

This chapter will focus specifically on the scalable production and approaches to achieving sustainability in manufacturing of cultured meat. Lessons will be drawn from other relevant industries. Although the technology for producing cultured meat has until now been piggy-backing on already existing biopharma technologies, more specifically on the upscaled cell culture for cell and gene therapies, the production of biologics from mammalian cells and the engineering of tissue constructs for regenerative medicine applications, significant breakthroughs need to happen to make cultured meat an industrial reality in the near future. Perhaps, these breakthroughs will be achieved by combining the existing biopharma-borrowed technologies with approaches and technologies from other relevant industries, particularly food science and manufacturing, given that cultured meat products are intended as foods.

2 Scale-up requirement of cells for production of cultured meat: scenario and timeline

Here we have developed a scenario for scale-up production of cultured meat through cell culture in order to get a better understanding of the challenges and opportunities in process development for manufacturing cultured meat and an understanding of the scale that is required (Fig. 1). For example, it was estimated that for the production of 1 tonne of cultured meat, 10^{14} cells would be needed (Post et al., 2020). Assuming a culture system where cell densities in the range of 5×10^6 cells/mL to 5×10^7 cells/mL could be achieved, the scale required to produce the equivalent biomass for 1 tonne would translate into

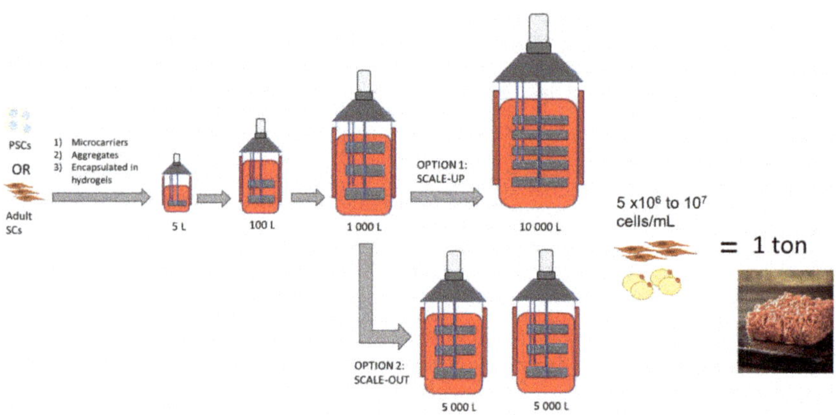

Figure 1 Schematic diagram of the scenario bioprocess for manufacturing cultured meat. PSCs: pluripotent stem cells. SCs: stem cells.

roughly a 5000–10 000 L expansion culture. Taking beef as an example, and Europe for reference, an average cattle farm occupies 34 hectares and hosts around 50 cattle which produce about 20-25 tonnes of beef meat per year (Animal Task Force, 2017). Therefore, to achieve a cultured beef production facility that could replace an average sized farm, a total biomass of 2.5×10^{15} cells would need to be produced, accounting to a total culture volume of 250 000 L.

Keeping in mind the structure of animal meat, which is to be replicated in cultured meat, depending on the cut, 80–90% constitutes muscle, while the remaining is fat and other connective tissue (Listrat et al., 2016). This means that the cell sources chosen for production of the component cells (muscle, fat and connective cells) will most likely need to be stem cells or progenitor cells with capacity to differentiate towards the component cell types (Zuk et al., 2001; Andersen et al., 2008; Post, 2014; Post and Hocquette, 2017). These choices will dictate important aspects of the bioprocess development such as if attachment substrates (e.g. microcarriers) are necessary, inoculation density to be used, feeding and mixing strategies that would be necessary to create a microenvironment suitable for cell proliferation.

Assuming the final production vessel to be the bottleneck (in terms of CAPEX investment) and an achievable final cell density of 10^7 cells/mL, a batch process would require roughly 4 days for proliferation in the larger vessel (i.e. 10 000 L). The 4 days of proliferation was calculated based on the reported doubling times for myogenic and adipogenic progenitors being in the range of 24–40 h (Kolkmann et al., 2020; Kolkmann et al., 2022; Hanga et al., 2021; Dohmen et al., 2022; Stout et al., 2022).

The seeding train would likely be completed with one or multiple 1000 L bioreactors, which in this scenario are assumed to not present the bottleneck in terms of CAPEX costs. However, a second stage of differentiation would also be necessary to produce the meat component cells. For myogenic differentiation, approximately 7 days would be needed, as previously reported (Kolkmann et al., 2022; Messmer et al., 2022), while adipogenic differentiation was found to be lengthier, lasting up to 4 weeks (Dohmen et al., 2022).

The differentiation could potentially take place in the same bioreactor vessel or in a separate, but similarly sized one. If the same bioreactor vessel is to be used for proliferation and subsequent differentiation, the 10 000 L bioreactor would need to be run approximately 25 times per year in order to produce the biomass equivalent to 1 tonne of muscle. This equates to approximately 275 days/year if taking into consideration the downtime between batches and the yearly facility maintenance. Of course, if the achieved cell density upon production would be half of the assumed one, or the process can only be scaled up to 5000 L, two production bioreactors would then be needed to achieve the same yearly yields. On the other hand, if similar size, but different vessels are to be used for proliferation and differentiation, the land footprint and the Capex costs would double, while the yearly yield would only increase by approximately 5 batches per year (33 batches), assuming the above-mentioned durations for proliferation and myogenesis. In Fig. 2, a schematic depicting the timeline for batches for the above-mentioned examples is presented. Especially in the case of adipogenesis, where the differentiation phase takes seven times longer than the proliferation phase (in the larger vessel), the proliferation bioreactor would remain idle for prolonged periods of time, and the use of a different vessel for differentiation would not increase the total yearly number of batches. Moreover, additional steps in bioprocessing would be required as the expanded cells would need to be harvested at the end of proliferation before introducing in the second bioreactor for differentiation. This scenario would most likely lead to a low techno-economic feasibility and to the myogenic differentiation process to become a production bottleneck.

The one-step expansion–differentiation process would be more feasible for the production of fat with adult mesenchymal stem cells as starting cell source, as they don't require protein coatings on the surface used for their expansion and differentiation whether this is a microcarrier or a planar tissue culture vessel (Rafiq et al., 2017; Rafiq et al., 2018; Hanga et al., 2020; Hanga et al., 2021; Ochs et al., 2022). However, if starting cell sources such as pluripotent stem cells would be used, for which protein coatings are necessary for both expansion and differentiation, regardless of the targeted lineage, the complexity of the bioprocess would increase significantly. Most often, the protein coating required for enhanced differentiation would be different than the protein coating necessary for expansion (Wilschut et al., 2010; Bodiou et al., 2020), thus making the one-step bioprocess difficult to achieve. This latter scenario is more likely for the production of muscle cells regardless of whether myosatellites or pluripotent stem cells are used as the starting cells (Post, 2014; Ding et al., 2018; Messmer et al., 2022).

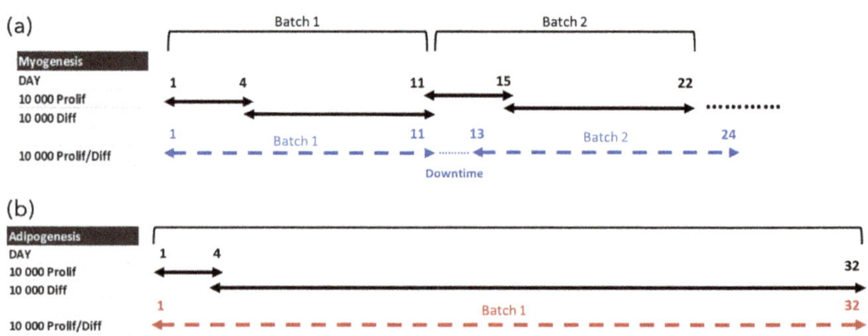

Figure 2 Batch timelines for different scenarios: when separate 10 000 L bioreactors are used for proliferation and differentiation or when one vessel bioreactor is used for a one-step bioprocess (proliferation/differentiation) for production of (a) muscle and (b) fat.

Cultured meat is a complex product and thus scaling up its production can be a challenging task that is highly dependent on the scenario to be employed. Every decision, from the cell source to the method of culture and bioreactor type and scale, will impact not only the final cell yield, but also the overall manufacturing costs. In Sections 3.1, 3.3, and 3.4, these decisions, which can result in bottlenecks, will be addressed in more detail.

3 Bottlenecks and challenges in scaling up production of cultured meat

There are many practical and scientific challenges pertaining to the industrialisation of cultured meat, and these will be discussed in the following sections.

3.1 Quality of starting cells

Unmodified precursor cells have a proliferative capacity that is theoretically limited by the Hayflick number of 50 doublings (Hayflick and Moorhead, 1961). For adult animals and *ex vivo* expansion conditions though, this number is practically much lower at approximately 30 population doublings (Hanga et al., 2020; Melzener et al., 2021). Even in this case, the proliferative capacity does not guarantee preservation of the ability to differentiate. To put things further into perspective, from a 0.5 g muscle tissue sourced through a biopsy, around 10 000 myosatellite cells are typically obtained (Post et al., 2020). Assuming a proliferation/differentiation capacity of 30 doublings, which is quite optimistic, 3-5 × 10^{12} cells could be produced from one biopsy (Melzener et al., 2021), equating to 50 kg of cultured muscle tissue. Because of the exponential nature of cell propagation, it becomes obvious that every doubling further that the cells can be pushed to will have a significant impact on the production yield from a single biopsy.

Therefore, generating a stable cell line becomes of utmost importance for scaling up cultured meat production. This could be achieved through cell immortalisation by serial passaging, similarly to the C2C12 mouse myoblast cells (Yaffe and Saxel, 1977) that are often used as a model cell line for the study of muscle tissue, or by using small molecules to extend the lifespan of primary cells and their differentiation capacity. Cell immortalisation typically bypasses senescence which can be of different types. Replicative senescence has been linked to two anti-proliferative pathways: p53/p21 and p16/Rb. When both of these pathways were suppressed, immortalisation was achieved in human cells (Shay et al., 1991; Ben-Porath and Weingberg, 2005; Wang et al., 2019) and more relevant in myoblasts (Ryu, 2017). However, this approach only results in a pre-immortal state as the short telomeres are still present and eventually the cells will reach the M2 cell cycle phase and cell death (Shay et al., 1991; Wang et al., 2019).

In some cell types, the abrogation of only one of these pathways has resulted in avoidance of senescence. For example, suppressing the p16 alone worked in human mesenchymal stem cells (Shibata et al., 2007; Wang et al., 2019). Other signalling pathways have also been linked to cellular senescence, and these include but are not limited to transforming growth factor beta/bone morphogenic protein (TGFβ/BMP), Wnt/β-catenin, mitogen-activated protein kinase (MAPK), hippo, Notch, fibroblast growth factor (FGF) and insulin-like growth factor (IGF) pathways (Wang et al., 2019). For example, it has been shown that partially inhibiting the p38 MAPK signalling pathway with a low concentration of the p38 inhibitor SB203580, paired box protein (PAX7) expression was maintained during prolonged *in vitro* culture and thus the proliferative capacity of satellite cells was prolonged while their differentiation ability was also preserved for longer (Bernet et al., 2014; Ding et al., 2018). Similarly, ectopic activation of fibroblast growth factor receptor 1 (FGFR1) signalling via transfection with insulin-like growth factor receptor (IFR1) expression vectors and treatment with Dimerizer B/B to activate iFR1 has been shown to increase PAX7 expression and to restore the proliferative capacity of aged satellite cells (Bernet et al., 2014). Overexpression of yes-associated protein 1 (YAP1) which is a major effector of the Hippo pathway was also shown to induce immortalisation of normal human keratinocytes (D'Addario et al., 2010; Wang et al., 2019). However, to the best of our knowledge, this has not been verified or reported for livestock cells. Other methods of cell immortalisation include more sophisticated approaches. For example, in 2016, Upside Foods, a cultured meat start-up submitted a patent for immortalisation of skeletal chicken muscle cells by combining overexpression of TERT and using CRISPR to knock out expressions of p15 and p16 (Thorley et al., 2016; Soice and Johnston, 2021). A recent study has also reported the spontaneous immortalisation and subsequent adaptation to suspension of primary chicken fibroblasts (Pasitka et al., 2023). The resulting cells though would have to be transdifferentiated into a cultured meat relevant tissue. More studies are needed into investigating methods of immortalising livestock primary cells or as a minimum prolonging the population doublings capacity while maintaining differentiation potential. It is worth noting that based on experience of immortalising human and other mammalian cell lines of interest in industrial biotechnology like human embryonic kidney cells (HEK293) or Chinese hamster ovary (CHO), particularly when using methods involving genetic engineering, abnormalities such as chromosomal translocations and copy number alternations, as well as genetic instability (Stepanenko and Dmitrenko, 2015; Vcelar et al., 2018) are a possibility. For these reasons, genetic modification is expected to raise questions amongst food regulators. As an example, the EU

Novel Food Regulation excludes genetically modified foods and as such, the use of genetically modified cells for food production automatically falls under the genetically modified organism (GMO) legislation (Post et al., 2020).

Another aspect to consider is that the starting cells for cultured meat are adherent dependent, meaning they typically require an attachment substrate which can be in various forms depending on the culture (e.g. tissue culture flasks, microcarriers, scaffolds). However, the necessity of an attachment substrate increases the complexity of the manufacturing process, as well as requires additional steps for cell recovery. Some attempts have been made at culturing cells as aggregates (Hosoyama et al., 2013; Westerman, 2015; Bodiou et al., 2020) which would bypass this necessity and result in a simplified process. Adherent cells can self-assemble to form small aggregates via cadherin binding (Duguay et al., 2003), which in theory also provides a 3D environment closer to the native cell environment. However, other challenges are associated with this type of culture. In such a scenario, control over the aggregate size and size distribution becomes an issue, typically resulting in gradients of nutrients and gases that lead to necrotic cores (Bodiou et al., 2020) and thus loss of cell viability alongside variability in cell quality as different cell layers are exposed to different culture conditions. Scalability of the method used for the production of aggregates is also of concern (Westerman, 2015; Hosoyama et al., 2013; Hosoyama et al., 2014; Bodiou et al., 2020; Manstein et al., 2021; Ma et al., 2022).

Ideally, a suspension cell line would ensure a simplified and streamlined manufacturing process. If cells no longer need cues from static tension via integrin binding to proliferate and differentiate, then attachment to a substrate is unnecessary, creating a suspension cell line. Multiple success stories of adapting cell lines such as CHO (Astley et al., 2007), HEK293 (Malm et al., 2020), baby hamster kidney cells (BHK) (Merten, 2015), green African monkey kidney cells (Vero) (Shen et al., 2019) and chicken fibroblasts (Pasitka et al., 2023) have been previously reported. However, adaptation to suspension cultures is typically a long process and success is not guaranteed as not all cell types are able to adapt. It is also worth noting that adaptation to suspension may also impact the cells' ability to differentiate and make tissue (Rubio et al., 2019).

3.2 Cell densities

The second challenge to be addressed is cell density. Achieving high cell densities is not a 'need to have' to realise cultured meat production but it can significantly reduce the production footprint and implicitly the CAPEX costs associated with the size of the bioreactors. To date, cell densities on microcarrier-based primary cell cultures were reported to be in a range of 2×10^5 cells/mL to 1×10^6 cells/mL (Hanga et al., 2020; Navaei et al., 2020; Hanga et al., 2021;

Dohmen et al., 2022; Ochs et al., 2022) and rarely up to 1×10^7 cells/mL, and only for established mammalian cell lines such as Vero, Madin–Darby canine kidney cells (MDCK) and adherent CHO (Trabelsi et al., 2006; Tharmalingam et al., 2015). These were achieved mostly in small-scale, semi-controlled spinner flask cultures and more rarely in bench-top, fully controlled bioreactors (Jung et al., 2012; Rafiq et al., 2013; Lawson et al., 2017; Yang et al., 2019; Hanga et al., 2021).

Although very high cell densities have been reported in aggregate-based cultures (3.5×10^7 cells/mL) (Manstein et al., 2021), there are only a handful of reports on aggregate-based cultures of myogenic cells (Hosoyama et al., 2013, 2014; Westerman, 2015), which are not geared towards maximising cell yields but rather towards the creation of organoids as a method for cell differentiation.

So far, the highest cell density in the context of cultured meat has been reported by Pasitka et al. (Pasitka et al., 2023) with single cell suspension of a spontaneously immortalised chicken fibroblast cell line, up to 1×10^8 cells/mL. Single-cell suspension culture is indeed the simplest and most scalable method for generating cell mass. Increased cell densities can also be achieved by using process intensification techniques, including medium/nutrient perfusion (Zhang et al., 2015; Zamani et al., 2018; Whitford, 2021), targeted medium supplementation (Yongky et al., 2019), concomitant medium and surface area addition at an optimal time point in culture (Hanga et al., 2021), continuous bioprocessing (Khanal and Lenhoff, 2021) or combinations of the above (Särnlund et al., 2021). It has to be noted though that continuous bioprocessing can mostly be applied to immortalised cell lines.

Although increased cell density is key to achieving realistically sized facilities for cultured meat production, contrary to what is generally thought, this would not necessarily save in medium related costs, since the metabolic profile of the cells (e.g. their requirement for nutrients) would remain the same. However, through process intensification techniques that bypass growth-limiting factors (e.g. running out of surface area), increased cell yield can be achieved with the same media volumes (Hanga et al., 2021). Medium associated costs can be tackled through a range of strategies, including metabolic engineering approaches, cell adaptation to less ammoniagenic nutrients, sophisticated medium customisation and feeding strategies, extension of medium lifespan, reduction of the raw materials/nutrients production costs through economies of scale and through switching to lower grade raw materials, as well as media recycling strategies that could be included in the manufacturing processes.

3.3 Suitable scaffolds

In their native environment, cells are surrounded by a naturally secreted nanofibrous and porous mesh called the extracellular matrix (ECM) composed of hundreds of proteins and proteoglycans (Molina et al., 2021). The ECM provides attachment points to anchorage dependent cells, releases biomechanical cues for tissue development and also has roles in tissue maintenance and repair (Walker et al., 2018). *In vitro*, the ECM is artificially constructed to mimic the tissue's native ECM structural and functional properties. The intention is to induce the properties in the scaffolds necessary for *in vitro* culture. These can take various forms depending on the process step when they are utilised. For example, for expansion and possibly even differentiation, these scaffolds would be in the form of microcarriers which are micrometre-sized beads on which cells grow attached (Bodiou et al., 2020) or as hydrogel capsules of different geometries where the cells are confined to the inside of the hydrogel (Hassan et al., 2013; Wollschlaeger et al., 2022). If structured cultured meats (e.g. steak and fillet) are intended as products, an additional step would be required in the manufacturing where the different cell types are combined and seeded onto a scaffold that would provide the structure of the tissue (Fig. 3).

Alternatively, cells can be grown without support materials, by growing them in two-dimensional monolayers and assembling the tissue via stacking of the intact cell sheets (layer-by-layer fabrication), in this way producing structured cultured meat (e.g. steaks, fillets) (Shahin-Shamsabadi and Selvaganapathy, 2022; Kang et al., 2021; Tanaka et al., 2022). In this approach, known as cell sheet engineering, synthetic polymers with stimuli-responsive properties such as poly(*N*-isopropylacrylamide) (pNIPAM) are typically used (Moschouris et al., 2016; De Pieri et al., 2021). By removing the stimuli, the properties of the synthetic substrate change, thus releasing the cell monolayer intact without the

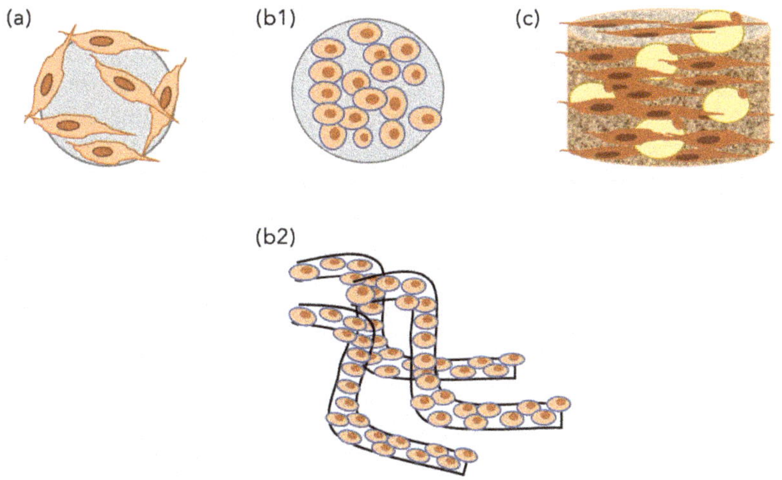

Figure 3 Illustrations of (a) cells cultured on microcarriers; (b) cells encapsulated in hydrogels with different geometries [(b1) spheres/capsules and (b2) fibres] and (c) scaffolds for co-culture of muscle and fat cells.

need for proteolytic enzymes to detach the cells. However, the intact cell sheet needs to be captured upon detachment and pre-layering, otherwise it will roll into a large aggregate under the elastic forces. Layer-by-layer fabrication of cultured meat would require a high number of intact cell sheets, making this approach difficult to scale-up. This approach was also translated to stimuli-responsive microcarriers, which are scalable platforms but with limited success in the cell detachment when very long exposure times to lower temperatures were required (Hanga and Holdich, 2014).

In tissue engineering, scaffolds have historically been composed of synthetic materials that have a higher plasticity and less batch variation relative to natural materials (Shivakumara and Demappa, 2019). However, synthetic materials could pose regulatory challenges for the production of cultured meat. In recent years there has been a push for the use of natural materials instead (Jahangirian et al., 2019). Nature offers an abundance of materials with potential for cultured meat, and these can be proteins (i.e. silk fibroin from spider webs, collagen from skin, bone and tendons, gelatine, elastin from elastic tissues, keratin from wool or hair, resilin from insect tendons), polysaccharides (i.e. cellulose, amylose, dextran, chitin and glycosaminoglycans) or even polynucleotides (i.e. DNA, RNA) (Yu et al., 2006; Silva et al., 2014; Ben-Arye et al., 2020; Bomkamp et al., 2022). Each of these biological materials has unique properties that are unmatched by synthetic materials. Protein-based materials are perhaps more suitable for cell culture applications as they mimic more closely the features of the ECM, and thus can promote cell migration, growth and even alignment. These characteristics make protein-based hydrogels suitable for applications such as cell encapsulation. Hydrogels in general have been predominantly used in cell encapsulation due to their ability to form mechanical stable, porous and hydrated 3D networks that are capable of facilitating the transport of nutrients and gases from the surrounding medium to the cells and also the transport of metabolic waste products from the cells to the surrounding medium (Nicodemus and Bryant, 2008; Wilson and McDevitt, 2013; Silva et al., 2014).

The success of hydrogel-based scaffolds is dependent on how closely they can mimic the desired properties of the native tissue ECM. These properties can be structural and topographical, physical, biochemical, mechanical and biological and can be influenced by the materials and the fabrication techniques. Structural and topographical properties affect the overall shape of the supports and can influence cell attachment, migration, infiltration and differentiation (Jenkins and Little, 2019). Fibre alignment or grooved patterns can promote the elongation of cells which can be useful for myogenic differentiation (Wang et al., 2015). Fibre diameter also affects pore size, making attachment along single fibres more likely, thus improving infiltration and O_2 diffusion (Zhang et al., 2010). Physical properties of scaffolds include surface charge (Hao et al., 2016) and degradation kinetics. Cells generally prefer slightly hydrophobic surfaces to attach to, although this is highly dependent on the cell type (Wang et al., 2016). Degradation rate is important too as it should match the proliferation time of the selected cell type. However, edible scaffold materials could allow for longer degradation times as the edible scaffold is incorporated into the final cultured meat product (Molina et al., 2021). Biochemical properties affect the function of the support provided. In their native tissues, cells attach to the ECM via integrin binding at specific receptor sites, made up of peptide sequencing motifs with the most common being RGD (Arg-Gly-Asp) found in fibronectin (Shin et al., 2003). Most animal-derived proteins such as collagen, fibrinogen or hyaluronic acid have excellent attachment properties but are extremely costly, although these can be produced recombinantly in fungi and bacteria (Post et al., 2020). Plant-based and synthetic materials can be coated with RGD peptides using click chemistry or sequential amino acid deposition; however, these are expensive, poorly scalable and can have low output efficiency. Other bioactive molecules and growth factors, such as basic fibroblast growth factor-2 (bFGF-2), can be coated on scaffolds during conditioning to improve cell attachment and to influence differentiation (Feng et al., 2017), but again this is not an optimal approach due to the associated costs.

Decellularised tissues are also a type of natural scaffold that uses the native ECM of animal, plant or fungus tissues. The cells are removed by washing the tissues typically with harsh detergents or enzymes. The pre-existing vasculature can also be used to help O_2 and nutrient diffusion across thicker tissues (Gui and Niklason, 2014). The ECM of animal tissues is most compatible, thanks to structural proteins like collagen and fibronectin which significantly enhance cell attachment (Somaiah et al., 2015). However, these are very expensive to produce and would increase reliance on the by-products of animal farming. Plant-based decellularised tissues are more cost efficient as they can be procured from plant waste products, allowing for a renewable supply in a circular economy (Cancelliere et al., 2021). Methods of decellularisation to strip out the immunogenic RNA and DNA while preserving the ECM and its native structures typically involves chemical treatment with detergents such as sodium dodecyl sulphate (SDS) and Triton. Other methods include detergent-free chemical treatment using bleach and sodium carbonate, and lyophilisation used in conjunction with DNAse I treatments (Kim et al., 2021). The application of supercritical fluids such as liquid carbon dioxide (CO_2) has also been shown to be advantageous relative to harsh detergent or acid-base treatments as it prevents denaturing of low molecular weight polymers and proteins. It may also prove more efficient at a large scale, as the simultaneous liquid-like solute solubility and gas-like diffusivity of supercritical CO_2 molecules increases the depth of tissue penetration (Kim et al., 2021; Wang et al., 2021).

The integrity of the supporting structure is dictated by its mechanical properties, which is crucial in bioreactor systems for withstanding shear stresses from impeller motion and fluid flow. The supports must stay intact and prevent cell detachment. Mechanical properties such as stiffness and Young's modulus (E) have also been shown to influence lineage-specific differentiation (Sun et al., 2016). For instance, soft tissue like fat (E = 2.5–5.0 kPa) and muscle (E = 8–17 kPa) require lower stiffness than harder tissues like bone (E = 25–40 kPa) (Li et al., 2011).

For cultured meat products, the materials used must be safe for human consumption, which disqualifies most synthetic materials. This means that ideally the materials used for scaffolds should be edible as that can allow the option of incorporating them in the final cultured meat product. In this case, choosing the right material is even more so important as it can influence the organoleptic properties (texture, colour, taste, etc.), and can also provide added nutritional benefits (Bomkamp et al., 2022; Ng and Kurisawa, 2021). The existing commercially available microcarriers are mostly composed of synthetic polymers such as polystyrene. Despite being useful for the purpose of cell attachment and proliferation (Rafiq et al., 2017; Rafiq et al., 2018; Hanga et al., 2020; Hanga et al., 2021), the cells require detachment through the means of a combination of proteolytic enzymes and agitation (Nienow et al., 2016). Many natural hydrogels, despite already being approved as food safe, may pose challenges in bioprocessing scale-up due to their batch variations and inflexible material properties (Catoira et al., 2019). Incorporation of hydrogel microcarriers into the end product would mean that continuous synthesis or purchase would be required, increasing overall production costs. Also, foreign materials that sound overly scientific or unnatural appearing on ingredients lists may influence consumer acceptance negatively (Pakseresht et al., 2022). Another option is to use either degradable or reusable scaffolds. Degradable scaffolds would allow for the formation of a continuous piece of muscle tissue, weakening supports as the tissue strength increases. Reusable scaffolds would be solid state, and cells would need to be detached through mechanical intervention, changes in surface chemical conformation or via enzymatic degradation of attachment ligands. Change in surface conformation via external stimuli could be achieved through functionalisation with thermo-responsive coatings, such as methylcellulose or poly (*N*-isopropylacrylamide) (Altomare et al., 2016; Akiyama, 2021). The building blocks of the degradable scaffolds could be filtered out of the media and recycled, meaning reusable scaffolds may be degradable, but degradable scaffolds may not always be reusable.

Degradation over time would allow for a more dynamic microenvironment, enabling targeting certain regions to form a patterned meat product reminiscent of muscle fibres and fatty tissue deposits. Ideally, the scaffolds could be degraded by a variety of enzymes, one example being matrix metalloproteinases (MMPs) which are secreted by myosatellite cells. However, of the 25 MMPs identified so far, many have been proven to act specifically on certain components of animal-derived ECM. For example, MMP-1, MMP-8, MMP-13 and MMP-18 cleave interstitial collagen types I, II and III, while MMP-2 and MMP-9 degrade denatured collagen types IV, VII and X (Chen and Li, 2009). This means for hydrogels derived from animal-free sources, MMPs secreted by myosatellites will be ineffective. Alternative degradation methods will be required for plant-derived polysaccharide hydrogels, such as ion chelation to break crosslinks. Chelating agents such as sodium citrate or ethylenediaminetetraacetic acid (EDTA) can sequester calcium to break down alginate, carrageenan etc. (Chueh et al., 2010). Specific lyase enzymes, for example alginate lyase (produced by bacteria that consume brown algae such as *Bacillus* sp. TAG8), can also be used (Tavafi et al., 2017).

3.4 Sustainability and techno-economic feasibility

For cultured meat to be successful in the long term, their production needs to be sustainable and economically feasible to achieve cost parity with traditional meat products. The concept of producing meat through cell culture is sustainable through multiple pathways when compared to traditional meat production through animal agriculture. Cultured meat would use land significantly more efficiently. For example, for poultry 60–300%, while for beef 2000–4000% land savings could be achieved (Tuomisto et al., 2014; Mattick et al., 2015). This would result in freeing up cropland that can be used to grow other foods like grains, fruits and vegetables. Cultured meat would also reduce the need for using synthetic fertilisers that have been associated with many detrimental effects on the soil, biodiversity and even human health (Ceballos et al., 2017).

Due to the closed and controlled nature of the production of cultured meat through cell cultivation in bioreactors, strategies can be employed to ensure the maximum benefits are achieved, including environmental ones. For example, the carbon dioxide produced by the cells through respiration could be captured and sequestered, thus leading to the possibility of producing carbon-negative meat (Mohareb et al., 2018). Cultured meat would not produce manure like animal farming, thus avoiding pollution of water and air and disease spreading as a result. The extent of this animal farming issue is significant as in the US alone, over 1 billion tonnes of manure are produced each year (EPA, 2013). Additionally, cultured meat doesn't require the use of antibiotics during its production, thus eliminating another danger to human health which is a consequence of animal farming. Animals are fed multiple antibiotics in industrialised animal farms typically at low doses when the antibiotics play a twofold role: to prevent disease and to promote growth (Hakim, 2018). However, the constant feeding of antibiotics allows bacteria to become resistant which not only poses significant health hazards, but it also comes at high healthcare-associated costs estimated to reach US$100 trillion by 2050 (O'Neill, 2016). Cultured meat bypasses all of these challenges and dangers.

Of course, cell cultivation does produce waste associated with the culture media and other components used in the process. However, these can be contained within the facility and could be treated at source. In this scenario, a financial incentive and perhaps opportunity would be to recover and recycle valuable unused nutrients and growth factors produced by the cells that otherwise would be discarded as waste. Overall, the cell culture media is perhaps the biggest challenge in achieving sustainability of cultured meat production, in addition to being the biggest waste and cost contributor. Finding methods to recover and recycle unused media components of value such as albumin, glucose, growth factors and amino acids, while removing products of metabolism (e.g. lactate, ammonia) that can inhibit cell growth when accumulated (Schop et al., 2009; Schop et al., 2010; Slivac et al., 2010; Hubalek et al., 2022), becomes of utmost importance for achieving economic feasibility and sustainability of cultured meat. Removing animal-derived components such as foetal bovine serum (FBS) and replacing them with cheaper, plant- and microbe-derived proteins, amino acids and growth factors would also significantly contribute to sustainability. Crude raw materials such as protein hydrolysates from plant, algal or even microbial sources could be potential candidates. Hydrolysates have been proven effective in promoting cell growth (Ng et al., 2020; Ho et al., 2021). However, they do exhibit batch-to-batch variability that can cause issues on manufacturing and cell performance.

Recycling of culture media is another aspect that should be considered in the manufacturing of these products to achieve cost parity. This could be achieved through several strategies: (1) chemical treatment of waste products of cell metabolism (e.g. lactate and ammonia) and (2) biological recycling where the waste products could serve as a feed to other biological entities with capacity to transform those waste products into nutrients for the cells. For example, for chemical recycling, ammonium ions in spent medium could be removed through ion exchange resins and membranes (Hubalek et al., 2022) and zeolite catalysis (Huang et al., 2014), with the latter being shown to decrease ammonia concentrations from 27.8 mg N/L to as low as 0.3 mg N/L within 1 hour (mg N/L is the mass of nitrate ions per litre of fluid). Once extracted from the waste medium, ammonia and other metabolic waste products could be commercialised (Chai et al., 2021). Ammonia, for example, could be used as fuel, to replace carbon-based ones, or it could be used as fertiliser for plants, even more so to grow plant-based feedstock for cultured meat. This would make sense, as the traditional production of ammonia-derived fertilisers contributes 1–2% to the global energy consumption (Hubalek et al., 2022). It could also be used to produce nitrogen gas for antioxidative packaging of cultured meat and other food products, or for gas-based pH/dO control in bioprocessing applications including cultured meat. Lactate, on the other hand, a by-product of glycolysis during cell proliferation, could be extracted and reused for inducing myogenic differentiation (Willkomm et al., 2014) or even in cell culture to maintain equilibrium (Freund and Croughan, 2018).

Lactate Supplementation and Adaptation (LSA) technology (Freund and Croughan, 2018) has been shown to lower lactate concentrations. Based on laws of mass action, supplementation of growth medium with lactate causes a net lactate production of zero via exploitation of the actual free energy of the oxidation of glucose to lactate and equilibrium balance of the solution. At the pyruvate oxidation step, there is a tipping point, which can be unbalanced through lactate addition, causing the reaction to flux towards the tricarboxylic acid (TCA) cycle for the continued production of ATP (adenosine triphosphate). Copper supplementation in medium also appeared to force a phenotypic shift in CHO cells, from net lactate production to lactate consumption, decreasing endpoint lactate levels in medium by driving up cell consumption of their own by-products (Luo et al., 2012). LSA is a technique that could be very useful in establishing efficient feeding regimes for cells (Konakovsky et al., 2016).

The biological recycling process would consist of either an at-line or offline secondary biological culture. For example, this could be a culture of microalgae. Algae consume ammonia as a source of energy, regulating its concentration in natural waterways (Chai et al., 2021). Indirect co-culture via addition of microalgae to the collected spent medium would allow algae to 'clean' the medium by removing ammonia (Haraguchi et al., 2021a,b). Algae are already used in the production of food ingredients such as omega-3 fatty acids or natural food colourants and dyes while also being used in fertiliser, bioplastic and pharmaceutical production. Algal waste from these industries could be fed back into the cultured meat industry, as enzymatic hydrolysis of algae biomass has been shown to generate L-amino acids (Romero Garcia et al., 2012). These amino acids could be used in peptide formation via sequential addition to build proteins and growth factors, which can then be used as less costly and sustainable medium supplements. Fermentation of nitrification-denitrification bacteria running in parallel bioreactors could also be used for stripping ammonium from media. Two novel strains of bacteria, *Alcaligenes* sp. S84S3 and *Proteus* sp. S19, have been shown to significantly reduce the concentration of ammonium ions; co-culture of these strains effectively sequestered 65.83% of the ammonium ions without producing of toxic nitrogen oxide intermediates over a 5-day incubation period at 30°C (Angar et al., 2016).

Water would also be a valuable component to recycle, but not necessarily a costly resource used in cultured meat production. It is the main diluent of cell culture medium components and typically MilliQ water or type I sterilised water (filtration, autoclaving) would be used for a mammalian cell-based manufacturing process. Although it is conceivable that some cell lines can also grow in sterile tap water, it yet remains to be seen if primary cells or culture meat relevant cell lines are able to grow and differentiate in tap water-based culture medium. Ultra-filtration is typically recommended for mammalian cell culture as organic (e.g. humic acids, tannins, pesticides and endocrine disrupters) and inorganic (e.g. mercury, lead, zinc, nickel, chromium, manganese and cadmium) compounds are quite common in tap water and most of them are known to have a negative impact on mammalian cell culture (Bánfalvi, 2011).

4 Other considerations

4.1 Analytical requirements for scalable production

An important contributor to the successful commercialisation of cultured meat is its compliance with food safety. Using comprehensive detection systems that can measure, analyse, monitor and control critical process parameters (CCPs) will not only satisfy regulators and customers but also allow deeper process knowledge, and enhance production efficiency at scale. Furthermore, automation and real-time monitoring can result in more sustainable processes through further optimisation and decreased production costs. The latter is especially relevant to cell culture medium, which is currently the most significant cost driver in cultured meat production, as it may enable some level of medium recycling and waste reduction (Specht, 2020; Djisalov et al., 2021).

Sensing key process variables in real time and imposing necessary changes ensures consistent manufacturing processes and consequently consistent product quality according to process analytical technology (PAT) philosophy (Glassey et al., 2011). PAT has already been successfully implemented in the chemical and pharmaceutical industries, while the biomanufacturing sector is struggling to catch up, mainly due to product and process complexity. Furthermore, many companies are still transitioning from manual to automated monitoring and control systems, in addition to slow turnaround time for the measurement of some key process parameters (Zhao et al., 2015; Emerson et al., 2020).

There are certain criteria that sensors used in commercial bioprocessing need to satisfy, and the main ones are presented next (Pohlscheidt et al., 2013; Biechele et al., 2015; Reyes et al., 2022; Esmonde-White et al., 2022). The sensors should have sufficient accuracy in the desired control range, linearity (i.e. the signal is directly proportional to concentration, so there is no need for dilution) with reliable performance over the long cell culture times, and ability to generate repeatable measurements of the same variable(s) under the same conditions during different times. High selectivity and sensitivity are also important sensor characteristics, because they allow determination of small amounts of the desired analyte(s) in the presence of other components. Sensors need to be robust to withstand different environmental conditions (e.g. steam sterilisation in stainless steel and glass bioreactors), and easy to validate, implement, use and maintain. Finally, short response time is extremely desirable, especially when the measurements are linked to a control action.

Monitoring data can be obtained either by taking a sample or not, and this can be done by four different methods (based on the physical connectivity of the analyser within the process), namely in-line (or *in situ*), at-line, online and offline (Callis et al., 1987; Rolfe, 2006; Dagge et al., 2009; EMA, 2016; Schwamb et al., 2015; Biechele et al., 2015; Minnich et al., 2016; Bellani et al., 2020). During in-line measurements, the sample is analysed within the bioreactor and not removed from it. At-line measurements involve removing the sample from the bioreactor (manually or via autosamplers) and analysing it in close proximity to it, whereas offline measurements require the sample to be taken to a different location (for example, a QC laboratory either on-site or off-site). Finally, online measurements involve diverting the sample from the bioreactor using a bypass, and all steps, including sampling, preparation, measurement and data transmission, are automated. In-line and online analysers allow for real-time monitoring, i.e. the response time of the sensor signal is shorter than the time in which detectable changes in the process occur (Minnich et al., 2016; Busse et al., 2017). At-line and especially offline measurements are associated with time delays and therefore do not contribute to the control of the dynamic process behaviour (Pohlscheidt et al., 2013; Gargalo et al., 2020). Based on their interaction with the sample, process analysers can also be classified as destructive or non-destructive (depending on whether any conditioning alters the sample or not) and invasive and non-invasive (depending on whether the probe is in direct contact with the sample or not) (Callis et al., 1987; Minnich et al., 2016; O'Mara et al., 2018; Djisalov et al., 2021).

In-line sensors for temperature, pH and dissolved oxygen (DO) and carbon dioxide (DCO_2) have been standard in production bioreactors for decades. Optical sensors (or optodes) for pH, DO and DCO_2 are preferred to electrochemical ones, because of their smaller size, long shelf life and availability as disposable devices, which allows them to be incorporated in single-use bioreactors (Biechele et al., 2015; Busse et al., 2017; Steiwedel et al., 2019; Reyes et al., 2022). A recent development in the detection of pH, temperature and oxygen in bioreactors involves the creation of a multiparametric sensor chip inside a capsule, which moves inside the reactor with the mechanical motion of the impeller and transmits information wirelessly to an external system control and data analysis (SCADA) (Busse et al., 2017; O'Mara et al.,

2018). The technology promises to offer continuous remote monitoring in a temporal and spatial manner irrespective of scale (multiple capsules can be added to one large reactor), whereas the electrochemical nature of the device enables its easy integration into instrumentation already existing in industry.

Biomass monitoring with reference to cell density and viability can be done using various direct and indirect methods (Djisalov et al., 2021). Indirect measurements include off-gas analysis, redox potential and osmolality of cell culture media, amongst others (Pohlscheidt et al., 2013; Allan et al., 2019), whereas offline manual cell count using a haemocytometer and microscope is a direct way to verify cell density and viability (Bellani et al., 2020). *In situ* microscopy (ISM) has the potential to provide real-time information on cell morphology, whether they are in suspension or on microcarriers (Zhao et al., 2015). However, the high optical density of some microcarriers prohibits their use for optical analysis, whereas others, such as Cytodex 1, are ideal candidates as they are transparent (Anton et al., 2007). Automatic image analysis of acquired pictures can also give data on cell concentration, viability and size distribution (aggregation) through image processing algorithms, although these algorithms must be tailored to specific cell species (Biechele et al., 2015; Schwamb et al., 2015). Furthermore, the technology has been tested in small volume reactors and not at production scale. Dielectric spectroscopy (DS) (also known as radio frequency impedance or biocapacitance) is another technology for the continuous non-invasive measurement of cell concentration, size and viability, whether they are on solid surfaces or in suspension (including on microcarriers) (Pohlscheidt et al., 2013; Steinwedel et al., 2019), thereby making it also very attractive for cultured meat bioprocess monitoring. One drawback could be that some microcarriers might become polarised by the electrical field as a result of their surface chemistry and presence of a surface charge on their surface, but this was not proven to be the case when Cytodex microcarriers were used in spinner flasks (Degouys et al., 1993). Successful online DS monitoring of cell cultures grown on microcarriers has been demonstrated in lab scale bioreactors, provided that the microcarriers are not fully covered by cells (El Wajgali et al., 2013), whereas Metze and co-workers (2020) have shown that the technology can be scaled up to 2000 L single used bioreactors. Online biocapacitance measurement also enables automatic determination of bolus feeding frequency in fed batch cultures through estimation of cell growth, as well as dynamic feeding strategies that calculate and adjust the feeding volume as needed (Reyes et al., 2022). Furthermore, the signal is independent of scale, as was demonstrated by Moore and co-workers (2019) for cultures of production cell lines from 5 L to 15 000 L, and therefore is a valuable tool for process development and scale-up.

Other spectroscopic sensors can also provide non-invasive measurements of cell density and viability, nutrients (i.e. glucose and glutamine) and metabolites (i.e. lactate, glutamate and ammonia). These include near-infrared (NIR), mid-infrared (MIR), UV/vis (including fluorescent) and Raman. MIR is a quantitative, sensitive and specific method used in process development for the identification of glucose, lactate and ammonia (Biechele et al., 2015; Schwamb et al., 2015). Strong water interference can mask important information in the MIR spectra; however, in-line measurements in aqueous solutions are possible with fibre optic probes that incorporate attenuated total reflection (ATR) technology and Fourier transformation (Biechele et al., 2015). Replacing the highly priced diamond ATR crystal with ATR loops of polycrystalline infrared fibres allows the use of MIR probes in single-use bioreactors (Busse et al., 2017; Steinwedel et al., 2019). NIR, which is also suitable for single-use bioreactors, has capacity to measure glucose and lactate and cell density, although it cannot adapt to cell morphology changes and the resulting peaks are broad and overlapping, making interpretation of spectra difficult (Schwamb et al., 2015; Biechele et al., 2015; Busse et al., 2017; Djisalov et al., 2021). Raman spectroscopy gives significantly weaker signals from water compared to MIR and is a leading PAT for in-line glucose monitoring and control (Buckley and Ryder, 2017; Kozma et al., 2017), while more recently it has been used for lactate, amino acids, ammonium, glutamine, glutamate, pH, cell viability and cell volume (Esmonde-White et al., 2022), albeit at laboratory and small bioreactor scales. Raman measurements can also reveal time-course molecular changes in live cells and provide information about molecular properties of stem cells during differentiation (Ghita et al., 2015; Lim et al., 2022). Electrochemical biosensors are employed mainly for glucose measurements, but also for lactate, glutamine and glutamate (Scheper et al., 1996; Biechele et al., 2015; Steinwedel et al., 2019; Djisalov et al., 2021). Steam sterilisation can damage the biological element of the biosensors and therefore they are used for online (flow injection analysis (FIA)-based systems) monitoring in steel and glass bioreactors, as well as at-line (Biechele et al., 2015). They can also tolerate γ-sterilisation and therefore single-use biosensors could be employed in disposable bioreactors (Busse et al., 2017; Steinwedel et al., 2019).

Although current sensing technologies can gather information for a number of process variables in addition to the 'traditional' ones, such as temperature and pressure, there are still challenges that need to be overcome to enable the efficient monitoring and control of the complex biological processes at scale. The manufacturing of 'new' bioproducts presents an excellent opportunity for the development and integration of modern sensor systems, which combine high sensitivity, stability, fast response times and preferably parallel analyses of CPPs, while remaining robust and cost-effective for industrial operations.

4.2 Digitalisation and automation

The efficiency of the manufacturing process together with the regulatory requirements and the main cost drivers are challenges that should be addressed before industrialisation of any product is established. One way of addressing these challenges is by employing automation to one or all steps of the manufacturing process. This has been the case for the 'classical' production industries, as well as cell and gene therapy, so the applicability to the cultured meat industry is also highly likely.

Automation has the ability to provide control over a bioprocess through better process monitoring through the use of sensors, while ensuring high-quality endpoint products (Rodrigues et al., 2011; Moutsatsou et al., 2019). Similar to cell therapy products, cultured meat products come with biological variation from the get-go, which is difficult to address due to the complexity of these products. Alongside the biological variation, in-process variation is also very likely as it is caused by human handling. Even if stringent protocols are applied, the multi-user handling which is almost always the case can significantly impact product quality due to minor imprecisions (e.g. variation in pipetting) adapted in protocols by each individual handler. Through automation, the in-process variation is eliminated as all tasks can be performed precisely and consistently every time. For example, for a medium change in the bioreactor, through automated pumps, the same consistent speed, force and accuracy can be applied, thus eliminating variability and increasing process reliability (Moutsatsou et al., 2019). Integration and automation of process analytics is another important strategy as it can eliminate the subjectivity of process decisions performed by users' judgement and instead it would make possible the use of more sophisticated and reliable processing rules (Schenk et al., 2015). Digitisation would allow for pattern recognition that otherwise would not be possible.

Spectroscopic and sometimes biosensor measurements can generate a large number of complex data. Using chemometrics (i.e. data manipulation utilising applied mathematics, multivariate statistics and computer science), it is possible to carry out real-time bioreactor monitoring and help fundamental understanding of bioprocesses (Biechele et al., 2015; Gargalo et al., 2020; Martynko and Kirsanov, 2020; Esmonde-White et al., 2022). Multivariate data analysis (MVDA) is used to extract qualitative and quantitative process information from the datasets, enabling continuous online prediction of process variables and/or estimation of the process status. Popular statistical tools such as principal component analysis (PCA) and partial least square regression (PLSR), together with advanced machine learning algorithms can analyse and model the large amount of data that are generated and therefore help to capture and interpret the complex and highly non-linear relationships between CPPs and critical quality attributes (CQAs) in bioprocesses (Narayanan et al., 2020).

Monitoring systems that are a combination of hardware devices and software-implemented models are termed soft sensors (Biechele et al., 2015; Gargalo et al., 2020; Reyes et al., 2022). The data can correspond to directly measurable variables or complex multivariable measurements, which, when analysed by mathematical algorithms in an online fashion, generate information about process parameters that are either difficult to measure directly or are measured at low sampling frequency. Soft sensors are perfect candidates for the PAT initiative and can improve processing times as well as manufacturing costs. They can be split into three categories – data-driven, model-driven and hybrid models – and the reader is referred to several comprehensive reviews for further reading (Sagmeister et al., 2013; Mandenius and Gustavsson, 2015; Reyes et al., 2022). Although soft sensors can play an integral role in process fingerprinting, process prediction and feedback control, there are still several barriers to their implementation into cultured meat production processes and bioindustry in general. For example, current at-line sensors offer higher precision and better calibration and stability than many *in situ* ones, and, therefore, there is a need to invest in better real-time, non-invasive analysers. Furthermore, soft sensors should be easily integrated into existing software environments and the devices should be adapted to standard process equipment within the production line (Randek and Mandenius, 2018).

It is clear that investing in automation, digital transformation and modelling approaches is the way forward in bioprocessing, including the cultured meat industry, and will bring it closer to realising the Industry 4.0 or Fourth Industrial Revolution concept, i.e. the integration of Internet of Things (IoT), cloud computing and analytics, artificial intelligence (AI) and machine learning into the manufacturing processes that enables predictive maintenance, better decision-making and improved efficiency (Narayanan et al., 2020).

4.3 Skills

For cultured meat production to achieve scale and outreach, the production facility would need to resemble more traditional food process plants, much like dairy processing plants and breweries. Bioprocessing of mammalian stem cells is currently somewhat of an art, requiring highly skilled and educated personnel to handle the cells, monitor the process and run the bioreactors. However, given the current shortage of skills particularly in the bioprocessing area, this approach is not sustainable and would become a bottleneck in the long term for the cultured meat industry moreover so as the same required skillsets are shared with other growing industries such as cell and gene therapy. Investment in automation could play a big role in ensuring that cultured meat plants can be run with high productivities and mostly by lower skilled personnel. Fully closed systems and processes would also help as a major CAPEX/OPEX cost in the production of biologics is typically the investment in heating, ventilation and air-conditioning (HVAC) and clean rooms. Fully closed and automated systems would eliminate the need for open processes and thus the background air quality level/grade of the facility can be reduced (Ochs et al., 2022).

5 Conclusion

Numerous technologies are available that could be utilised to scale up the manufacture of cultured meat. These technologies have been borrowed from industries such as cell and gene therapy and tissue engineering. However, for cultured meat production, in order to be economically feasible and to reach cost parity with traditional meat, a hybrid approach between tissue engineering and food science technology would be most likely to succeed. Cultured meat production is more complex than any other biopharmaceutical product as it involves multiple stem cell types, expansion, proliferation, possibly co-culture of different cell types and the use of scaffolds in various forms depending on the stage of the process. When designing such a complex manufacturing process and the facility around it, many inter-dependent variables come into play, and these are both of a biological and an engineering nature. Biological variables include cell type, cell density, cell metabolism, proliferation and differentiation capacity, while engineering variables are related to the bioprocess itself if a one-step expansion–subsequent differentiation bioprocess would be possible, bioreactor scale, mixing and feeding strategy, media recycling, facility design and operation skills. For cultured meat to rip the maximum benefits, its production needs to be designed to be sustainable. This means that all components of the bioprocess, including scaffolds, cells, media components (e.g. proteins and growth factors) as well as other resources used, such as energy and water, need to be sustainable. This challenge requires careful considerations when designing such a bioprocess. Perhaps, this is the time when innovative approaches and novel concepts of out-of-the-box technologies could be tested in the production of cultured meat.

6 List of abbreviations

ATP	adenosine triphosphate
ATR	attenuated total reflection
bFGF-2	basic fibroblast growth factor-2

BHK	baby hamster kidney cells
CCPs	control critical process parameters
CHO	Chinese hamster ovary
CQAs	critical quality attributes
DS	dielectric spectroscopy
ECM	extracellular matrix
EDTA	ethylenediaminetetraacetic acid
FGF	fibroblast growth factor
FGFR1	fibroblast growth factor receptor 1
FIA	flow injection analysis
GMO	genetically modified organism
HEK293	human embryonic kidney cells
HVAC	heating, ventilation and air-conditioning
IFR1	insulin-like growth factor receptor
IGF	insulin-like growth factor
IoT	Internet of Things
ISM	*in situ* microscopy
LSA	lactate supplementation and adaptation
MAPK	mitogen-activated protein kinase
MDCK	Madin–Darby canine kidney cells
MIR	mid-infrared
MMPs	matrix metalloproteinases
MVDA	multivariate data analysis
NIR	near-infrared
PAT	process analytical technology
PAX7	paired box protein
PCA	principal component analysis
PLSR	partial least square regression
pNIPAM	poly(*N*-isopropylacrylamide)
SCADA	system control and data analysis
SDS	sodium dodecyl sulphate
TCA	tricarboxylic acid
TGFβ/BMP	transforming growth factor beta / bone morphogenic protein
YAP1	yes-associated protein 1

7 Where to look for further information

A very good source of information on all aspects of cultured meat including considerations for manufacturing is the US-based charity devoted to promoting and supporting alternative proteins worldwide - The Good Food Institute (gfi.org). Of particular importance, on their website, databases of newly published research can be found. On their YouTube channel, recorded scientific presentations can also be accessed covering all relevant topics including scaled-up production of alternative proteins.

Conferences of particular interest in understanding all aspects of manufacturing of biological products in general include Bioprocessing Summit, Bioprocess UK, while conferences with a focus on cultured meat include the International Scientific Conference on Cultured Meat (ISCCM) organised every year in Maastricht (The Netherlands), Cultivate UK, Good Food Conference organised every year in San Francisco.

Further introductory information can be found on dedicated websites such as www.whatiscultivatedmeat.com or on the websites of various associations such as Cellular Agriculture Europe (www.cellularagriculture.eu), CellAgri Portugal (cellagri.pt), Cellular Agriculture Greece (cellag.gr), and Cellular Agriculture UK (cellag.uk).

Recently, the UK research councils have funded the Cellular Agriculture Manufacturing Hub (CARMA) led by University of Bath which will run for 7 years and will investigate how to manufacture cultivated meat at scale. More information on the CARMA can be found at https://www.ukri.org/news/vaccine-and-food-manufacturing-hubs-will-save-lives-and-cut-carbon/ and https://gfieurope.org/blog/12-million-cultivated-meat-investment-a-seismic-move-for-uk/.

8 References

Akiyama, Y. (2021). Design of temperature-responsive cell culture surfaces for cell sheet engineering. *Cyborg Bionic Syst.* 2021, 5738457. https://doi.org/10.34133/2021/5738457.

Allan, S. J., De Bank, P. A. and Ellis, M. J. (2019). Bioprocess design considerations for cultured meat production with a focus on the expansion bioreactor. *Front. Sustain. Food Syst.* 3, 44. https://doi.org/10.3389/fsufs.2019.00044.

Altomare, L., Cochis, A., Carletta, A., Rimondini, L. and Farè, S. (2016). Thermo-responsive methylcellulose hydrogels as temporary substrate for cell sheet biofabrication. *J. Mater. Sci. Mater. Med.* 27(5), 95. https://doi.org/10.1007/S10856-016-5703-8.

Andersen, D. C., Schrøder, H. D. and Jensen, C. H. (2008). Non-cultured adipose-derived CD45− side population cells are enriched for progenitors that give rise to myofibres in vivo. *Exp. Cell Res.* 314(16), 2951–2964. https://doi.org/10.1016/J.YEXCR.2008.06.018.

Angar, Y., Kebbouche-Gana, S., Djelali, N. E. and Khemili-Talbi, S. (2016). Novel approach for the ammonium removal by simultaneous heterotrophic nitrification and denitrification using a novel bacterial species co-culture. *World J. Microbiol. Biotechnol.* 32(3), 36. https://doi.org/10.1007/S11274-015-2007-Y.

Animal Task Force - 'Facts and figures: Why is European animal livestock production important today?' (2017). Available at: http://animaltaskforce.eu/Portals/0/ATF

/Downloads/Facts%20and%20figures%20sustainable%20and%20competitive%20livestock%20sector%20in%20EU_FINAL.pdf. Accessed on 8th of August 2023.

Anton, F., Burzlaff, A., Kasper, C., Brückerhoff, T. and Scheper, T. (2007). Preliminary study towards the use of in-situ microscopy for the online analysis of microcarrier cultivations. *Eng. Life Sci.* 7(1), 91–96.

Astley, K., Naciri, M., Racher, A. and Al-Rubeai, M. (2007). The role of p21 cip1 in adaptation of CHO cells to suspension and protein-free culture. *J. Biotechnol.* 130(3), 282–290.

Bánfalvi, G. (2011). Heavy metals, trace elements and their cellular effects. In: Banfalvi, G. (Ed.). *Cellular Effects of Heavy Metals*. Dordrecht: Springer.

BBC (2020). https://doi.org/10.1007/978-94-007-0428-2_1. Available at: https://www.bbc.co.uk/news/business-55155741. Accessed on 27th of July 2022.

Bellani, C. F., Ajeian, J., Duffy, L., Miotto, M., Groenewegen, L. and Connon, C. J. (2020). Scale-up technologies for the manufacture of adherent cells. *Front. Nutr.* 7, 575146. https://doi.org/10.3389/fnut.2020.575146.

Ben-Arye, T., Shandalov, Y., Ben-Shaul, S., Landau, S., Zagury, Y., Ianovici, I., Lavon, N. and Levenberg, S. (2020). Textured soy protein scaffolds enable the generation of three-dimensional bovine skeletal muscle tissue for cell-based meat. *Nat. Food* 1(4), 210–220. https://doi.org/10.1038/s43016-020-0046-5.

Ben-Porath, I. and Weinberg, R. A. (2005). The signals and pathways activating cellular senescence. *Int. J. Biochem. Cell Biol.* 37(5), 961–976.

Bernet, J. D., Doles, J. D., Hall, J. K., Kelly-Tanaka, K., Carter, T. A. and Olwin, B. B. (2014). P38 MAPK signaling underlies a cell autonomous loss of stem cell self-renewal in aged skeletal muscle. *Nat. Med.* 20(3), 265–271.

Biechele, P., Busse, C., Solle, D., Scheper, T. and Reardon, K. (2015). Sensor systems for bioprocess monitoring. *Eng. Life Sci.* 15(5), 469–488. https://doi.org/10.1002/elsc.201500014.

Bodiou, V., Moutsatsou, P. and Post, M. J. (2020). Microcarriers for upscaling cultured meat production. *Front. Nutr.* 7, 10.

Bomkamp, C., Skaalure, S. C., Fernando, G. F., Swartz, E. W., Specht, E. A. (2022). Scaffolding biomaterials for 3D cultivated meat: prospects and challenges. *Adv. Sci. (Weinh)* 9(3), e2102908. https://doi.org/10.1002/advs.202102908.

Buckley, K. and Ryder, A. G. (2017). Applications of Raman spectroscopy in biopharmaceutical manufacturing: A short review. *Appl. Spectrosc.* 71(6), 1085–1116. https://doi.org/10.1177/0003702817703270.

Busse, C., Biechele, P., de Vries, I., Reardon, K. F., Solle, D. and Scheper, T. (2017). Sensors for disposable bioreactors. *Eng. Life Sci.* 17(8), 940–952. https://doi.org/10.1002/elsc.201700049.

Callis, J. B., Illman, D. L. and Kowalski, B. R. (1987). Process analytical chemistry. *Anal. Chem.* 59(9), 624A–637A. https://doi.org/10.1021/ac00136a001.

Cancelliere, R., Zurlo, F., Micheli, L. and Melino, S. (2021). Vegetable waste scaffolds for 3D-stem cell proliferating systems and low-cost biosensors. *Talanta* 223(1), 121671. https://doi.org/10.1016/J.TALANTA.2020.121671.

Catoira, M. C., Fusaro, L., di Francesco, D., Ramella, M. and Boccafoschi, F. (2019). Overview of natural hydrogels for regenerative medicine applications. *J. Mater. Sci. Mater. Med.* 30(10), 115. https://doi.org/10.1007/S10856-019-6318-7.

Ceballos, G., Ehrlich, P. R. and Dirzo, R. (2017). Biological annihilation via the ongoing sixth mass extinction signaled by vertebrate population losses and declines. *Proc. Natl Acad. Sci. U. S. A.* 114(30), E6089–E6096.

Chai, W. S., Chew, C. H., Munawaroh, H. S. H., Ashokkumar, V., Cheng, C. K., Park, Y. K. and Show, P. L. (2021). Microalgae and ammonia: a review on inter-relationship. *Fuel* 303, 121303. https://doi.org/10.1016/J.FUEL.2021.121303.

Chen, X. and Li, Y. (2009). Role of matrix metalloproteinases in skeletal muscle: migration, differentiation, regeneration and fibrosis. *Cell Adh. Migr.* 3(4), 337–341. https://doi.org/10.4161/CAM.3.4.9338.

Chueh, B. H., Zheng, Y., Torisawa, Y. S., Hsiao, A. Y., Ge, C., Hsiong, S., Huebsch, N., Franceschi, R., Mooney, D. J. and Takayama, S. (2010). Patterning alginate hydrogels using light-directed release of caged calcium in a microfluidic device. *Biomed. Microdevices* 12(1), 145–151. https://doi.org/10.1007/s10544-009-9369-6.

D'Addario, I., Abbruzzese, C., Lo Iacono, M., Teson, M., Golisano, O. and Barone, V. (2010). Overexpression of YAP1 induces immortalization of normal human keratinocytes by blocking clonal evolution. *Histochem. Cell Biol.* 134(3), 265–276.

Dagge, L., Harr, K., Paul, M. and Schnedl, G. (2009). Classification of process analysis: Offline, atline, online, inline. *Cem. Int.* 7, 72–81.

De Pieri, A., Rochev, Y. and Zeugolis, D. I. (2021). Scaffold-free cell-based tissue engineering therapies: Advances, shortfalls and forecast. *npj Regen. Med.* 6(1), 18. https://doi.org/10.1038/s41536-021-00133-3.

Degouys, V., Cerckel, I., Garcia, A., Harfield, J., Dubois, D., Fabry, L. and Miller, A. O. A. (1993). Dielectric spectroscopy of mammalian cells - 2. simultaneous in situ evaluation by aperture impedance pulse spectroscopy and low frequency dielectric spectroscopy of the biomass of HTC cells on cytodex 3. *Cytotechnology* 13(3), 195–202. https://doi.org/10.1007/BF00749815.

Ding, S., Swennen, G. N. M., Messmer, T., Gagliardi, M., Molin, D. G. M., Li, C., Zhou, G. and Post, M. J. (2018). Maintaining bovine satellite cells stemness through p38 pathway. *Sci. Rep.* 8(1), 10808.

Djisalov, M., Knežić, T., Podunavac, I., Živojević, K., Radonic, V., Knežević, N. Ž., Bobrinetskiy, I. and Gadjanski, I. (2021). Cultivating multidisciplinarity: manufacturing and sensing challenges in cultured meat production. *Biology (Basel)* 10(3), 204. https://doi.org/10.3390/biology10030204.

Dohmen, R. G. J., Hubalek, S., Melke, J., Messmer, T., Cantoni, F., Mei, A., Hueber, R., Mitic, R., Remmers, D., Moutsatsou, P., Post, M. J., Jackisch, L. and Flack, J. E. (2022). Muscle-derived fibro-adipogenic progenitor cells for production of cultured bovine adipose tissue. *NPJ Sci. Food* 6(1), 6. https://doi.org/10.1038/s41538-021-00122-2.

Duguay, D., Foty, R. A. and Steinberg, M. S. (2003). Cadherin-mediated cell adhesion and tissue segregation: Qualitative and quantitative determinants. *Dev. Biol.* 253(2), 309–323. https://doi.org/10.1016/S0012-1606(02)00016-7.

El Wajgali, A., Esteban, G., Fournier, F., Pinton, H. and Marc, A. (2013). Impact of microcarrier coverage on using permittivity for on-line monitoring high adherent Vero cell densities in perfusion bioreactors. *Biochem. Eng. J.* 70, 173–179. https://doi.org/10.1016/j.bej.2012.11.001.

Emerson, J., Kara, B. and Glassey, J. (2020). Multivariate data analysis in cell gene therapy manufacturing. *Biotechnol. Adv.* 45, 107637.

EPA (2013). *Literature Review of Livestock and Poultry Manure and Implications for Water Quality*. Washington, DC: Office of Water.

Esmonde-White, K. A., Cuellar, M. and Lewis, I. R. (2022). The role of Raman spectroscopy in biopharmaceuticals from development to manufacturing. *Anal. Bioanal. Chem.* 414(2), 969–991. https://doi.org/10.1007/s00216-021-03727-4.

European Medicines Agency (2016). *Guideline on Process Validation for Finished Products - Information and Data to Be Provided in Regulatory Submissions* (1st revision).

Feng, L., Li, Y., Zeng, W., Xia, B., Zhou, D. and Zhou, J. (2017). Enhancing effects of basic fibroblast growth factor and fibronectin on osteoblast adhesion to bone scaffolds for bone tissue engineering through extracellular matrix-integrin pathway. *Exp. Ther. Med.* 14(6), 6087–6092. https://doi.org/10.3892/etm.2017.5320.

Freund, N. W. and Croughan, M. S. (2018). A simple method to reduce both lactic acid and ammonium production in industrial animal cell culture. *Int. J. Mol. Sci.* 19(2), 385, https://doi.org/10.3390/ijms19020385.

Gargalo, C. L., Udugama, I., Pontius, K., Lopez, P. C., Nielsen, R. F., Hasanzadeh, A., Mansouri, S. S., Bayer, C., Junicke, H. and Gernaey, K. V. (2020). Towards smart biomanufacturing: A perspective on recent developments in industrial measurement and monitoring technologies for bio-based production processes. *J. Ind. Microbiol. Biotechnol.* 47(11), 947–964. https://doi.org/10.1007/s10295-020-02308-1.

Ghita, A., Pascut, F. C., Sottile, V., Denning, C. and Notingher, I. (2015). Applications of Raman micro-spectroscopy to stem cell technology: Label-free molecular discrimination and monitoring cell differentiation. *EPJ Tech. Instrum.* 2(1), 6. https://doi.org/10.1140/epjti/s40485-015-0016-8.

Glassey, J., Gernaey, K. V., Clemens, C., Schulz, T. W., Oliveira, R., Striedner, G. and Mandenius, C. F. (2011). Process analytical technology (PAT) for biopharmaceuticals. *Biotechnol. J.* 6(4), 369–377. https://doi.org/10.1002/biot.201000356.

Gui, L. and Niklason, L. E. (2014). Vascular tissue engineering: building perfusable vasculature for implantation. *Curr. Opin. Chem. Eng.* 3, 68–74. https://doi.org/10.1016/J.COCHE.2013.11.004.

Hakim, D. (2018). At hamburger central, antibiotics for cattle that aren't sick. *New York Times*. Available at: https://www.nytimes.Com/2018/03/23/business/cattle-antiobiotics.html.

Hao, L., Fu, X., Li, T., Zhao, N., Shi, X., Cui, F., Du, C. and Wang, Y. (2016). Surface chemistry from wettability and charge for the control of mesenchymal stem cell fate through self-assembled monolayers. *Colloids Surf. B Biointerfaces* 148, 549–556. https://doi.org/10.1016/j.colsurfb.2016.09.027.

Hanga, M. P. and Holdich, R. G. (2014). Membrane emulsification for the production of uniform poly-N-isopropylacrylamide-coated alginate particles using internal gelation. *Chemical Engineering Research and Design* 92(9), 1664–1673.

Hanga, M. P., Ali, J., Moutsatsou, P., de la Raga, F. A., Hewitt, C. J., Nienow, A. W. and Wall, I. (2020). Bioprocess development for scalable production of cultivated meat. *Biotechnol. Bioeng.* 117(10), 3029–3039.

Hanga, M. P., de la Raga, F. A., Moutsatsou, P., Hewitt, C. J., Nienow, A. W. and Wall, I. (2021). Scale-up of an intensified bioprocess for the expansion of bovine adipose-derived stem cells (bASCs) in stirred tank bioreactors. *Biotechnol. Bioeng.* 118(8), 3175–3186.

Haraguchi, Y. and Shimizu, T. (2021a). Microalgal culture in animal cell waste medium for sustainable 'cultured food' production. *Arch. Microbiol.* 203(9), 5525–5532. https://doi.org/10.1007/S00203-021-02509-X.

Haraguchi, Y. and Shimizu, T. (2021b). Three-dimensional tissue fabrication system by co-culture of microalgae and animal cells for production of thicker and healthy cultured food. *Biotechnol. Lett.* 43(6), 1117–1129. https://doi.org/10.1007/S10529-021-03106-0.

Hassan, W., Dong, Y. and Wang, W. (2013). Encapsulation and 3D culture of human adipose-derived stem cells in an *in-situ* crosslinked hybrid hydrogel composed of PEG-based hyperbranched copolymer and hyaluronic acid. *Stem Cell Res. Ther.* 4(2), 32. https://doi.org/10.1186/scrt182.

Hayflick, L. and Moorhead, P. S. (1961). The serial cultivation of human diploid cell strains. *Exp. Cell Res.* 25, 585–621.

Ho, Y. Y., Lu, H. K., Lin, Z. F. S., Lin, H. W., Ho, Y. S. and Ng, S. K. (2021). Applications and analysis of hydrolysates in animal cell culture. *Bioresour. Bioprocess.* 8(1), 93.

Hosoyama, T., Meyer, M. G., Krakora, D. and Suzuki, M. (2013). Isolation and in vitro propagation of human skeletal muscle progenitor cells from fetal muscle. *Cell Biol. Int.* 37(2), 191–196.

Hosoyama, T., McGivern, J. V., Van Dyke, J. M., Ebert, A. D. and Suzuki, M. (2014). Derivation of myogenic progenitors directly from human pluripotent stem cells using a sphere-based culture. *Stem Cells Transl. Med.* 3(5), 564–574.

Huang, Y., Song, C., Li, L. and Zhou, Y. (2014). The mechanism and performance of zeolites for ammonia removal in the zeolite packed electrolysis reactor. *Electrochemistry* 82(7), 557–560.

Hubalek, S., Post, M. J. and Moutsatsou, P. (2022). Towards resource-efficient and cost-efficient cultured meat. *Curr. Opin. Food Sci.* 47, 100885.

Jahangirian, H., Azizi, S., Rafiee-Moghaddam, R., Baratvand, B. and Webster, T. J. (2019). Status of plant protein-based green scaffolds for regenerative medicine applications. *Biomolecules* 9(10), 619. https://doi.org/10.3390/biom9100619.

Jenkins, T. L. and Little, D. (2019). Synthetic scaffolds for musculoskeletal tissue engineering: Cellular responses to fiber parameters. *NPJ Regen. Med.* 4, 15 https://doi.org/10.1038/s41536-019-0076-5.

Jung, S., Panchalingam, K. M., Wuerth, R. D., Rosenberg, L. and Behie, L. A. (2012). Large-scale production of human mesenchymal stem cells for clinical applications. *Biotechnol. Appl. Biochem.* 59(2), 106–120. https://doi.org/10.1002/BAB.1006.

Kang, D. H., Louis, F., Liu, H., Shimoda, H., Nishiyama, Y., Nozawa, H., Kakitani, M., Takagi, D., Kasa, D., Nagamori, E., Irie, S., Kitano, S. and Matsusaki, M. (2021). Engineered whole cut meat-like tissue by the assembly of cell fibers using tendon-gel integrated bioprinting. *Nat. Commun.* 12(1), 5059. https://doi.org/10.1038/s41467-021-25236-9.

Khanal, O. and Lenhoff, A. M. (2021). Developments and opportunities in continuous biopharmaceutical manufacturing. *mAbs* 13(1), 1903664, https://doi.org/10.1080/19420862.2021.1903664.

Kim, B. S., Kim, J. U., So, K. H. and Hwang, N. S. (2021). Supercritical fluid-based decellularization technologies for regenerative medicine applications. *Macromol. Biosci.* 21(8), e2100160. https://doi.org/10.1002/MABI.202100160.

Kolkmann, A. M., Van Essen, A., Post, M. J. and Moutsatsou, P. (2022). Development of a chemically defined medium for in vitro expansion of primary bovine satellite cells. *Front. Bioeng. Biotechnol.* 10, 895289. https://doi.org/10.3389/fbioe.2022.895289.

Kolkmann, A. M., Post, M. J., Rutjens, M. A. M., van Essen, A. L. M. and Moutsatsou, P. (2020). Serum-free media for the growth of primary bovine myoblasts. *Cytotechnology* 72(1), 111–120.

Konakovsky, V., Clemens, C., Müller, M. M., Bechmann, J., Berger, M., Schlatter, S. and Herwig, C. (2016). Metabolic control in mammalian fed-batch cell cultures for reduced lactic acid accumulation and improved process robustness. *Bioengineering (Basel)* 3(1), 5.

Kozma, B., Hirsch, E., Gergely, S., Párta, L., Pataki, H. and Salgó, A. (2017). On-line prediction of the glucose concentration of CHO cell cultivations by NIR and Raman spectroscopy: comparative scalability test with a shake flask model system. *J. Pharm. Biomed. Anal.* 145, 346–355. https://doi.org/10.1016/j.jpba.2017.06.070.

Lawson, T., Kehoe, D. E., Schnitzler, A. C., Rapiejko, P. J., Der, K. A., Philbrick, K., Punreddy, S., Rigby, S., Smith, R., Feng, Q., Murrell, J. R. and Rook, M. S. (2017). Process development for expansion of human mesenchymal stromal cells in a 50 L single-use stirred tank bioreactor. *Biochem. Eng. J.* 120, 49–62.

Li, D., Zhou, J., Chowdhury, F., Cheng, J., Wang, N. and Wang, F. (2011). Role of mechanical factors in fate decisions of stem cells. *Regen. Med.* 6(2), 229–240. https://doi.org/10.2217/RME.11.2.

Lim, D., Renteria, E. S., Sime, D. S., Ju, Y. M., Kim, J. H., Criswell, T., Shupe, T. D., Atala, A., Marini, F. C., Gurcan, M. N., Soker, S., Hunsberger, J. and Yoo, J. J. (2022). Bioreactor design and validation for manufacturing strategies in tissue engineering. *Biodes Manuf.* 5(1), 43–63. https://doi.org/10.1007/s42242-021-00154-3.

Listrat, A., Lebret, B., Louveau, I., Astruc, T., Bonnet, M., Lefaucheur, L., Picard, B. and Bugeon, J. (2016). How muscle structure and composition influence meat and flesh quality. *TheScientificWorldJOURNAL* 2016, 3182746, https://doi.org/10.1155/2016/3182746.

Luo, J., Vijayasankaran, N., Autsen, J., Santuray, R., Hudson, T., Amanullah, A. and Li, F. (2012). Comparative metabolite analysis to understand lactate metabolism shift in Chinese hamster ovary cell culture process. *Biotechnol. Bioeng.* 109(1), 146–156.

Ma, J.-Y., Wang, Y.-X., Huang, Y., Zhang, Y., Cui, Y.-X. and Kong, D.-M. (2022). Chemical-biological approaches for the direct regulation of cell-cell aggregation. *Aggregate* 3(2), e166.

Malm, M., Saghaleyni, R., Lundqvist, M., Giudici, M., Chotteau, V., Field, R., Varley, P. G., Hatton, D., Grassi, L., Svensson, T., Nielsen, J. and Rockberg, J. (2020). Evolution from adherent to suspension: Systems biology of HEK293 cell line development. *Sci. Rep.* 10(1), 18996. https://doi.org/10.1038/s41598-020-76137-8.

Mandenius, C.-F. and Gustavsson, R. (2015). Mini-review: Soft sensors as means for PAT in the manufacture of bio-therapeutics. *J. Chem. Technol. Biotechnol.* 90(2), 215–227. https://doi.org/10.1002/jctb.4477.

Manstein, F., Ullmann, K., Kropp, C., Halloin, C., Triebert, W., Franke, A., Farr, C. M., Sahabian, A., Haase, A., Breitkreuz, Y., Peitz, M., Brüstle, O., Kalies, S., Martin, U., Olmer, R. and Zweigerdt, R. (2021). High density bioprocessing of human pluripotent stem cells by metabolic control and in silico modelling. *Stem Cells Transl. Med.* 10(7), 1063–1080. https://doi.org/10.1002/sctm.20-0453.

Martynko, E. and Kirsanov, D. (2020). Application of chemometrics in biosensing: a review. *Biosensors (Basel)* 10(8), 100. https://doi.org/10.3390/bios10080100.

Mattick, C. S., Landis, A. E., Allenby, B. R. and Genovese, N. J. (2015). Anticipatory life cycle analysis of in vitro biomass cultivation for cultured meat production in the United States. *Environ. Sci. Technol.* 49(19), 11941–11949.

Melzener, L., Verzijden, K. E., Buijs, A. J., Post, M. J. and Flack, J. E. (2021). Cultured beef: From small biopsy to substantial quantity. *J. Sci. Food Agri.* 101(1), 7–14.

Merten, O. W. (2015). Advances in cell culture: Anchorage dependence. *Philos. Trans. R. Soc. Lond. B Biol. Sci.* 370(1661), 20140040. https://doi.org/10.1098/rstb.2014.0040.

Messmer, T., Klevernic, I., Furquim, C., Ovchinnikova, E., Dogan, A., Cruz, H., Post, M. J. and Flack, J. E. (2022). A serum-free media formulation for cultured meat production supports bovine satellite cell differentiation in the absence of serum starvation. *Nat. Food* 3(1), 74–85; https://doi.org/10.1038/s43016-021-00419-1.

Metze, S., Ruhl, S., Greller, G., Grimm, C. and Scholz, J. (2020). Monitoring online biomass with a capacitance sensor during scale-up of industrially relevant CHO cell culture fed-batch processes in single-use bioreactors. *Bioprocess Biosyst. Eng.* 43(2), 193–205. https://doi.org/10.1007/s00449-019-02216-4.

Minnich, C., Hardy, S. and Krämer, S. (2016). Stopping the Babylonian confusion: an updated nomenclature for process analyzers in PAT applications. *Chem. Ing. Tech.* 88(6), 694–697. https://doi.org/10.1002/cite.201500188.

Mohareb, E. A., Heller, M. C. and Guthrie, P. M. (2018). Cities' role in mitigating United States food system greenhouse gas emissions. *Environ. Sci. Technol.* 52(10), 5545–5554. https://doi.org/10.1021/acs.est.7b02600.

Molina, M. I., Malollari, K. G. and Komvopoulos, K. (2021). Design challenges in polymeric scaffolds for tissue engineering. *Front. Bioeng. Biotechnol.* 9, 617141. https://doi.org/10.3389/fbioe.2021.617141.

Moore, B., Sanford, R. and Zhang, A. (2019). Case study: the characterization and implementation of dielectric spectroscopy (biocapacitance) for process control in a commercial GMP CHO manufacturing process. *Biotechnol. Prog.* 35(3), e2782. https://doi.org/10.1002/btpr.2782.

Moschouris, K., Firoozi, N. and Kang, Y. (2016). The application of cell sheet engineering in the vascularization of tissue regeneration. *Regen. Med.* 11(6), 559–570; https://doi.org/10.2217/rme-2016-0059.

Moutsatsou, P., Ochs, J., Schmitt, R. H., Hewitt, C. J. and Hanga, M. P. (2019). Automation in cell and gene therapy manufacturing: From past to future. *Biotechnol. Lett.* 41(11), 1245–1253.

Narayanan, H., Luna, M. F., von, M., Cruz, M. N., Polotti, G., Morbidelli, M., Butté, A. and Sokolov, M. (2020). Bioprocessing in the digital age: the role of process models. *Biotechnol. J.* 15, 1900172. https://doi.org/10.1002/biot.201900172.

Navaei, A., Uth, N., Sargent, A., Levinson, Y. and Ramaswamy, S. (2020). Important considerations for cell therapy manufacturing of mesenchymal stem cells. *Cytotherapy* 22(5), S92. https://doi.org/10.1016/j.jcyt.2020.03.158.

Nicodemus, G. D. and Bryant, S. J. (2008). Cell encapsulation in biodegradable hydrogels for tissue engineering applications. *Tissue Eng. Part B Rev.* 14(2), 149–165. https://doi.org/10.1089/ten.teb.2007.0332.

Nienow, A. W., Hewitt, C. J., Heathman, T. R. J., Glyn, V. A. M., Fonte, G. N., Hanga, M. P., Coopman, K. and Rafiq, Q. A. (2016). Agitation conditions for the culture and detachment of hMSCs from microcarriers in multiple bioreactor platforms. *Biochem. Eng. J.* 108, 24–29.

Ng, J. Y., Chua, M. L., Zhang, C., Hong, S., Kumar, Y., Gokhale, R. and Ee, P. L. R. (2020). *Chlorella vulgaris* extract as a serum replacement that enhances mammalian cell growth and protein expression. *Front. Bioeng. Biotechnol.* 8, 564667.

Ng, S. and Kurisawa, M. (2021). Integrating biomaterials and food Biopolymers for cultured meat production. *Acta Biomater.* 124, 108–129. https://doi.org/10.1016/j.actbio.2021.01.017.

Ochs, J., Hanga, M. P., Shaw, G., Duffy, N., Kulik, M., Tissin, N., Reibert, D., Biermann, F., Moutsatsou, P., Ratnayake, S., Nienow, A., Koenig, N., Schmitt, R., Rafiq, Q., Hewitt, C. J., Barry, F. and Murphy, J. M. (2022). Needle-to-needle robot-assisted manufacture of cell therapy products. *Bioeng. Transl. Med.* 7(3), e10387. https://doi.org/10.1002/btm2.10387.

O'Mara, P., Farrell, A., Bones, J. and Twomey, K. (2018). Staying alive! Sensors used for monitoring cell health in bioreactors. *Talanta* 176, 130–139. https://doi.org/10.1016/j.talanta.2017.07.088.

O'Neill, J. (2016). "Tackling drug-resistant infections globally: Final report and recommendations." London, UK: Review on Antimicrobial Resistance. Available at: https://amr-review.org/sites/default/files/160518_Final%20paper_with%20cover.pdf.

Pakseresht, A., Ahmadi Kaliji, S. and Canavari, M. (2022). Review of factors affecting consumer acceptance of cultured meat. *Appetite* 170, 105829. https://doi.org/10.1016/J.APPET.2021.105829.

Pasitka, L., Cohen, M., Ehrlich, A., Gildor, B., Reuveni, E., Ayyash, M., Wissotsky, G., Herscovici, A., Kaminker, R., Niv, A., Bitcover, R., Dadia, O., Rudik, A., Voloschin, A., Shimoni, M., Cinnamon, Y. and Nahmias, Y. (2023). Spontaneous immortalization of chicken fibroblasts generates stable, high-yield cell lines for serum-free production of cultured meat. *Nat. Food* 4(1), 35–50. https://doi.org/10.1038/s43016-022-00658-w.

Pohlscheidt, M., Charaniya, S., Bork, C., Jenzsch, M., Noetzel, T. L. and Luebbert, A. (2013). Bioprocess and fermentation monitoring. In: Flickinger, M. C. (Ed.). *Upstream Industrial Biotechnology: Equipment, Process Design, Sensing, Control, and cGMP Operations* (1st edn.) (vol. 2). John Wiley & Sons, Inc., 1471–1491.

Post, M. J. (2014). Cultured beef: Medical technology to produce food. *J. Sci. Food Agric.* 94(6), 1039–1041.

Post, M. J. and Hocquette, J. F. (2017). Chapter 16. New sources of animal proteins in vitro meat. In: Purslow, P. P. (Ed.). *New Aspects of Meat Quality*. Elsevier Ltd: Cambridge, 425–441.

Post, M. J., Levenberg, S., Kaplan, D. L., Genovese, N., Fu, J., Bryant, C. J., Negowetti, N., Verzijden, K. and Moutsatsou, P. (2020). Scientific, sustainability and regulatory challenges of cultured meat. *Nat. Food* 1, 403–415. https://doi.org/10.1038/s43016-020-0112-z.

Rafiq, Q. A., Brosnan, K. M., Coopman, K., Nienow, A. W. and Hewitt, C. J. (2013). Culture of human mesenchymal stem cells on microcarriers in a 5L stirred-tank bioreactor. *Biotechnol. Lett.* 35(8), 1233–1245.

Rafiq, Q. A., Hanga, M. P., Heathman, T. R. J., Coopman, K., Nienow, A. W., Williams, D. J. and Hewitt, C. J. (2017). Process development of human multipotent stromal cell microcarrier culture using an automated high-throughput microbioreactor. *Biotechnol. Bioeng.* 114(10), 2253–2266.

Rafiq, Q. A., Ruck, S., Hanga, M. P., Heathman, T. R. J., Coopman, K., Nienow, A. W., Williams, D. J. and Hewitt, C. J. (2018). Qualitative and quantitative demonstration of bead-to-bead transfer with bone marrow-derived human mesenchymal stem cells on microcarriers: Utilising the phenomenon to improve culture performance. *Biochem. Eng. J.* 135, 11–21.

Randek, J. and Mandenius, C. F. (2018). On-line soft sensing in upstream bioprocessing. *Crit. Rev. Biotechnol.* 38(1), 106–121, https://doi.org/10.1080/07388551.2017.1312271.

Reyes, S. J., Durocher, Y., Pham, P. L. and Henry, O. (2022). Modern sensor tools and techniques for monitoring, controlling, and improving cell culture processes. *Processes* 10(2), 189. https://doi.org/10.3390/pr10020189.

Rodrigues, C. A. V., Fernandes, T. G., Diogo, M. M., da Silva, C. L. and Cabral, J. M. S. (2011). Stem cell cultivation in bioreactors. *Biotechnol. Adv.* 29(6), 815–829.

Rolfe, P. (2006). Sensing in tissue bioreactors. *Meas. Sci. Technol.* 17(3), 578–583. https://doi.org/10.1088/0957-0233/17/3/S20.

Romero García, J. M., Acién Fernández, F. G. and Fernández Sevilla, J. M. (2012). Development of a process for the production of l-amino-acids concentrates from microalgae by enzymatic hydrolysis. *Bioresour. Technol.* 112, 164–170. https://doi.org/10.1016/J.BIORTECH.2012.02.094.

Rubio, N. R., Fish, K. D., Trimmer, B. A. and Kaplan, D. L. (2019). In Vitro Insect Muscle for Tissue Engineering Applications. *ACS Biomater. Sci. Eng.* 5(2), 1071–1082, https://doi.org/10.1021/acsbiomaterials.8b01261.

Ryu, W. (2017). *Molecular Virology of Human Pathogenic Viruses*. Amsterdam, The Netherlands: Elsevier.

Sagmeister, P., Wechselberger, P., Jazini, M., Meitz, A., Langemann, T. and Herwig, C. (2013). Soft sensor assisted dynamic bioprocess control: efficient tools for bioprocess development. *Chem. Eng. Sci.* 96, 190–198, https://doi.org/10.1016/j.ces.2013.02.069.

Särnlund, S., Jiang, Y. and Chotteau, V. (2021). Process intensification to produce a difficult-to-express therapeutic enzyme by high cell density perfusion or enhanced fed-batch. *Biotechnol. Bioeng.* 118(9), 3533–3544.

Schenk, F. W., Kulik, M. and Schmitt, R. (2015). Metrology-based quality and process control in automated stem cell production. *Tech. Mess.* 82(6), 309–316. https://doi.org/10.1515/teme-2015-0036.

Scheper, T. H., Hilmer, J. M., Lammers, F., Müller, C. and Reinecke, M. (1996). Biosensors in bioprocess monitoring. *J. Chromatogr. A* 725(1), 3–12. https://doi.org/10.1016/0021-9673(95)00998-1.

Schop, D., Janssen, F. W., van Rijn, L. D., Fernandes, H., Bloem, R. M., de Bruijn, J. D. and van Dijkhuizen-Radersma, R. (2009). Growth, metabolism and growth inhibitors of mesenchymal stem cells. *Tissue Eng. Part A* 15(8), 1877–1886.

Schop, D., van Dijkhuizen-Radersma, R., Borgart, E., Janssen, F. W., Rozemuller, H., Prins, H. J. and de Bruijn, J. D. (2010). Expansion of human mesenchymal stromal cells on microcarriers: Growth and metabolism. *J. Tissue Eng. Regen. Med.* 4(2), 131–140.

Schwamb, S., Puskeiler, R. and Wiedemann, P. (2015). Monitoring of cell culture. In: Al-Rubeai, M. (Ed.). *Animal Cell Culture*. Springer, 185–221.

Shahin-Shamsabadi, A. and Selvaganapathy, P. R. (2022). Engineering murine adipocytes and skeletal muscle cells in meat-like constructs using self-assembled layer-by-layer biofabrication: A platform for development of cultivated meat. *Cells Tissues Organs* 211(3), 304–312. https://doi.org/10.1159/000511764.

Shay, J. W., Pereira-Smith, O. M. and Wright, W. E. (1991). A role for both RB and p53 in the regulation of human cellular senescence. *Exp. Cell Res.* 196(1), 33-39.

Shen, C. F., Guilbault, C., Li, X., Elahi, S. M., Ansorge, S., Kamen, A. and Gilbert, R. (2019). Development of suspension-adapted Vero cell culture process technology for production of viral vaccines. *Vaccine* 37(47), 6996–7002.

Shibata, K. R., Aoyama, T., Shima, Y., Fukiage, K., Otsuka, S., Furu, M., Kohno, Y., Ito, K., Fujibayashi, S., Neo, M., Nakayama, T., Nakamura, T. and Toguchida, J. (2007). Expression of the p161NK4A gene is associated closely with senescence of human mesenchymal stem cells and is potentially silenced by DNA methylation during in vitro expansion. *Stem Cells* 25(9), 2371–2382.

Shin, H., Jo, S. and Mikos, A. G. (2003). Biomimetic materials for tissue engineering. *Biomaterials* 24(24), 4353–4364. https://doi.org/10.1016/S0142-9612(03)00339-9.

Shivakumara, L. R. and Demappa, T. (2019). Synthesis and Swelling Behavior of Sodium Alginate/poly(vinyl alcohol) hydrogels. *Turk. J. Pharm. Sci.* 16(3), 252–260. https://doi.org/10.4274/TJPS.GALENOS.2018.92408.

Silva, R., Fabry, B. and Boccaccini, A. R. (2014). Fibrous protein-based hydrogels for cell encapsulation. *Biomaterials* 35(25), 6727–6738.

Slivac, I., Blajić, V., Radošević, K., Kniewald, Z. and Gaurina Srček, V. (2010). Influence of different ammonium, lactate and glutamine concentrations on CHO cell growth. *Cytotechnology* 62(6), 585–594. https://doi.org/10.1007/S10616-010-9312-Y.

Soice, E. and Johnston, J. (2021). Immortalizing cells for human consumption. *Int. J. Mol. Sci.* 22(21), 11660. https://doi.org/10.3390/ijms222111660.

Somaiah, C., Kumar, A., Mawrie, D., Sharma, A., Patil, S. D., Bhattacharyya, J., Swaminathan, R. and Jaganathan, B. G. (2015). Collagen promotes higher adhesion, survival and proliferation of mesenchymal stem cells. *PLoS ONE* 10(12), e0145068. https://doi.org/10.1371/JOURNAL.PONE.0145068.

Specht, L. (2020). *An Analysis of Culture Medium Costs and Production Volumes for Cultivated Meat*. Washington, DC: The Good Food Institute.

Steinwedel, T., Dahlmann, K., Solle, D., Scheper, T., Reardon, K. F. and Lammers, F. (2019). Sensors for disposable bioreactor systems. In: Eibl, R. and Eibl, D. (Eds). *Single-Use Technology in Biopharmaceutical Manufacture*, 69–82. https://doi.org/10.1002/9781119477891.ch6.

Stepanenko, A. A. and Dmitrenko, V. V. (2015). HEK293 in cell biology and cancer research: phenotype, karyotype, tumorigenicity, and stress-induced genome-phenotype evolution. *Gene* 569(2), 182–190.

Stephens, N., Di Silvio, L., Dunsford, I., Ellis, M., Glencross, A. and Sexton, A. (2018). Bringing cultured meat to market: Technical, socio-political, and regulatory challenges in cellular agriculture. *Trends Food Sci. Technol.* 78, 155–166.

Stephens, N. and Ellis, M. (2020). Cellular agriculture in the UK: A review. *Wellcome Open Res.* 5, 12.

Stout, A. J., Mirliani, A. B., Rittenberg, M. L., Shub, M., White, E. C., Yuen, J. S. K. and Kaplan, D. L. (2022). Simple and effective serum-free medium for sustained expansion of bovine satellite cells for cell cultured meat. *Commun. Biol.* 5(1), 466. https://doi.org/10.1038/s42003-022-03423-8.

Sun, Z., Guo, S. S. and Fässler, R. (2016). Integrin-mediated mechanotransduction. *J. Cell Biol.* 215(4), 445–456. https://doi.org/10.1083/jcb.201609037.

Tanaka, R. I., Sakaguchi, K., Yoshida, A., Takahashi, H., Haraguchi, Y. and Shimizu, T. (2022). Production of scaffold-free cell-based meat using cell sheet technology. *NPJ Sci. Food* 3;6(1):41. https://doi.org/10.1038/s41538-022-00155-1.Tavafi, Tavafi, H., Abdi-Ali, A., Ghadam, P. and Gharavi, S. (2017). Screening of alginate lyase-producing bacteria and optimization of media compositions for extracellular alginate lyase production. *Iran. Biomed. J.* 21(1), 48–56. https://doi.org/10.6091/.21.1.48.

Tharmalingam, T., Wu, C. H., Callahan, S. and Goudar, C. T. (2015). A framework for real-time glycosylation monitoring (RT-GM) in mammalian cell culture. *Biotechnol. Bioeng.* 112(6), 1146–1154.

Thorley, M., Duguez, S., Mazza, E. M. C., Valsoni, S., Bigot, A., Mamchaoui, K., Harmon, B., Voit, T., Mouly, V. and Duddy, W. (2016). Skeletal muscle Characteristics Are Preserved in hTERT/cdk4 human myogenic cell lines. *Skelet. Muscle* 6(1), 43.

Trabelsi, K., Rourou, S., Loukil, H., Majou, S. and Kallel, H. (2006). Optimisation of virus yield as a strategy to improve rabies vaccine production by Vero cells in a bioreactor. *J. Biotechnol.* 121(2), 261–271.

Tuomisto, H. L., Ellis, M. J. and Haastrup, P. (2014). Environmental impacts of cultured meat: alternative production scenarios. In: Schenck, R. and Huizenga, D. (Eds). *Proceedings of the 9th International Conference on Life Cycle Assessment in the Agri-Food Sector*. ACLCA: San Francisco.

Vcelar, S., Jadhav, V., Melcher, M., Auer, N., Hrdina, A., Sagmeister, R., Heffner, K., Puklowski, A., Betenbaugh, M., Wenger, T., Leisch, F., Baumann, M. and Borth, N. (2018). Karyotype variation of CHO host cell lines over time in culture characterized by chromosome counting and chromosome painting. *Biotechnol. Bioeng.* 115(1), 165–173.

Walker, C., Mojares, E. and del Río Hernández, A. (2018). Role of extracellular matrix in development and cancer progression. *Int. J. Mol. Sci.* 19(10). https://doi.org/10.3390/IJMS19103028.

Wang, L., Wu, Y., Guo, B. and Ma, P. X. (2015). Nanofiber yarn/hydrogel core-shell scaffolds mimicking native skeletal muscle tissue for guiding 3D myoblast alignment, elongation, and differentiation. *ACS Nano* 9(9), 9167–9179. https://doi.org/10.1021/acsnano.5b03644.

Wang, L., Zhang, C., Gao, F. and Pan, G. (2016). Needleless electrospinning for scaled-up production of ultrafine chitosan hybrid nanofibers used for air filtration. *RSC Adv.* 6(107), 105988–105995.

Wang, Y., Chen, S., Yan, Z. and Pei, M. (2019). A prospect of cell immortalization combined with matrix microenvironmental optimization strategy for tissue engineering and regeneration. *Cell Biosci.* 9, 7. https://doi.org/10.1186/s13578-018-0264-9.

Wang, D., Cai, Z., Huang, X. and Wang, L. (2021). Study on the dissolution and diffusion of supercritical carbon dioxide in polystyrene melts based on adsorption and diffusion mechanism. *ACS Omega* 6(3), 1971–1984. https://doi.org/10.1021/acsomega.0c04751.

Westerman, K. A. (2015). Myospheres are composed of two cell types: one that is myogenic and a second that is mesenchymal. *PLOS ONE* 10(2), e0116956. https://doi.org/10.1371/journal.pone.0116956.

Whitford, W. G. (2021). Bioprocess intensification: Technologies and goals. In: *Process Control, Intensification and Digitalisation in Continuous Biomanufacturing* (1st edn.). Wiley-VCH Verlag GmBH.

Willkomm, L., Schubert, S., Jung, R., Elsen, M., Borde, J., Gehlert, S., Suhr, F. and Bloch, W. (2014). Lactate regulates myogenesis in C2C12 myoblasts in vitro. *Stem Cell Res.* 12(3), 742–753, https://doi.org/10.1016/j.scr.2014.03.004.

Wilschut, K. J., Haagsman, H. P. and Roelen, B. A. J. (2010). Extracellular matrix components direct porcine muscle stem cell behavior. *Exp. Cell Res.* 316(3), 341–352. https://doi.org/10.1016/j.yexcr.2009.10.014.

Wilson, J. L. and McDevitt, T. C. (2013). Stem cell microencapsulation for phenotypic control, bioprocessing, and transplantation. *Biotechnol. Bioeng.* 110(3), 667–682.

Wollschlaeger, J. O., Maatz, R., Albrecht, F. B., Klatt, A., Heine, S., Blaeser, A. and Kluger, P. J. (2022). Scaffolds for cultured meat on the basis of polysaccharide hydrogels enriched with plant-based proteins. *Gels* 8(2), 94.

Yaffe, D. and Saxel, O. (1977). Serial passaging and differentiation of myogenic cells isolated from dystrophic mouse muscle. *Nature* 270(5639), 725–727. https://doi.org/10.1038/270725a0.

Yang, J., Gurtin, P., Jia, G., Lv, Z., Yang, H. and Ju, D. (2019). Large-scale microcarrier culture of HEK293T cells and Vero cells in single-use bioreactors. *AMB Express* 9(1), 70.

Yongky, A., Xu, J., Tian, J., Oliveira, C., Zhao, J., McFarland, K., Borys, M. C. and Li, Z. J. (2019). Process intensification in fed-batch production bioreactors using non-perfusion seed cultures. *mAbs* 11(8), 1502–1514. https://doi.org/10.1080/19420862.2019.1652075.

Yu, L., Dean, K. and Li, L. (2006). Polymer blends and composites from renewable sources. *Prog. Polym. Sci.* 31(6), 576–602.

Zamani, L., Lundqvist, M., Zhang, Y., Aberg, M., Edfors, F., Bidkhori, G., Lindahl, A., Mie, A., Mardinoglu, A., Field, R., Turner, R., Rockberg, J. and Chotteau, V. (2018). High cell density perfusion culture has a maintained exoproteome and metabolome. *Biotechnol. J.* 13(10), 1800036, 11 pages.

Zhang, Y., Fan, W., Ma, Z., Wu, C., Fang, W., Liu, G. and Xiao, Y. (2010). The effects of pore architecture in silk fibroin scaffolds on the growth and differentiation of mesenchymal stem cells expressing BMP7. *Acta Biomater.* 6(8), 3021–3028. https://doi.org/10.1016/j.actbio.2010.02.030.

Zhang, Y., Stobbe, P., Silvander, C. O. and Chotteau, V. (2015). Very high cell density perfusion of CHO cells anchored in a non-woven matrix-based bioreactor. *J. Biotechnol.* 213, 28–41.

Zhao, L., Fu, H.-Y., Zhou, W. and Hu, W.-S. (2015). Advances in process monitoring tools for cell culture bioprocesses. *Eng. Life Sci.* 15(5), 459–468. https://doi.org/10.1002/elsc.201500006.

Zuk, P. A., Zhu, M., Mizuno, H., Huang, J., Futrell, J. W., Katz, A. J., Benhaim, P., Lorenz, H. P. and Hedrick, M. H. (2001). Multilineage cells from human adipose tissue: implications for cell-based therapies. *Tissue Eng.* 7(2), 211–228. http://doi.org/10.1089/107632701300062859.

Part 3

Quality and sustainability issues

Chapter 10

Developing cultured meat as a food product

J. F. Young, A. Abraham*, M. K. Rasmussen, S. Skrivergaard and M. Therkildsen, Aarhus University, Denmark*

1 Introduction

Research in and development of cultured meat has focused primarily on technical issues such as tissue engineering and upscaling production, since these are prerequisites to developing the product as a food commodity. However, since the aim is to create a food product for consumption, it is critical to address food quality at an early stage in order to develop an end product of high nutritional and sensory quality that consumers want (Fraeye et al., 2020). Aspects of food quality include an increased lipid content. This could be increased in culture through trans-differentiation of myocytes to adipocytes or through co-culturing adipocytes with myocytes. Alternatively, lipids could be mixed with the muscle cells post-harvest by adding droplets of lipids or by adding another cultured product derived from adipocytes. However, there is currently little research on this and other quality aspects of cultured meat as a food product (Fraeye et al., 2020; Dohmen et al., 2022).

* Contributed equally to this chapter.

http://dx.doi.org/10.19103/AS.2023.0130.13

This chapter will address the most important quality aspects of conventionally produced meat to identify those elements that need to be introduced into the production of cultured meat products. Additives to control texture and taste are widely available in the food industry but this chapter will focus on exploiting inherent product characteristics and processes. The raw material (the tissue from the animal or cultured cells) is critical for the final quality of the food product and consumer acceptance.

When discussing the conversion of muscle cells in culture into food, compared to the postmortem conversion of muscle tissue into meat, several differences must be considered. In the case of animals, the processes from muscle to meat involve systemic events, like reactions in the outer layer of the muscle as well as in capillarisation within muscle tissue. Differences in tissue layers and capillarisation are elements that are not present *per se* in cultured meat, although some research into scaffolding seeks to mimic capillarisation using a hollow fibre reactor (Post et al., 2020). The diffusion of nutrients and waste products in and out of single muscle cells in the whole animal versus cells in culture could thus be different and ultimately affect the quality of the product. An example is the addition of the B12 vitamin to cultured meat; in animals, this is naturally produced in the rumen or from the environment and transported via the blood flow to muscle tissue.

While many features of the whole animal may be desirable, there are also negative elements we do not have to consider, e.g. stress which can cause certain systemic off-flavours. Animals can get stressed during transport and slaughter where systemic hormones can significantly affect cellular metabolism and thus meat quality (Støier et al., 2001). The same will not be the case for cultured meat as extracellular conditions can be controlled in the bioreactor. Another systemic element in the entire animal is microbial fermentation in the gut. An example is the fermentation-derived tryptophan metabolite skatole (3-methylindole), which is one of the major contributors to the undesirable boar-taint flavour in pork (Rasmussen and Zamaratskaia, 2014).

In a balanced diet, meat is the main contributor of lipids, high-quality protein and bioavailable iron as well as a highly desired umami flavour (Young et al., 2013). Meat derived from conventional farm animals consists of muscle, fat and connective tissue. The relative amounts vary with species, breed and the specific cut since the amount of intra-muscular fat and connective tissue differs between different carcass cuts. Many parameters including housing conditions, type of feed, stress levels, slaughter process and handling after slaughter determine the quality of the meat. From a consumer perspective, visual quality (and thus the decision to purchase) is determined by colour and visible intramuscular fat (IMF)/leanness, whilst eating quality is determined by characteristics such as aroma, tenderness, juiciness and flavour (Grunert, 1997).

In conventional slaughtered carcass muscle, nutrient and oxygen deprivation and temperature reduction initiate changes in pH and induce biochemical reactions which generate texture change and changes in taste and flavour components. The most important processes are hydrolysis of proteins, oxidation of lipids and other metabolic processes. These generate smaller molecules from which the characteristic meaty flavour appears after

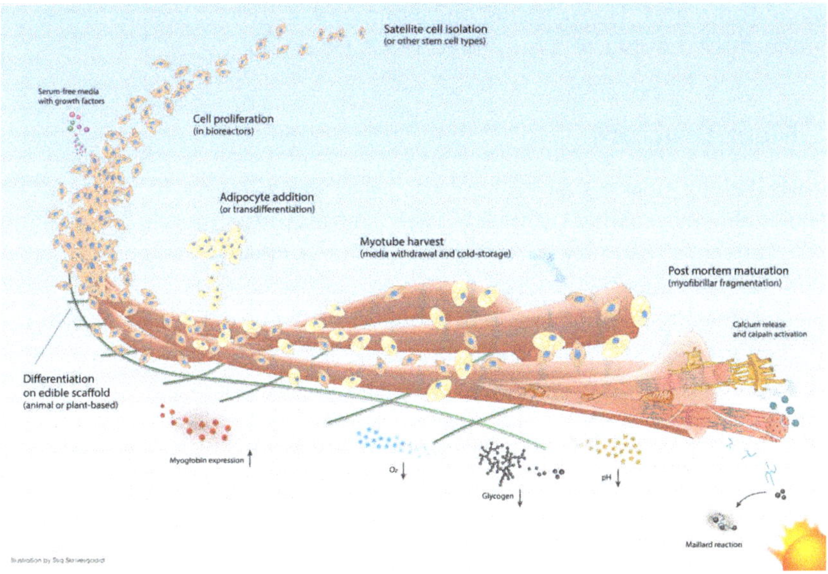

Figure 1 Illustration of the main processes from cell isolation to a matured cultivated meat product. After initial isolation of an animal muscle-derived satellite cell, or another stem cell type (e.g. ESCs and iPSCs) with myogenic capacity, the cells must undergo a high degree of proliferation in a large-scale bioreactor with a sustainable low-cost serum-free media. The media must contain all required nutrients, such as sugars, amino acids, lipids, minerals, metal ions and growth factors. After achieving high cell numbers, the cells can be seeded unto an edible 3D scaffold (also in a bioreactor setting) and begin their myogenic differentiation, fusion and formation of aligned myotubes. In this process, proteins such as myoglobin will be highly expressed, while highly ordered myofibrillar protein structures will be formed. At this step, adipocytes could also be introduced or the transdifferentiation of myogenic cells could be initiated to enhance downstream taste and nutrition through lipid accumulation. The formed myotubes within the scaffold will then be harvested and stored in a temperature-controlled environment. This withdrawal from oxygenated nutrient-rich media and decrease in temperature will set in motion key elements of canonical post-mortem maturation processes. Oxygen concentration will drop within the structures, as will glycogen when converted to glucose now being used in anaerobic respiration which in turn decreases the pH. As the cellular structures break down during these steps, calcium will be released mainly from the sarcoplasmic reticulum and activate the calpain system which will degrade the myofibrillar proteins and fragment the ordered structures. After maturation, the cultivated meat- product is processed and cooked, in which flavourful compounds can be created through protein-sugar Maillard-type reactions.

interactions and Maillard reactions initiated by hot temperatures during cooking (Dashdorj et al., 2015). From a technical and processing perspective, water-holding capacity and lipid texture are parameters affecting both juiciness and processability. Last, but not least, the product should contain not only the expected nutritional value but also be safe to consume.

Which of these elements are relevant when considering a completely different type of meat production such as cultured meat? To what extent can they be controlled? These are the questions we will address in this chapter. The complexity of these factors is summarised in Fig. 1.

2 Meat structure

Meat is skeletal muscle, consisting of approximately 90% muscle fibres and 10% fat and connective tissue (Listrat et al., 2016). An individual muscle is surrounded by a connective tissue layer (epimysium) which is inextensible and tough and holds multiple muscle bundles. A muscle bundle consists of 30–40 muscle fibres encased by another connective tissue layer (perimysium). Each individual muscle fibre is also encased by a connective tissue layer (endomysium) which is in proximity to the cellular membrane (sarcolemma) of the myofiber. These connective tissue structures give the raw or uncooked meat its texture (Astruc, 2014), whilst less abundant tissues like blood vessels, capillaries and nerves play a minor role. Collagen and elastin are the major proteins, with collagen being prominent in intramuscular connective tissues, acting as a scaffold for the muscle. Connective tissue has been shown to contribute to meat flavour, and meat from older animals has a more desirable meaty flavour due to higher glycosylation of collagen (Weir, 1992). The addition of beef collagen to a chicken product has been shown to have higher sensory acceptance than the addition of vegetable proteins (Prestes et al., 2013), highlighting the role of collagen in meat flavour.

A muscle fibre is multi-nucleated, encompassing myofibrils along with other organelles such as sarcoplasmic reticulum and mitochondria. Myofibrils have a series of sarcomeres which are the basic contractile unit of a muscle. Each sarcomere is formed by thin and thick filaments mainly consisting of the proteins actin and myosin, respectively. Titin is another filamentous protein that runs along the myofibril, supporting the structure of the sarcomeres (Guo and Greaser, 2017). Other major cytoskeletal proteins include filamin, vinculin, desmin, talin, synemin and dystrophin (Guo and Greaser, 2017). Overall, the firmness of a meat slice can be attributed to the degree of degradation of myofibrillar proteins and to the background toughness associated with collagen.

Researchers are looking into building tissue based on muscle cells and a lipid source, either as adipocytes, accumulated lipids in muscle cells or added after muscle cell culturing. Research is also considering connective tissue as an extracellular matrix in supporting muscle cell adhesion and signalling (though not as an important part of the edible commodity). The structure of cultured meat is determined by the muscle fibres which, as differentiated myotubes, contain myofibrillar proteins as well as other cytoskeletal proteins and organelles. However, myofibers are not encased by connective tissue nor are they organised and encased in larger muscle bundles. Collagen could form part of scaffolding for tissue, be added to the product or form part of a heterogeneous cell culture. Some fibroblasts produce collagen and elastin though age-related glycosylation does not occur in a cultured product.

As a result, tissue engineered from muscle cells and adipocytes will have a very loose structure and may have to be supported by scaffolds (e.g. plant based or synthetic) to mimic the texture of conventionally grown meat. This can happen either as an integral part of the culturing process as a subsequent addition or by crosslinking, e.g., with transglutaminase (Buchert et al., 2010). Scaffolds may not closely mimic connective tissue but, to meet consumer expectations of products other than minced meat, it will be important to ensure the correct rheological properties, mouthfeel and taste of the scaffold. The tenderness of cultured meat is not focused on the degradation of myofibrillar structures and solubility of connective tissue as for conventionally grown tissue but on introducing a firmer structure through choice of scaffolding. The major function of this scaffold is to align differentiating myocytes and create a three-dimensional framework for a highly structured myotube formation. However, depending on the degradability of the scaffolding material, the majority of the material will likely end up in the finished meat product, ultimately making it a hybrid product if the scaffold is non-animal derived. This makes it important to consider the organoleptic, nutritional and functional properties of the scaffold as a significant part of a final product. Scaffolding materials need to be edible, low cost and capable of large-scale production.

Compatible scaffold biomaterials could be of plant origin such as pectin, starch, cellulose and protein isolates from soy or zein or could be derived from algae or fungi, e.g. alginate and chitosan, respectively (Ng and Kurisawa, 2021). Animal-derived gelatin is also a possibility and, as a co-product from meat processing, supports the concept of a circular bioeconomy using side streams. However, the use of animal materials might affect marketing and consumer acceptance due to the 'animal-free' stance taken by most of the cultivated meat industry. Several functional properties of these materials are known, such as the stability and moisture of hydrogels (Akbar and Ahmed, 2018). However, future research needs to address their scaffolding capabilities, as well as their maturation (post-harvest degradation) and cooking behaviour in the context of cultivated meat.

A major part of water-holding capacity could still be maintained in cultured meat, as water-holding capacity and the characteristic juiciness of meat arise from the myofibrillar structure in which water is bound within the complex organisation of actin and myosin filaments. In a conventional piece of meat, approximately 85% of the water is held within the intra-myofibrillar space, while the remainder is held in the space between myofibrils, between the myofibrils and the sarcolemma and between muscle fibre bundles (Bowker, 2017). The proportion of water held in the intra-myofibrillar space depends on capillary forces created by myosin and actin filaments. As the muscle cells enter rigor mortis, the space between the myosin and actin filaments is compromised by a reduction in pH because of intra-cellular anaerobic metabolism. Due to the drop in pH and proximity to the isoelectric point of the major proteins myosin and actin, the repulsion between the filaments is reduced, lowering the amount of water that can be held within the myofibrillar space. During rigor mortis, the sarcomeres are shortened, also leading to loss of water. The water-holding capacity of meat can increase again during ageing as cytoskeletal proteins are degraded, which results in intracellular translocation of water into the partly fragmented myofibrillar structure (Kristensen and Purslow, 2001). In relation to cultured meat, this emphasises the importance of ensuring the differentiation of myocytes into fused myotubes with a distinct myofibrillar structure, controlling pH decline post-harvest as well as ageing conditions this is critical to the juiciness of the final product.

3 Appearance

Consumers associate the colour of meat with wholesomeness which means colour has an important effect on purchasing decisions (Mancini and Hunt, 2005). Muscle ultrastructure has an impact on the appearance of fresh meat as it modulates the way light is absorbed or scattered. The amount and composition of fat can also affect appearance since more IMF makes the meat appear lighter. The iron-containing heme-protein myoglobin is the major contributor to meat colour, along with other heme-proteins such as cytochrome C and haemoglobin. Meat from species with a higher myoglobin content such as beef will appear redder than pork or chicken, which has a lower amount of myoglobin.

Proliferating *in-vitro* muscle cells produce lower quantities of myoglobin compared to *in-vivo* cells, meaning total myoglobin content is low in cultured muscle cells (Graber and Woodworth, 1986).

However, myoglobin is upregulated during differentiation (Skrivergaard et al., 2021; Kanatous and Mammen, 2010). It is therefore important to ensure available iron in cultured cell media to support final colour development. Simsa et al. (2019) reported that bioartificial muscle supplemented with heme proteins during proliferation more closely simulated the colour of beef compared to the cultured muscle without heme-protein supplementation. These findings emphasise the importance of heme proteins. It may be challenging in cultured meat production to meet consumer expectations of red meat since it is potentially difficult to match the colour of the conventional counterpart.

Various redox forms of myoglobin have a significant impact on the colour of fresh and cooked meat. The iron atom of myoglobin can exist in three different redox forms, depending on oxidation status and the ligand to which iron is bound. When the iron is in the reduced form, and with no ligands, it exists as deoxymyoglobin which is purple-red in colour. When oxygen is bound to the iron, it becomes oxymyoglobin, characterised by the bright cherry red colour of fresh meat. The iron can become oxidised and turn into a brown metmyoglobin form which is less appealing to the consumer. During cooking, denaturation will result in the unfolding of myoglobin, which makes it oxidise easily and contributes to the cooked colour of meat. In cultured meat products, it may be possible to alter the form of the myoglobin in the harvested product.

After slaughter, when blood is drained from the carcass of the animal, oxygen is lost and metabolism becomes anaerobic, resulting in acidification. The speed at which pH declines after slaughter and the ultimate pH of the carcass is a result of the muscle glycogen content and the cooling profile of the carcass. If muscle glycogen is too low, meat may end up as dark firm and dry (DFD). High muscle glycogen, combined with high-temperature post-mortem, may cause pale soft and exudative (PSE) meat. Both DFD and PSE are undesirable meat qualities (Gonzalez-Rivas et al., 2020). This means that pH decline is of major importance for the final quality and functionality of a cultured meat product.

Once cultured meat is produced in a bioreactor, it should be harvested in a controlled manner. Intracellular glycogen content can be controlled through the glucose concentration in the culturing medium prior to harvest (Skurat et al., 2017). Oxygen level and temperature can be controlled in the bioreactor to manage intracellular pH decline in the product. These aspects will be important future research topics for ensuring a cultured meat product of high quality.

4 Flavour

Flavour refers to a sensory perception of food based on both taste and smell. Taste components of the meat directly affect taste buds in the mouth and tongue. Flavour components (pre-coursers) are volatiles and are only released when the meat is heated during the cooking process. The overall flavour from meat is the result of lipid oxidation and the Maillard reaction during heating (Aaslyng and Meinert, 2017) when free fatty acids, peptides and free amino acids, reducing sugars, nucleotides and vitamins are available (Dashdorj et al.,

2015). The flavour of conventional meat develops during ageing because of the release of free fatty acids, peptides/free amino acids and reducing carbohydrates as a result of lipolysis, proteolysis and glycolysis induced by endogenous enzymes in the muscle tissue during the ageing period (Gorraiz et al., 2002; Dashdorj et al., 2015).

4.1 Protein degradation/maturation

Protein degradation is primarily induced by endogenous proteolytic enzyme processes, breaking down both sarcoplasmic and myofibrillar proteins in muscle tissue. The longer the ageing time, the more peptides are released (Fu et al., 2017). The initial degradation of structural proteins plays a major role in the texture of meat, whilst further proteolysis of protein fragments and peptides contributes to flavour generation. The breakdown of proteins occurs in a sequential way (Houbak et al., 2008). The calpain system initiates the process of protein degradation, followed by other proteolytic enzymes (such as the proteasome complex which is responsible for hydrolysis) through to smaller peptides. These peptides are then degraded to free amino acids by tri-, di and aminopeptidases.

The calpain system involved in proteolytic degradation post-mortem involves the ubiquitous iso-forms calpain-1 and calpain-2 and possibly the muscle-specific iso-form calpain-3, together with their inhibitor calpastatin. Calpain-1 and -2 are both calcium dependent with a calcium requirement of half-maximal activity of calpain of 3–50 μM and 400–800 μM, respectively (Goll et al., 2003). However, this concentration is much higher than the physiological calcium concentration of 50–300 nM in the sarcoplasm of *in-vivo* muscle fibres (Goll et al., 2003). For the calpain system to be active in the initial degradation of key myofibrillar proteins post-mortem, the calcium concentration needs to increase, at least locally. An increase in calcium concentration occurs due to physiological changes in the tissue post-mortem caused by temperature and oxygen reduction, resulting in decreased pH. The sarcoplasmic reticulum and mitochondria become less stable and eventually calcium leaks out in the sarcoplasm (Cornforth et al., 1980; Ji and Takahashi, 2006; Whiting, 1980). Recently, Lyu and Ertbjerg (2022) demonstrated that calpains gradually translocate from the sarcoplasm to the myofibrils post-mortem, coinciding with the degradation of structural proteins. When calpains are bound to myofibrils, calpastatin is unable to interact with these enzymes and inhibits their activity (Lyu and Ertbjerg, 2022).

Studies on porcine satellite cells have shown that calpain-1, -2 and -3 and calpastatin are expressed during proliferation, fusion and differentiation (Theil et al., 2006). Similar results are seen in bovine satellite cells (Yang et al., 2012;

Wang et al., 2017). Calpain-1, calpain-3 and calpastatin m-RNA expression increases in the days following differentiation (Theil et al., 2006), suggesting that these enzymes are expressed in the myotubes and may play a role during maturation. The increased expression of both calpain-1 and calpastatin also suggests that a similar mechanism occurs in conventional muscle tissue post-mortem. Translocation to myofibrils is necessary for the enzymes to be proteolytically active and not inhibited by calpastatin.

Upon conventional slaughter of animals, a situation similar to apoptosis arises due to oxygen depletion, pH decrease and calcium release into the cytosol of the muscle fibre (Park et al., 2010). This activates the proteolytic caspases, leading to the release of protein fragments (Laville et al., 2009; Kemp and Wheeler, 2011), which can act as substrates for other proteolytic enzymes. A similar situation as apoptosis can be imagined in harvesting myotubes, as they will most probably be exposed to starvation (removal of nutrients) and hypoxia. Members of the caspase family increase during myogenesis. Caspases 7 and 12 were found to be involved in apoptotic cell death under hypoxia conditions in bovine satellite cells (Yang et al., 2012).

4.2 Lipid incorporation and content

In conventional meat cuts, lipids are generally present in three ways: subcutaneous, intermuscular and intramuscular. Variations between different breeds, age of the animal, sex (incl. castration), feeding amount and feed composition generally explain variations in the content and composition of fat in conventional meat (Therkildsen et al., 2017; Park et al., 2018). Lipids appearing as IMF play a crucial role in the flavour of meat, emphasising the importance of including fat in a cultured meat product to secure the desired flavour. IMF represents both fats between muscle fibres and within muscle fibres, whereas, in cultured muscle, it would only be intra-myocellular fat (IMCF) that contributes to the flavour in muscle cell cultures.

Some of the factors affecting IMF in conventional meat production may also be relevant for IMCF content and composition of cultured meat. The effect of sex and breed on IMF may to some extent be maintained in *in-vitro* cultured meat. It has recently been demonstrated, e.g., that satellite cells from different donor breeds maintain growth and differentiation characteristics *in vitro* (Melzener et al., 2022). It remains to be explored if the same is true for the ability to accumulate IMCF. This indicates a potential donor effect in relation to cellular genetics. Some of the effect of sex and breeding *in vivo* is also related to the presence of steroid hormones. Female and castrated animals display larger IMF content compared to non-castrated male animals. In cultured meat production, favourable conditions for IMCF accumulation without specific steroid hormones can easily be obtained in other ways.

The fatty acid profile of feed is to some extent reflected in the lipid stores of animals as reviewed for both ruminants (Arshad et al., 2018) and monogastric animals (Young et al., 2003). Generally, manipulation of the fatty acid profile of the meat through feeding regimes is more efficient for monogastric animals. In the case of cultured meat, the cellular incorporation of fatty acids is reflected by culture media composition (Belal et al., 2019). However, the turnover rate and desired composition need investigation. Examples of preferred flavours are the increased content of oleic acid from grain-fed ruminant meat, which is denoted a positive 'ruminant fat', as opposed to increased linoleic acid in forage which can cause a 'fishy' and 'grassy' off-flavour (Melton et al., 1982; Campo et al., 2003; Therkildsen et al., 2017). Feed-related fatty acids co-occur with other components in feed, the effect of which on flavour is less well investigated.

Whether the impact of fatty acid composition on conventional meat taste is similar to that of the fatty acid composition of cultured cells remains to be explored. Cultured meat does not contain the equivalent of subcutaneous fat, but this is often not requested by the consumer. IMF causing 'marbling' in conventional meat is important for mouthfeel and juiciness and could be part of the cultured product as co-cultured adipocytes or as lipids added after muscle culturing. However, fat may also be incorporated into cells of the cultured product. The addition of fatty acids to media is seen as a relatively easy way to provide satellite cells with more substrate to increase IMCF content.

Previous studies have shown that adding oleic or palmitoleic acid to media increases the accumulation of lipid droplets in bovine satellite cultures (Li et al., 2019; Zhang et al., 2021). However, the addition of large quantities of fatty acids to cell media complicates the maintenance of pH. This can be solved by conjugating fatty acids [e.g. bovine serum albumin (BSA)] which also facilitates uptake. To further improve IMCF content, satellite cell cultures can be stimulated to increase the uptake and accumulation of lipids. By activating the transcription factors peroxisome proliferator-activated receptor gamma (PPARγ) and CCAAT/enhancer binding protein-dependent pathways in muscle cell models, IMCF content can be increased (Park et al., 2018).

The way in which lipid is incorporated affects the consumed product as the cooking process will lead to a loss of subcutaneous and inter-muscular fat from traditional meat (Gerber et al., 2009). For cultured meat, this may be the case if a lipid source is added or mixed in after the culturing stage. If fat is deposited as IMCF, this will be retained after cooking.

4.3 Flavour development from Maillard reactions and minor components

Animal feeding regimes affect the incorporation of nutrients as well as post-mortem meat metabolism, which then affect quality, as reviewed by Andersen et al. (2005). As discussed in Section 3.2, lipids are incorporated into cells over time, whilst carbohydrate storage in the form of glycogen is more dynamic. Sugar feeding and energy-consuming stress of animals before slaughter affects their glycogen stores (Rosenvold et al., 2001) and hence acidification potential post-mortem with an impact on pH decrease and meat quality. When culturing cells, it is possible to control glycogen load in cells at harvest (Skurat et al., 2017). Post-harvest conditions will then determine how much is converted to energy through metabolism and how much will remain as carbohydrates.

As reviewed by Khan et al. (2015) and Arshad et al. (2018), flavour develops partly from lipid oxidation and partly from precursors in muscle cells when reducing carbohydrates in the muscle cell interact with free amino acids/peptides after protein degradation. When cooking the product under high temperatures, the reducing carbohydrates and amino acids interact. Complex Maillard reactions are initiated, resulting in the formation of e.g. pyrroles, furans, carbonyls and pyrazines (Jousse et al., 2002). Lipid oxidation and thiamine degradation products also contribute flavour components and interact with Maillard reaction products, giving rise to the complex flavour notes which create a meaty flavour (Dashdorj et al., 2015). Myoglobin contributes to the characteristic metallic taste of meat due to heme content. Myoglobin has been explored as a flavour additive to plant-based meat analogues, albeit produced recombinantly from soy leghaemoglobin protein (https://www.canada.ca/en/health-canada/services/food-nutrition/genetically-modified-foods- other-novel-foods/approved-products/soy-leghemoglobin/document.html).

As reviewed by Dashdorj et al. (2015), many minor components contribute flavour notes. Hypoxanthine, anserine, carnosine and some specific amino acids contribute bitterness, while other amino and organic acids contribute sweetness and sourness. The 'Umami' taste is described as savoury, brothy or beefy. It originates from components such as 5-inosine monophosphate, 5-guanosine monophosphate and the amino acids aspartic acid and glutamic acid (Dashdorj et al., 2015). Minor components and precursors for Maillard reactions can be present in muscle cells and hence develop in the product during maturation and heat treatment. To obtain the best possible meaty flavour, uptake from the culture medium is critical for some components. For others, the maturation process is critical for flavour development.

5 Nutritional quality

Meat is a nutrient-dense food with a highly variable lipid content, with the subcutaneous fraction varying particularly between breeds and feeding status (Therkildsen et al., 2017; Park et al., 2018). The lean part consists of approximately 20–25% protein (Williams, 2007), around 70% water, 5% lipids, 1% ash and 1% carbohydrates (Kauffman, 2012). As meat is relatively low in carbohydrates, it has a low glycaemic index (Biesalski and Nohr, 2009). However, glycogen content is more important for structure (Section 2) and flavour (Section 4.3). The composition of micronutrients (e.g. vitamins A, E and B12 and iron) in meat can vary depending on the type of muscle, age, breed, sex, type of feed and the production system.

5.1 Protein

Meat proteins have a high Protein Digestibility Corrected Amino Acid Score (PDCAAS) (Schaafsma, 2000) of 0.92 (Pereira and Vicente, 2013) and are a source of all essential amino acids as well as some non-essential amino acids (Ahmad et al., 2018). Cultured meat is often framed as an alternative source of protein alongside plant-based meat alternatives. However, proteins from cultured meat are expected to be comparable to conventional meat in terms of biological value, content of essential amino acids and high protein bioavailability. As many cultured meat products will contain less connective tissue and less crosslinking than tissue from conventional production, bioavailability may be higher but protein content at a cellular level may be less on a weight basis if the myofibril content of cells is less dense. When conventional meat is at the consumption stage, the protein percentage will be around 30% (Lofgren, 2013) due to water and potentially some lipid loss.

5.2 Lipids

Meat is a contributor of fatty acids to the diet. Total fat content can vary depending on the type of muscle and degree of marbling (Reig et al., 2013). Ruminant meat has higher amounts of saturated fatty acids (more than 40% of fatty acids) compared to pork with less in chicken (Wood and Enser, 2017). Polyunsaturated fatty acids from animal feed are converted to monounsaturated fatty acids and saturated fatty acids in the rumen. Stearic and palmitic acid forms the bulk of saturated fatty acids situated mainly in the subcutaneous fraction (Wood et al., 1989). Ruminant meat also contains trans-fatty acids (conjugated linoleic acid) formed by ruminant microbes (Wood, 2017).

The fatty acid composition can be modified by using different feed ingredients, e.g. fish oils or algae, to increase eicosapentaenoic acids (EPAs) or docosapentaenoic acids (DHAs). Feed containing high amounts of saturated fats such as tallow increase palmitic acids and stearic oleic acids (Juárez et al., 2021).

The subcutaneous, saturated fat segment of conventional cuts is not relevant when culturing meat, whereas intramuscular and more unsaturated lipids are desirable. Fatty acid profiles of cellular lipids are more closely affected by media composition than through feed composition in conventional production. However, the extent to which content and composition can be controlled in cells requires further study.

5.3 Bioactive components, vitamins and minerals

Meat has been classified as a functional food because of naturally occurring compounds that can improve physiological functions (Gehring, 2017). Anserine and carnitine not only promote wound healing, recovery from fatigue and stress relief but also function as antioxidants together with glutathione and ubiquinone. Conjugated linoleic acid (CLA) has been found to promote weight loss and improve cardiovascular health (Gehring, 2017). Taurine improves eye health, reduces heart diseases and promotes cognitive function. Meat contains micronutrients such as copper, phosphorous, magnesium, selenium, zinc and iron (Lombardi-Boccia et al., 2013; Bohrer, 2017) in a highly bioavailable form (Tetens et al., 2007; Bohrer, 2017). Meat is a valuable source of many B vitamins such as thiamine, riboflavin, niacin (Lombardi-Boccia et al., 2013) and vitamin B12 (Bohrer, 2017).

The high bioavailability of heme-iron from conventional meat is a valuable nutrient in a mixed diet which should be mimicked in cultured meat production. As mentioned in Sections 3 and 4.3, myoglobin is important for the appearance and taste of the meat. The amount and bioavailability of iron from cultured meat may be improved through culture medium composition. Vitamin B12 is another potential limiting nutritional factor in a diet with less animal-based products. It is important to investigate how to incorporate B12 in cultured meat via the culture medium. B vitamins and minerals should be present in cultured meat as part of muscle metabolism but the amounts need to be investigated in comparison to conventional meat.

6 Microbial safety

Conventional muscle/meat is theoretically sterile until slaughter. At this point skin is disrupted, allowing microorganisms access to the highly nutritious substrate. Increasing efforts are made to reduce contamination and growth of microorganisms during slaughter and immediate post-slaughter stages by reducing contact with contaminating agents, scalding surfaces and reducing the temperature in the cutting area. However, there will always be an endogenous flora on the cut. Cultured meat is produced in bioreactors and is sterile until it leaves the bioreactor. After this point, care should be taken to avoid contamination and maintain a low temperature (e.g. under 5°C). However, a sterile product is also at elevated risk because there is no endogenous flora to compete with a potential pathogen. It is important to avoid both pathogenic and spoilage microorganisms, although some microorganisms can positively contribute to flavour development (e.g. through lipase secretion). This line of research will become important once commercial production lines are established. Since production conditions are different, there will need to be dedicated safety legislation for cultured meat. Safety may include the addition of protective or even flavour-contributing (starter) cultures commonly used in the meat industry. From a comprehensive review of preservation principles (Elsser-Gravesen and Elsser-Gravesen, 2014) it should be possible to develop appropriate safety strategies for individual products.

7 Key research questions

As production processes for cultured meat are relatively well established, product quality should become more of a priority. From our perspective, the most pressing research question for the scientific community to address is how to achieve high nutritional and sensory quality. Research needs to focus on the incorporation of specific nutrients into myotubes and how they can be manipulated during the production process, e.g. when to introduce PUFA-like EPA and DHA and B12 in media to ensure optimum content in the final product. Sensory quality is very dependent on the incorporation of fat and post-harvest changes in myotubes. Research needs to establish the right balance between fat and myofibrillar proteins during production processes such as trans- differentiation of myocytes or co-culture of adipocytes. There is a need to understand the changes needed in the myotubes during processing and post-harvest which lead to pH decrease, fragmented proteins, peptides and free amino acids important for flavour development and texture. The process requires nutrient removal, oxygen reduction and decreased temperature. However, the timing of these steps is critical and requires research and development for upscaling.

8 Conclusion

Cultured meat is a food product. Concomitant development must include all facets for consumer acceptance - appearance, taste, nutrition and safety. A wealth of knowledge to tap into from research on conventional meat products regarding colour, texture and flavour development exists. However, this new production concept calls for multidisciplinary fundamental knowledge generation on how to meet the nutritional expectations and required safety, and not least provide evidence to support new legislation. It is pivotal for consumer acceptance of cultured meat that the product quality is high and highly prioritised prior to market entrance.

9 Acknowledgement

The authors acknowledge the grants that supported the work for this chapter: The Danish Agricultural Agency (33010-NIFA-19-716), Ministry of Food, Agriculture and Fisheries of Denmark, The Graduate School of Technical Sciences (GSTS), Centre of Innovative Food Research (CiFOOD), Aarhus University Research Fund (AUFF- E-2020-9-20), Aarhus University, Denmark. Affiliated Associate Professor Niall W. G. Young, Aarhus University, is acknowledged for comments to the chapter.

10 Where to look for further information

Cultured meat product quality is a research area in ascendance. Scientific literature is scarce but will surely increase. The environments discussing and picking up news and trends on cultured meat in general are NGOs such as "New Harvest" https://new-harvest.org/, "Good Food Institute" https://gfi.org/ and local and regional branches of "Cellular Agriculture" (e.g. https://www.cellag.org/ and Australia, Canada, Europe, UK).

11 References

Aaslyng, M. D. and Meinert, L. 2017. Meat flavour in pork and beef – from animal to meal. *Meat Science* 132, 112–117.

Ahmad, R. S., Imran, A. and Hussain, M. B. 2018. Nutritional composition of meat. In Arshad, M.S. (ed.), *Meat Science and Nutrition*. London: IntechOpen.

Akbar, A. and Ahmed, S. 2018. Development of hydrogels from edible polymers. *Polymers for Food Applications*, January, 551–589.

Andersen, H. J., Oksbjerg, N., Young, J. F. and Therkildsen, M. 2005. Feeding and meat quality – a future approach. *Meat Science* 70(3), 543–554.

Arshad, M. S., Sohaib, M., Ahmad, R. S., Nadeem, M. T., Imran, A., Arshad, M. U., Kwon, J. H. and Amjad, Z. 2018. Ruminant meat flavor influenced by different factors with special reference to fatty acids. *Lipids in Health and Disease* 17(1), 223.

Astruc, T. 2014. Connective tissue: structure, function, and influence on meat quality. In: Dikeman, M., and Devine, C. (Eds.), *Encyclopedia of Meat Sciences* (2nd edn.). Oxford: Academic Press.

Belal, S. A., Kang, D. R., Sivakumar, A. S., Choe, H. S. and Shim, K. S. 2019. Effect of long chain fatty acids on triacylglycerol accumulation, fatty acid composition and related gene expression in primary cultured bovine satellite cells. *Animal Biotechnology* 30(4), 323–331. doi: 10.1080/10495398.2018.1496925.

Biesalski, H. K. and Nohr, D. 2009. The nutritional quality of meat. In: Kerry, J. P. and Ledward, D. (Eds). *Improving the Sensory and Nutritional Quality of Fresh Meat*. Cambridge: Woodhead Publishing Ltd., 161–177.

Bohrer, B. M. 2017. Review: Nutritent density and nutritional value of meat products and non-meat foods high in protein. *Trends in Food Science and Technology* 65, 103–112.

Bowker, B. 2017. Chapter 4. Developments in our understanding of water-holding capacity. In: Petracci, M. and Berri, C. (Eds). *Poultry Quality Evaluation*. Woodhead Publishing.

Buchert, J., Cura, D. E., Ma, H., Gasparetti, C., Monogioudi, E., Faccio, G., Mattinen, M., Boer, H., Partanen, R., Selinheimo, E., Lantto, R. and Kruus, K. 2010. Crosslinking food proteins for improved functionality. *Annual Review of Food Science and Technology*, 1, 113–138.

Campo, M. M., Nute, G. R., Wood, J. D., Elmore, S. J., Mottram, D. S. and Enser, M. 2003. Modelling the effect of fatty acids in odour development of cooked meat in vitro: part I—sensory perception. *Meat Science* 63(3), 367–375.

Cornforth, D. P., Pearson, A. M. and Merkel, R. A. 1980. Relationship of mitochondria and sarcoplasmic reticulum to cold shortening. *Meat Science* 4(2), 103–121.

Dashdorj, D., Amna, T. and Hwang, I. 2015. Influence of specific taste-active components on meat flavor as affected by intrinsic and extrinsic factors: an overview. *European Food Research and Technology* 241(2), 157–171.

Dohmen, R. G. J., Hubalek, S., Melke, J., Messmer, T., Cantoni, F., Mei, A., Hueber, R., Mitic, R., Remmers, D., Moutsatsou, P., Post, M. J., Jackisch, L. and Flack, J. E. 2022. Muscle-derived fibro-adipogenic progenitor cells for production of cultured bovine adipose tissue. *NPJ Science of Food* 6(1), 6.

Elsser-Gravesen, D. and Elsser-Gravesen, A. 2014. Biopreservatives. *Advances in Biochemistry and Engineering Biotechnology* 143, 29–49.

Fraeye, I., Kratka, M., Vandenburgh, H. and Thorrez, L. 2020. Sensorial and nutritional aspects of cultured meat in comparison to traditional meat: much to be inferred. *Frontiers in Nutrition* 7, 35. doi: 10.3389/fnut.2020.00035.

Fu, Y., Young, J. F. and Therkildsen, M. 2017. Bioactive peptides in beef: endogenous generation through postmortem aging. *Meat Science* 123, 134–142.

Gehring, K. 2017. Chapter 21. Meat and health. In: Toldrà, F. (Ed.), *Lawrie's Meat Science. Woodhead Publishing Series in Food Science, Technology and Nutrition*, 661–678. https://www.sciencedirect.com/book/9780081006948/lawries-meat-science#book-description.

Gerber, N., Scheeder, M. R. L. and Wenk, C. 2009. The influence of cooking and fat trimming on the actual nutrient intake from meat. *Meat Science* 81(1), 148–154.

Goll, D. E., Thompson, V. F., Li, H., Wei, W. and Cong, J. 2003. The calpain system. *Physiological Reviews* 83(3), 731–801.

Gonzalez-Rivas, P. A., Chauhan, S. S., Ha, M., Fegan, N., Dunshea, F. R. and Warner, R. D. 2020. Effects of heat stress on animal physiology, metabolism, and meat quality: a review. *Meat Science* 162, 108025.

Gorraiz, C., Beriain, M. J., Chasco, J. and Insausti, K. 2002. Effect of aging time on volatile compounds, odor, and flavor of cooked beef from pirenaica and Friesian bulls and heifers. *Journal of Food Science* 67(3), 916–922.

Graber, S. G. and Woodworth, R. C. 1986. Myoglobin expression in L6 muscle cells. Role of differentiation and heme. *Journal of Biological Chemistry* 261(20), 9150–9154.

Grunert, K. G. 1997. What's in a steak? A cross-cultural study on the quality perception of beef. *Food Quality and Preference* 8(3), 157–174.

Guo, W. and Greaser, M. L. 2017. Chapter 2. Muscle structure, proteins, and meat quality. In: Purslow, P. P. (Ed.), *New Aspects of Meat Quality*. Woodhead Publishing.

Houbak, M. B., Ertbjerg, P. and Therkildsen, M. 2008. In vitro study to evaluate the degradation of bovine muscle proteins post-mortem by proteasome and μ-calpain. *Meat Science* 79(1), 77–85.

Ji, J. R. and Takahashi, K. 2006. Changes in concentration of sarcoplasmic free calcium during post- mortem ageing of meat. *Meat Science* 73(3), 395–403.

Jousse, F., Jongen, T., Agterof, W., Russell, S. and Braat, P. 2002. Simplified kinetic scheme of flavor formation by the Maillard reaction. *Journal of Food Science* 67(7), 2534–2542.

Juárez, M., Lam, S., Bohrer, B. M., Dugan, M. E. R., Vahmani, P., Aalhus, J., Juárez, A., López-Campos, O., Prieto, N. and Segura, J. 2021. Enhancing the nutritional value of red meat through genetic and feeding strategies. *Foods (Basel, Switzerland)* 10(4), 872.

Kanatous, S. B. and Mammen, P. P. A. 2010. Regulation of myoglobin expression. *The Journal of Experimental Biology* 213(16), 2741-2747.

Kauffman, R. G. 2012. *Meat Composition from: Handbook of Meat and Meat Processing*. CRC Press, 45–61.

Kemp, C. M. and Wheeler, T. L. 2011. Effects of manipulation of the caspase system on myofibrillar protein degradation in vitro. *Journal of Animal Science* 89(10), 3262-3271.

Khan, M. I., Jo, C. and Tariq, M. R. 2015. Meat flavor precursors and factors influencing flavor precursors--A systematic review. *Meat Science* 110, 278–284.

Kristensen, L. and Purslow, P. P. 2001. The effect of ageing on the water-holding capacity of pork: role of cytoskeletal proteins. *Meat Science* 58(1), 17–23.

Laville, E., Sayd, T., Morzel, M., Blinet, S., Chambon, C., Lepetit, J., Renand, G. and Hocquette, J. F. 2009. Proteome changes during meat aging in tough and tender beef suggest the importance of apoptosis and protein solubility for beef aging and tenderization. *Journal of Agricultural and Food Chemistry* 57(22), 10755–10764.

Li, X. Z., Yan, Y., Zhang, J. F., Sun, J. F., Sun, B., Yan, C. G., Choi, S. H., Johnson, B. J., Kim, J. K. and Smith, S. B. 2019. Oleic acid in the absence of a PPARγ agonist increases adipogenic gene expression in bovine muscle satellite cells1. *Journal of Animal Science* 97(10), 4114–4123.

Listrat, A., Lebret, B., Louveau, I., Astruc, T., Bonnet, M., Lefaucheur, L., Picard, B. and Bugeon, J. 2016. How muscle structure and composition influence meat and flesh quality. *TheScientificWorldJournal* 2016, 3182746.

Lofgren, P. A. 2013. Meat, poultry, and meat products: nutritional value. In: Caballero, B. (Ed.), *Encyclopedia of Human Nutrition* (3rd edn.). Waltham: Academic Press.

Lombardi-Boccia, G., Lanzi, S., Lucarinbi, M. and Di Lullo, G. 2013. Meat and Meat Products Consumption in Italy: contribution to Trace Elements, heme Iron and selected B Vitamins supply. *International Journal for Vitamin and Nutrition Research* 74, 4.

Lyu, J. and Ertbjerg, P. 2022. Sarcoplasmic and myofibril-bound calpains during storage of pork longissimus muscle: new insights on protein degradation. *Food Chemistry* 372, 131347.

Mancini, R. A. and Hunt, M. C. 2005. Current research in meat color. *Meat Science* 71(1), 100-121.

Melton, S. L., Black, J. M., Davis, G. W. and Backus, W. R. 1982. Flavor and selected chemical components of ground beef from steers backgrounded on pasture and fed corn up to 140 days. *Journal of Food Science* 47(3), 699–704.

Melzener, L., Ding, S., Hueber, R., Messmer, T., Zhou, G., Post, M. J. and Flack, J. E. 2022. Comparative analysis of cattle breeds as satellite cell donors for cultured beef. *bioRxiv*, 2022.01.14.476358.

Ng, S. and Kurisawa, M. 2021. Integrating biomaterials and food Biopolymers for cultured meat production. *Acta Biomaterialia* 124, 108–129. doi: 10.1016/j.actbio.2021.01.017.

Park, S. J., Beak, S. H., Jung, D. J. S., Kim, S. Y., Jeong, I. H., Piao, M. Y., Kang, H. J. and Fassah, D. M., Na, Park, S. J., Beak, S. H., Jung, D. J. S., Kim, S. Y., Jeong, I. H., Piao, M. Y., Kang, H. J., Fassah, D. M., Na, S. W., Yoo, S. P. and Baik, M. 2018. Genetic, management, and nutritional factors affecting intramuscular fat deposition in beef cattle - a review. *Asian-Australasian Journal of Animal Sciences* 31(7), 1043–1061.

Park, S. Y., Kim, H. Y., Lee, J. H., Yoon, K. H., Chang, M. S. and Park, S. K. 2010. The age-dependent induction of apoptosis-inducing factor (AIF) in the human semitendinosus skeletal muscle. *Cellular and Molecular Biology Letters* 15(1), 1–12.

Pereira, P. M. and Vicente, A. F. 2013. Meat nutritional composition and nutritive role in the human diet. *Meat Science* 93(3), 586–592.

Post, M. J., Levenberg, S., Kaplan, D. L., Genovese, N., Fu, J., Negowetti, C. N., Verzijden, K. and Moutsasou, P. 2020. Sceinetific, sustainability and regularory challenges of cultured meat. *Nature Food* 1, 403–415.

Prestes, R. C., Graboski, A., Roman, S. S., Kempka, A. P., Toniazzo, G., Demiate, I. M. and Di Luccio, M. 2013. Effects of the addition of collagen and degree of comminution in the quality of chicken ham. *Journal of Applied Poultry Research* 22(4), 885–903.

Rasmussen, M. K. and Zamaratskaia, G. 2014. Regulation of porcine hepatic cytochrome P450 – implication for boar taint. *Computational and Structural Biotechnology Journal* 11(19), 106–112.

Reig, M., Aristoy, M. C. and Toldrá, F. 2013. Variability in the contents of pork meat nutrients and how it may affect food composition databases. *Food Chemistry* 140(3), 478–482.

Rosenvold, K., Lærke, H. N., Jensen, S. K., Karlsson, A. H., Lundström, K. and Andersen, H. J. 2001. Strategic finishing feeding as a tool in the control of pork quality. *Meat Science* 59(4), 397–406.

Schaafsma, G. 2000. The protein digestibility-corrected amino acid score. *Journal of Nutrition* 130(7), 1865S–1867S. doi: 10.1093/jn/130.7.1865S.

Simsa, R., Yuen, J., Stout, A., Rubio, N., Fogelstrand, P. and Kaplan, D. L. 2019. Extracellular heme proteins influence bovine myosatellite cell proliferation and the color of cell-based meat. *Foods* 8(10) doi: 10.3390/foods8100521.

Skrivergaard, S., Rasmussen, M. K., Therkildsen, M. and Young, J. F. 2021. Bovine satellite cells isolated after 2 and 5 days of tissue storage maintain the proliferative and myogenic capacity needed for cultured meat production. *International Journal of Molecular Sciences* 22(16), 8376.

Skurat, A. V., Segvich, D. M., DePaoli-Roach, A. A. and Roach, P. J. 2017. Novel method for detection of glycogen in cells. *Glycobiology* 27(5), 416–424.

Støier, S., Aaslyng, M. D., Olsen, E. V. and Henckel, P. 2001. The effect of stress during lairage and stunning on muscle metabolism and drip loss in Danish pork. *Meat Science* 59(2), 127–131.

Tetens, I., Bendtsen, K. M., Henriksen, M., Ersboll, A. K. and Milman, N. 2007. The impact of a meat- versus a vegetable-based diet on iron status in women of childbearing age with small iron stores. *European Journal of Nutrition* 46(8), 439–445.

Theil, P. K., Sørensen, I. L., Therkildsen, M. and Oksbjerg, N. 2006. Changes in proteolytic enzyme mRNAs relevant for meat quality during myogenesis of primary porcine satellite cells. *Meat Science* 73(2), 335–343.

Therkildsen, M., Spleth, P., Lange, E.-M. and Hedelund, P. I. 2017. The flavor of high-quality beef - a review. *Acta Agriculturae Scandinavica, Section A – Animal Science* 67(3–4), 85–95.

Wang, W., Yang, Y. B., Ma, X. Y., Yu, X. L. and Hwang, I. 2017. Changes in calpain and caspase gene expression at the mRNA level during bovine muscle satellite cell myogenesis and the correlation between the cell model and the muscle tissue. *Russian Journal of Bioorganic Chemistry* 43(3), 270–277.

Weir, G. S. D. 1992. Proteins as a source of flavour. In: Hudson, B. J. F. (Ed.). *Biochemistry of Food Proteins*. Boston, MA: Springer US.

Whiting, R. C. 1980. Calcium uptake by bovine muscle mitochondria and sarcoplasmic reticulum. *Journal of Food Science* 45(2), 288–292.

Williams, P. 2007. Nutritional composition of red meat. *Nutrition and Dietetics* 64(s4), S113–S119.

Wood, J. D. 2017. Chapter 20. Meat composition and nutritional value. In: Toldrà, F. (Ed.). *Lawrie's Meat Science. Woodhead Publishing Series in Food Science, Technology and Nutrition*, 661–678.

Wood, J. D. and Enser, M. 2017. Manipulating the fatty acid composition of meat to improve nutritional value and meat quality. In: Purslow, P. (Eds.), *New Aspects of Meat Quality - from Genes to Ethics*, 1st edition. Woodhead Publishing, GB (vol. 2017), 501–535.

Wood, J. D., Enser, M., Whittington, F. M., Moncrieff, C. B. and Kempster, A. J. 1989. Backfat composition in pigs - differences between fat thickness groups and sexes. *Livestock Production Science* 22(3–4), 351–362. doi: 10.1016/0301-6226(89)90066-3.

Yang, Y. B., Pandurangan, M. and Hwang, I. 2012. Changes in proteolytic enzymes mRNAs and proteins relevant for meat quality during myogenesis and hypoxia of primary bovine satellite cells. *In Vitro Cellular and Developmental Biology. Animal* 48(6), 359–368.

Young, J. F., Rosenvold, K., Stagsted, J., Steffensen, C. L., Nielsen, J. H. and Andersen, H. J. 2003. Significance of preslaughter stress and different tissue PUFA levels on the oxidative status and stability of porcine muscle and meat. *Journal of Agricultural and Food Chemistry* 51(23), 6877–6881.

Young, J. F., Therkildsen, M., Ekstrand, B., Che, B. N., Larsen, M. K., Oksbjerg, N. and Stagsted, J. 2013. Novel aspects of health promoting compounds in meat. *Meat Science* 95(4), 904–911.

Zhang, J., Li, Q., Nogoy, K. M. C., Sun, J., Sun, B., Wang, Y., Tang, L., Yu, J., Jin, X., Li, X. and Choi, S. H. 2021. Effect of palmitoleic acid on the differentiation of bovine skeletal muscle satellite cells. *Journal of Animal Science and Technology* 63(4), 919–933.

Chapter 11

Assessing the environmental impact of commercial-scale cultivated meat production

Hanna L. Tuomisto, University of Helsinki and Natural Resources Institute Finland (Luke), Finland

1 Introduction

Agriculture and food systems are major contributors to environmental change. Agriculture is responsible for over 80% of global biodiversity loss, nutrient losses and land use change (Campbell et al., 2017). Food systems contribute around a third of all human-induced greenhouse gas (GHG) emissions (Crippa et al., 2021), of which 60% are from livestock production (Xu et al., 2021). As the consumption of livestock products is predicted to increase due to growing global population size and affluence (WRI, 2019), the environmental impacts of food production are likely to increase unless there are major changes in the production and consumption of livestock products.

Major transformative changes are urgently required in food production and diets in order to reduce the pressure to expand and intensify livestock production. Numerous studies have shown that switching to plant-based diets would substantially reduce pressure on the environment. Estimates show that global transformation to veganism could lead to 80% lower greenhouse emissions from food production compared to business-as-usual in 2050 (Willet et al., 2019). Changes in food production and consumption are also crucial for achieving climate targets of limiting the increase in global temperature to 1.5° above preindustrial levels (Clark et al., 2020). However, a complete conversion to veganism is unrealistic due to low consumer acceptance.

http://dx.doi.org/10.19103/AS.2023.0130.14

Plant-based meat substitutes that mimic the appearance, texture and taste of meat are already widely available, but despite increased interest, the market share of these alternative protein sources remains low compared to livestock products. The question is whether cultivated meat could have the potential to improve the sustainability of food production by offering a more appealing and environmentally sustainable alternative to livestock meat.

Cellular agriculture utilises cell-culturing technologies to cultivate microbial, plant and animal cells in bioreactors with the aim of producing alternatives to livestock products with lower environmental impacts (Rischer et al., 2020). Microbes can be used in food products directly or programmed to produce recombinant proteins to replace, for instance, milk or egg proteins. Environmental life cycle assessment (LCA) studies have indicated that cell-cultured alternatives could have a lower environmental impact compared to meat (Mattick et al., 2015; Tuomisto et al., 2022; Sinke and Odegaard, 2021), milk (Behm et al., 2022) and eggs (Järviö et al., 2021b).

This chapter focuses on the environmental assessment of cultivated meat. Section 2 provides an overview of the environmental issues related to livestock production. Section 3 introduces the LCA method commonly used to estimate environmental impacts of products. Section 4 reviews the LCA studies of cultivated meat, and Section 5 draws conclusions regarding the current understanding of the environmental performance of cultivated meat.

2 Environmental impacts of livestock production

2.1 Land use and biodiversity

Grasslands and croplands used for livestock feed production occupy a third of the world's total land area and 80% of all agricultural land area (WRI, 2019). Increased consumption of livestock products is the main driver of land use change, and 67% of deforestation for agriculture is caused by feed production for livestock (Poore and Nemecek, 2018). The deforestation of the Amazon rainforest is largely driven by the expansion of agricultural land for the production of soybean feed and beef, mainly for export. Due to the global feed and meat markets, it can be argued that the consumption of livestock meat in most places around the world has at least an indirect contribution to the deforestation of tropical forests. Tropical forests support two-thirds of the world's biodiversity, despite covering just 10% of the earth's land area (Giam, 2017). The current global deforestation rates of 10 million hectares per year threaten to result in major terminal losses of biodiversity (FAO and UNEP, 2020).

Deforestation for food production is not the only contributor to biodiversity loss. Changes in agricultural practices have also caused major losses of agro-biodiversity (Dudley and Alexander, 2017), especially changes in livestock grazing practices, which have led to a decline in some wild species, such as butterflies and beetles that are dependent on extensive grazing (Filazzola et al., 2020). The intensification of farming through increased use of fertilisers and pesticides, mono-cropping and reduced numbers of crop varieties have all contributed to the decline in biodiversity. Furthermore, the simplification of agricultural landscapes due to larger field sizes and fewer non-crop habitats (such as field boundaries, set-aside lands, wildflower meadows and on-farm trees) have caused wild species extinction and increased the number of endangered species.

Changes in land use and intensification of agriculture are driven mainly by the need to produce more and cheaper food for the world's growing and increasingly affluent population. Statistics show that gross domestic product per capita correlates with the consumption of livestock products per capita to a certain point (Tilman and Clark, 2014). The consumption of livestock products per person has been predicted to increase the most in developing economies where consumption levels are currently low (WRI, 2019). If livestock product consumption trends increase as predicted and food production practices do not change, the global agricultural land area will need to expand by an area equal to approximately the size of all current forests and savannahs on Earth by 2050 (WRI, 2019).

2.2 Climate change

Livestock production has been estimated to account for approximately 20% of global human-induced GHG emissions when both direct emissions and indirect emissions of livestock production are included (Xu et al., 2021; Crippa et al., 2021). Direct emissions include methane emissions from the digestive process of ruminants (e.g. cows and sheep) and from manure management, as well as nitrous oxide emissions from manure and the use of nitrogen fertilisers. Indirect emissions include those related to the production of farming inputs (e.g. fertilisers, pesticides and fuels), transportation of livestock, fuel use on farms and emissions from land use change, such as carbon emissions from converting forests to agricultural land. Direct emissions from agriculture (including crop and livestock production) are estimated to account for approximately 11% globally, but when indirect emissions are included, the total contribution is approximately 25% of all human-induced GHG emissions (WRI, 2019). Over 80% of the GHG emissions of livestock products are related to land use change and farming practices, whilst processing, packaging and retail activities contribute to a minor degree (Fig. 1).

Cattle production contributes 65% of all GHG emissions of global livestock production (Xu et al., 2021). Beef emissions are approximately six times higher per unit of product when compared to pork and eight times

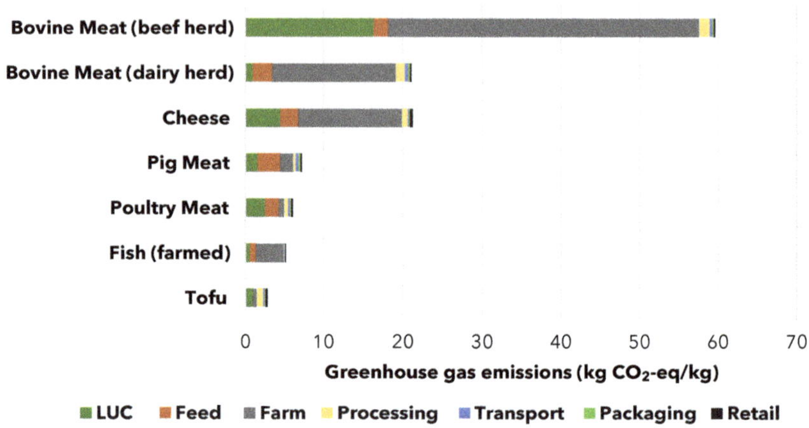

Figure 1 Greenhouse gas emissions of livestock products, fish and tofu from different life cycle stages. LUC, land use change. Source: Adapted from: Poore and Nemecek (2018).

higher than poultry meat (Poore and Nemecek, 2018). This is due to methane emissions emitted during the digestive process of ruminants, as well as larger feed requirements. The production of 1 kg of beef meat requires an average of 25 kg of feed, whereas pork needs only 6–7 kg and poultry 3 kg (Alexander et al., 2016).

2.3 Water use and emissions to waterways

Estimates of the contribution of livestock production to global water use depend on the methods used. Some methods include only 'blue water', which includes freshwater withdrawals from rivers, reservoirs, lakes and aquifers, whilst the water footprint method also includes 'green water', which includes rainwater and 'grey water', which is the amount of water needed to dilute wastewater to a level that meets the water quality standards of a particular waterway. The contribution of livestock production to total water use has been estimated at 30% of all freshwater use globally (Mekonnen and Hoekstra, 2012) and 8% of blue water use (FAO, 2006). The relative differences in the water footprint of livestock products also depend on the method used. When green, blue and grey water are included, ruminants have a relatively high water footprint due to the larger land area requirements for feed production and therefore higher green water footprint compared to monogastric animals. If only the blue water footprint is considered, the differences between livestock products are low. Extra caution, however, needs to be applied when interpreting the results of studies including only blue water footprint, as some methods use country-specific scarcity factors to describe the impact of water use on the depletion of water resources in the specific location (Boulay et al., 2019). Therefore, using the same amount of water in a water-scarce area has a much higher impact than using the same amount of water in a water-abundant region.

Livestock production is also a major contributor to the pollution of waterways due to phosphorus and nitrogen losses from fields and manure management systems (Willett et al., 2019). The increase in the size of livestock farms and the concentration of livestock production in certain areas and crop production in others have led to a situation in which manure cannot be efficiently utilised as a fertiliser on fields, which used to be a common practice in traditional farming systems and produced all the feed for the farm's own livestock. Modern industrial livestock farms rely heavily on imported feed that can be sourced overseas. Therefore, livestock farms generate excess manure as the field area near the farm is insufficient to spread the manure in moderate amounts. This leads to a situation in which manure is applied to fields in substantially higher amounts than the crops can use or is stored outdoors in unprotected systems. In both cases, nutrients end up leaching into waterways, causing eutrophication, which leads to increased growth in water plants and algae, which disturbs the balance of the ecosystem and causes negative consequences to aquatic species. The eutrophication impact of ruminants is higher compared to pork, poultry and plant-based protein sources (Poore and Nemecek, 2018).

3 Life cycle assessment

3.1 An overview of life cycle assessment

LCA is commonly used to estimate the environmental impact of food products (Clark et al., 2022; Poore and Nemecek, 2018) and has been standardised by the International Organisation for Standardisation (ISO, 2006). As ISO standards provide flexibility regarding methodological choices, various harmonisation projects are ongoing with an aim of improving the comparability of LCA studies. For instance, the European Commission's Environmental Footprint project aims to develop detailed guidelines for assessing the environmental impacts of products and organisations for marketing purposes.

Most available LCA studies are attributional LCAs (ALCAs) that account for the emissions and resource use along the life cycle of a product and use allocation methods to share the impacts between the co-products (Poore and Nemecek, 2018). Consequential LCA (CLCA) is a less commonly used approach that aims to assess the marginal impacts of changes in a product system or the production of an additional product (Smetana et al., 2019). This means that, instead of quantifying the impacts of the production of inputs needed for the product system, CLCA quantifies the impacts of changes in all affected product systems. CLCA does not involve the use of allocation methods, but the system has been expanded to include all co-products. The remainder of this chapter focuses on ALCA, as all currently available LCA studies of cultivated meat utilise the ALCA approach.

LCA studies consist of four main steps: (1) goal and scope definition, (2) inventory analysis, (3) impact assessment and (4) interpretation (Fig. 2). The goal and scope definition step defines the aims of the study, the functional unit (FU), system boundaries, data quality requirements and allocation procedures. Inventory analysis involves the collection of data for the study. Impact assessment converts the data to the environmental impact results. The interpretation phase includes the assessment of the robustness of the analysis and draws conclusions and recommendations. These steps are further discussed in the following sections.

3.2 Goal and scope definition

3.2.1 Goals of the study

The goal of the study determines the selection of methodological choices. The first step is to define the intended application of the LCA study, which can be, for instance, policy design, hotspot analysis, product development, eco-design, marketing and research. The target audience is also a key consideration. Some studies can be targeted to a specific group, for instance, the public, policymakers, researchers, business to business or for internal purposes. In some cases, LCAs may target a wide range of different audiences. The choice of method is also dependent on whether the study is comparative.

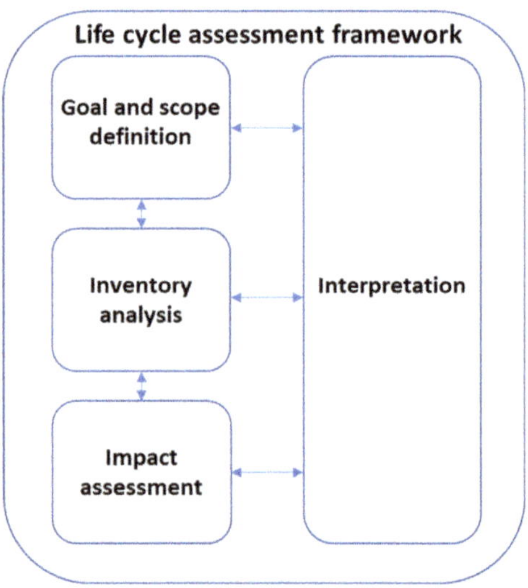

Figure 2 Framework of the life cycle assessment (LCA) method. Source: ISO (2006).

The aim of the study also defines the quality of data required from each step of the life cycle. Published LCA studies of cultivated meat are prospective and aim to estimate the impact of future technologies. This means that the data quality requirements are different compared to studies carried out for existing processes.

3.2.2 Functional unit

The FU is the unit of analysis towards which environmental impacts are allocated. The definition of the FU needs to determine what, how much, how well and how long. For instance, in the case of a cultivated meat product, the FU can be 1 kg of unpackaged cultivated meat (with 20% protein content) at the factory gate. Instead of using a weight unit as an FU, more complex indexes that consider the nutritional content of the foods have been developed (McAuliffe et al., 2020). These indexes consider both health-promoting and health-reducing nutrients. Nutritional indexes can be useful when comparing, for instance, protein-rich foods that have different types of nutritional profiles (e.g. plant- and animal-source foods). However, the use of nutritional indexes as FU is not straightforward as diets consist of many different food items, and therefore the nutritional content of the whole diet is more important than the nutrients obtained from single food items. Therefore, the unit of protein is commonly used as an FU when the environmental impacts of protein-rich foods are compared (Poore and Nemecek, 2018).

3.2.3 System boundaries

System boundaries define included and excluded processes in an LCA study. The aim of the study partly determines the choice of boundaries. For instance, if the aim is to compare different production methods of food products, it is not necessary to include packaging, retail and consumption steps. This type of assessment is commonly known as a cradle-to-gate assessment. If the aim of the study is to identify hotspots along the whole life cycle of a product, a cradle-to-grave assessment can be used (Fig. 3).

The LCA of studies of cultivated meat have used system boundaries including processes from the extraction of natural resources up to the harvesting of the cultured cell biomass from the bioreactors. None of the studies include processes related to product formulation or packaging. Some LCA studies of cultivated meat include the production of bioreactors, but the manufacturing of other infrastructure (such as buildings) is excluded. When the impacts of cultivated meat are compared with livestock meat, the limitation of the system boundaries to the cradle-to-gate is reasonable as most environmental impacts of livestock production originate within that boundary (Fig. 1).

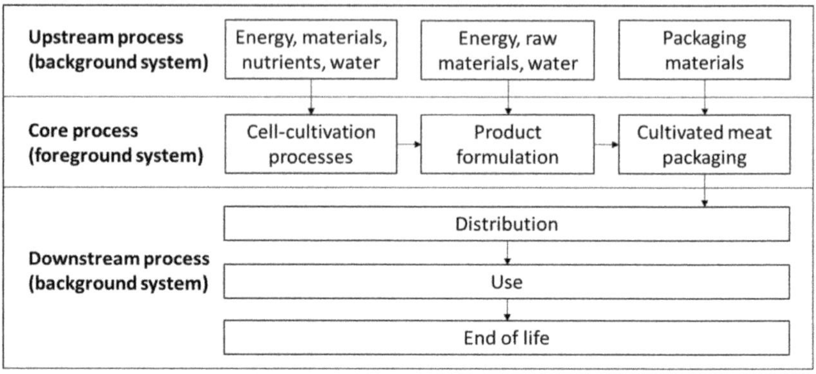

Figure 3 Example of simplified system boundary diagram for cultivated meat production for a cradle-to-grave assessment.

3.3 Inventory analysis

3.3.1 Data gathering

The inventory analysis step of an LCA study involves the collection of data regarding material and energy inputs and emission flows along the defined system boundaries. The data quality requirements depend on the purpose of the study. If the aim is to carry out a screening analysis to understand hotspots of impacts in the product system and estimate the rough level of the impacts at the order of magnitude level, the data quality requirements are not as high as for a more detailed LCA study. Data quality requirements may also differ between product phases. Higher quality data may be required for core processes (e.g. cell-culturing processes) than for upstream processes (e.g. production of inputs). LCA databases provide life cycle inventory (LCI) data for thousands of products. Therefore, common practice is to collect activity data for the core processes regarding the material and energy input requirements and outputs and use the existing databases as a source of LCI data for the inputs.

3.3.2 Co-product allocation

In ALCAs, allocation methods are used to share the impact of the whole production process between the co-products. As the first option, allocation should be avoided by creating sub-processes, but that is not usually possible in the case of food products. As a second option, system expansion can be used, which means that co-products are assigned the impact that alternative production methods would have. This means that, in the case of dairy production, meat would receive the same impacts as producing meat in beef herd systems. However, system expansion has been argued to belong to the CLCA approach and therefore is rarely used in the context of ALCA studies.

Instead, ALCA studies commonly use allocation methods based on the dry or wet mass, energy content or economic value of the co-products (Kyttä et al., 2022). The choice of the allocation method is critical as it can have a major impact on the results.

In the case of cultivated meat production, lactate and ammonia are produced as co-products. If these products are used as inputs in other product systems, some of the impacts of the whole cultured meat production process can be allocated to them. Tuomisto et al. (2022) compared alternative allocation methods, including product substitution that used lactate to replace lactic acid production and dry and wet mass-based allocations. The substitution-based allocation resulted in 5–48% of the impacts of the whole production process being allocated to lactate depending on the impact category, whilst dry mass-based allocation resulted in 85% and wet mass allocation 63% of the impacts allocated to lactate. Economic allocation would be the most reasonable choice when price information is available.

3.4 Life cycle impact assessment

The life cycle impact assessment (LCIA) phase involves converting the LCI data to selected environmental impacts. As many LCIA methods are available, the results of different LCA studies cannot be directly compared when methods are not aligned. Some LCIA methods include up to 20 different environmental impact categories, such as climate change, land use, water resource depletion, eutrophication potential etc. It is common practice to choose only the most relevant environmental impact categories in each study. LCA software is generally used for carrying out the LCIAs.

The first step of the LCIA is the grouping and characterisation of resource and emission flows to form the environmental impact results for each impact category. It is common to end the LCIA with this step. However, when a study includes many environmental impact categories, normalisation and weighting steps may be used for aggregating the results to a single score. The weighting factors aim to describe the relative importance of different environmental impact categories, for instance, based on the opinions of stakeholders and experts or based on the distance to targets (Tuomisto et al., 2012).

3.5 Interpretation

Interpretation involves the evaluation of the robustness of the results in terms of data quality. The Monte Carlo assessment is commonly used for the uncertainty assessment in LCA studies. Sensitivity assessments are used to test how specific assumptions or uncertainties in the data affect the results. Hotspots are identified by using contribution assessments that show the impacts of each life cycle stage (Fig. 1). The limitations of the study are carefully analysed.

Conclusions and recommendations are formulated based on consideration of the results together with the uncertainty assessments and limitations.

4 Environmental impacts of cultivated meat

4.1 Overview of life cycle assessment studies of cultivated meat

Few prospective LCA studies have been carried out to estimate the environmental impacts of cultivated meat production (Table 1). All the studies have had to use assumptive modelling as cultivated meat technologies are still under development. Therefore, LCA studies have high uncertainties. For instance, it is not known whether the system designs modelled in the studies would be able to yield the level of product outputs assumed. There are also major uncertainties related to the ingredients of the culture medium, the design of the bioreactors, the length of the cultivation process and the nutritional content of the products.

The first LCA of cultivated meat (Tuomisto and Teixeira de Mattos, 2011) can be regarded as a screening analysis aimed at providing rough estimates of the possible scale of the impacts at the level of the order of magnitudes. The study assumed a hypothetical production scenario that used cyanobacterial hydrolysate as the main source of amino acids and sugars. The cyanobacteria were assumed to be produced in open ponds in commercial production facilities. The production of other inputs for the culture medium (e.g. vitamins, minerals and growth factors) was excluded. A stirred tank bioreactor was assumed to be used, and the energy use requirements of the bioreactor were based on simplified calculations. Three different locations were included – Spain, Thailand and California in the USA. Due to the use of cyanobacteria as the main source of nutrients, this study had the lowest land use estimates compared to the other LCA studies carried out (Table 1). The low climate impact results are partly not only caused by the cyanobacteria-based culture medium but also likely due to underestimated energy requirements for bioreactor processes. Cleaning of the bioreactors, for example, was not included.

Table 1 Comparison of the climate impact and land use of cultivated meat based on published life cycle assessments

	Climate impact (kg CO_2-eq/kg product)	Land use (m^2/kg product)
Tuomisto et al. (2022)	4.9-25.2	1.8-6.9
Sinke and Odegaard (2021)	2.5-13.5	1.7-1.8
Mattick et al. (2015)	3.0-25.5	1.5-9.5
Smetana et al. (2015)	23.9-24.6	0.4-0.8
Tuomisto et al. (2011)	1.9-2.2	0.2

Smetana et al. (2015) used the study by Tuomisto and Teixeira de Mattos (2011) as a basis but modelled the production of cyanobacteria in bioreactors instead of open ponds. This increased energy use requirements and therefore climate impact, whilst the land use requirements remained relatively low (Table 1).

Mattick et al. (2015) modelled the environmental impacts of cultivated meat production in the US. The study assumed the use of Chinese hamster ovarian (CHO) cells cultivated in a stirred-tank bioreactor using serum-free culture medium, including standard basal medium (synthetic amino acids, vitamins and minerals) and soybean hydrolysate. The study also included cleaning the bioreactors and the buildings' energy requirements. The results showed higher environmental impacts for cultivated meat compared to the study by Tuomisto and Teixeira de Mattos (2011), but the uncertainty ranges were wide (Table 1).

Sinke and Odegaard (2021) carried out an LCA of cultivated meat based on the data collected from start-up companies supplemented with modelling and data from the literature. The data from the companies included various cell types and cultured media options. The aim was to estimate the potential ranges of the impacts rather than quantifying detailed impacts of any specific product type. The study also modelled future scenarios for 2050, assuming improvements in both cultivated meat and livestock systems. Climate impacts and land use were at the lower end of the uncertainty ranges found in Mattick et al. (2015).

Tuomisto et al. (2022) estimated the environmental impacts of cultivated meat production in hollow-fibre bioreactors in the United Kingdom (UK). The study aimed to understand how different process designs affect environmental impacts. The baseline scenario assumed the use of immortalised murine myoblast cell line C2C12 cultivated in a standard culture medium consisting of basal medium and foetal bovine serum (FBS) and using an average UK electricity mix as the energy source. The alternative scenarios included serum-free medium, CHO cells, renewable energy use and different assumptions regarding cell metabolism. The climate and land use impact ranges were in line with the uncertainty ranges in Mattick et al. (2015).

4.2 Land use

Currently available estimates of the land use of cultivated meat (Mattick et al., 2015; Tuomisto et al., 2022; Sinke and Odegaard, 2021) show lower land area requirements compared to conventionally produced beef although not necessarily poultry (Fig. 4). These studies assumed relatively short differentiation and maturation periods, meaning the muscle tissue is not yet well developed. When the aim is to produce cultivated meat with a whole meat-type texture, a longer cell-culturing time is required. This increases the use of

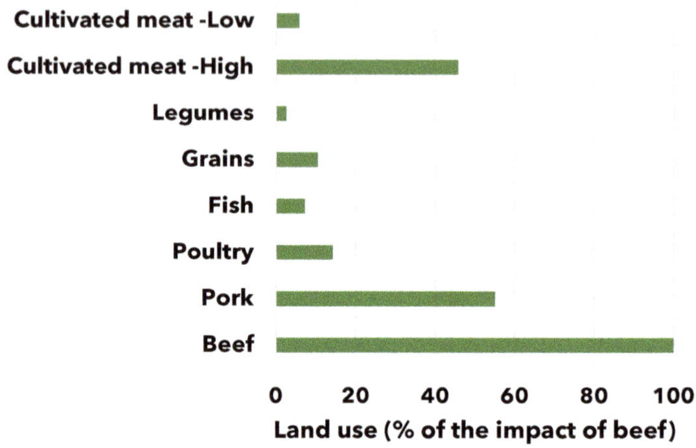

Figure 4 Relative differences in land use of cultivated meat (low and high indicate the lowest and highest impacts, respectively, found in existing LCA studies) and plant- and animal-based foods normalised based on the protein content.

nutrients and associated land to produce them when nutrients from agricultural origin are used. As beef production requires around ten times more land than poultry production (Poore and Nemecek, 2018), it is likely that cultivated meat production would have substantially lower land use requirements compared with beef even though nutrient use requirements would be higher than in the published papers.

The impact of cultivated meat on land use requirements depends on the source of nutrients used for feeding the cells (i.e. the culture medium). The culture mediums traditionally used in mammalian cell cultures consist of basal medium – including glucose, synthetic amino acids, vitamins and minerals – and FBS. This medium, however, is not suitable for commercial-scale cultivated meat production due to its high price and inclusion of animal-based ingredients. The development of suitable alternative culture media for cultivated meat production is ongoing, and one of the main challenges is to replace FBS with non-animal-based sources of growth factors. Possible sources include microalgae, yeast, fungi or plant extracts. Tuomisto et al. (2022) showed that serum-free mediums would reduce the environmental impacts of cultivated meat only if the cell yields were at the same level as the standard medium. Alternative sources of amino acids, vitamins and minerals have also been considered as synthetic production processes are energy intensive and require relatively high amounts of glucose obtained from agricultural crops, such as maize (Mattick et al., 2015). Side streams of agriculture and food processing could provide potential sources of recycled nutrients (Upcraft et al., 2021).

The amount of nutrients needed per unit of cultivated meat produced is also a key factor affecting land area requirements. Knowledge of nutrient use requirements of cultivated mammalian cells is currently lacking. Mammalian cell culturing practices for medical research purposes have not yet been optimised in terms of medium use as nutrient costs have not been a critical factor in the process. Therefore, further primary research on the nutrient requirements of cultivated mammalian cells is needed before improved estimates of land use requirements for cultivated meat production can be produced.

Total nutrient use requirements also depend on the potential to recycle unused nutrients in the process or alternatively utilise leftover nutrients for other production processes, for instance, as an ingredient for fertilisers. In advanced bioreactor systems, it would be possible to use filtration to remove the unused nutrients from the outflow medium and feed them back to the bioreactor or collect them to be used for other products. When nutrients are utilised for other products and do not end up in waste management, part of the impacts of producing those nutrients can be allocated to the co-products, which reduces the impact of cultivated meat.

The development and discovery of new sources of feedstock ingredients may further reduce the land use requirements of cultivated meat. Hydrogen oxidising bacteria (HOB), for example, is an example of a potential source of nutrients for cultivated meat production. HOB can use carbon dioxide as a source of carbon, and hydrogen extracted from water molecules as a source of hydrogen. Therefore, it does not require any agricultural land for the production process and has minimal land area requirements, as only indirect land use for the production of energy and other inputs is needed (Sillman et al., 2020; Järviö et al., 2021a).

4.3 Energy use

In cultivated meat production processes, direct energy use consists of temperature control of bioreactors, pumping, mixing, aeration, cleaning and drying processes. Indirect energy use includes the production of culture medium ingredients, materials and infrastructure. The total energy requirements of cultivated meat production depend on the ingredients used for the culture medium and the design of the bioreactor. The production of synthetic medium ingredients is especially an energy-intensive process (Tuomisto et al., 2022). The energy requirements of the bioreactor processes depend on the level of heat integration (e.g. the use of heat exchangers), insulation and potential to utilise, for instance, seawater for cooling.

Current estimates of industrial energy requirements (e.g. electricity, heat or gas) show that the energy used for cultivated meat production is higher than that for livestock production and even higher than that needed for beef production (Mattick et al., 2015; Tuomisto et al., 2022). However, the focus on industrial energy usage alone is not fair as livestock receive energy from feed that they use

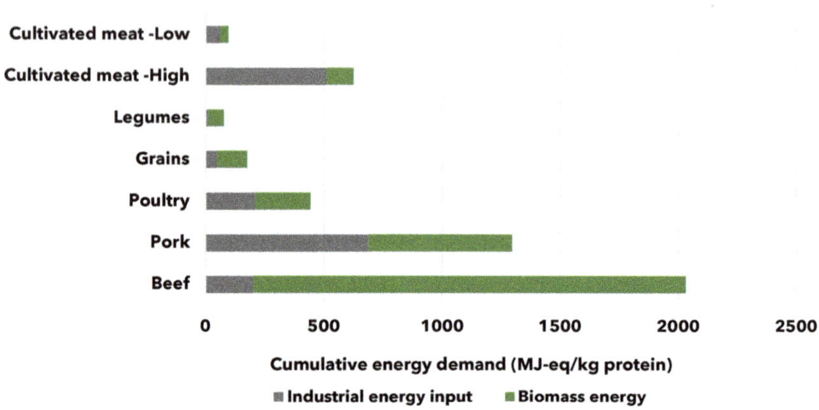

Figure 5 Cumulative energy demands of cultivated meat (low and high indicate the lowest and highest impacts, respectively, found in the existing LCA studies) and plant- and animal-based foods normalised based on protein content. Industrial energy input means electricity, oil and gas, whilst biomass energy includes the energy content of feed.

to maintain bodily functions. In cultivated meat production processes, energy in the feed has to be replaced with industrial energy. Therefore, if the total energy use of both industrial and feed energy are compared, cultivated meat would require a third of the energy than that is needed for beef production (Fig. 5).

Increased demand for industrial energy would require higher capacity to produce sustainable energy. As demand for energy is increasing in areas other than the food sector, the challenge of securing clean energy for all uses is high. In that sense, the utilisation of direct sunlight by plants and using them as feed for livestock has its own advantages, as the system does not require complex industrial processes for converting energy from one type to another. This is especially the case in the context of livestock grazing systems in which energy is not needed for harvesting, processing and transportation of feed. Due to capacity limitations in the clean and sustainable supply of industrial energy, cultivated meat production processes need to be as energy efficient as possible.

4.4 Climate impact

The majority of GHG emissions generated from cultivated meat production are associated with energy consumption. Therefore, energy source is an important factor when determining the climatic impact of the product. Due to their relatively high energy requirements, the use of fossil fuels in cultivated meat production can lead to higher climate impact when compared to pork and poultry but still lower than beef (Fig. 6) (Mattick et al., 2015; Tuomisto et al., 2022). The use of renewable energy in cultivated meat production processes

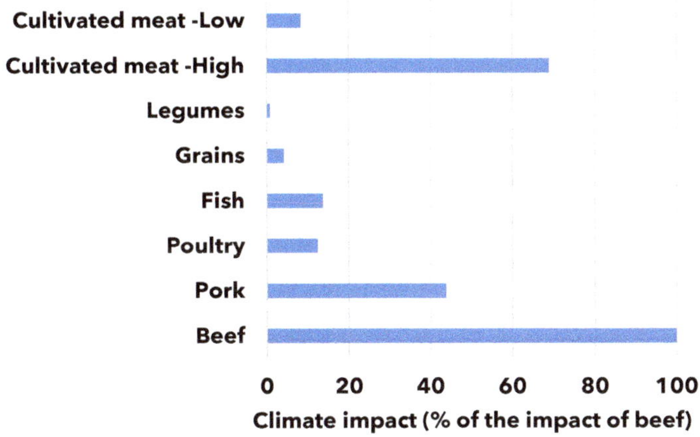

Figure 6 Relative differences in the climate impact of cultivated meat (low and high indicate the lowest and highest impacts, respectively, found in the existing LCA studies) and plant- and animal-based foods normalised based on the protein content.

can substantially reduce climate impacts (Sinke and Odegaard, 2021; Tuomisto et al., 2022).

The climate impact of cultivated meat production also depends on the type of culture medium ingredients used. The production of ingredients originating from agriculture or requiring agricultural crops in the production process (e.g. synthetic amino acids) causes GHG emissions from agricultural activities, such as nitrous oxide emissions from the use of nitrogen fertilisers. Therefore, emissions from cultivated meat production are not related to energy generation alone.

When considering the climate impact of cultivated meat production from a more holistic perspective, the potential benefits of lower land area requirements should be considered. If cultivated meat production requires substantially less land than livestock meat, agricultural land could potentially be converted to forests, which would increase the carbon storage in soil and vegetation and therefore reduce the total amount of carbon in the atmosphere, mitigating against climate change.

Commonly used methods for reporting climate impacts convert all GHG emissions to GHG equivalents that reflect the climate impact of the gases during a 100-year timeframe as a comparison to carbon dioxide. Methane is 28 times stronger than carbon dioxide and nitrous oxide 265 times stronger. Lynch and Pierrehumbert (2019) argued that the commonly used methods for reporting climate impacts give overly high emphasis to methane emissions as methane has an atmospheric lifetime of just 12 years, whilst carbon dioxide persists over a much longer time period. The study used a 1000-year timeframe for comparing the climate impacts of cultivated meat with beef and found that in the longer-term analysis, some of the cultivated meat systems resulted in higher climate impacts than some of the beef systems. However, the reason why methane has a higher climate impact during shorter periods is that in a 20-year period, it traps 84 times more heat per mass unit than carbon dioxide and 28 times more during a 100-year period. As the importance of limiting climate change grows ever more urgent, the avoidance of emissions with a high climate impact in the short term is crucial. Furthermore, carbon dioxide can be removed from the atmosphere by growing more vegetation, whilst methane cannot be removed during its lifetime. As Intergovernmental Panel on Climate Change (IPCC) has highlighted the importance of the emissions generated by 2030 in terms of determining the severity of climate change we will face (IPCC, 2018), it might even be reasonable to switch from using the 100-year timeframe to the 20-year timeframe, which would give even more importance to methane emissions.

4.5 Water use and emissions to water

Water use in cultivated meat production consists of direct and indirect water inputs. Direct water use includes water used in the culture medium, temperature control and cleaning of the bioreactors. Indirect water use is that used for the production of inputs, such as culture medium ingredients and energy. The water use of cultivated meat production depends on factors such as the type of medium ingredients and energy sources used, the potential to recycle water used in the culture medium and the potential to utilise seawater in cooling processes.

The relative differences in water use between cultivated meat and livestock meat depend on the methods used. When using the water footprint approach, which includes green, blue and grey water (as explained in Section 1), the water use in livestock meat production is relatively higher (Tuomisto and Teixeira de Mattos, 2011) than in methods that include only blue water (Tuomisto et al., 2022). This is due to the fact that livestock production requires more agricultural land, and thus green water (i.e. rainwater) use is higher.

Emissions to waterways from cultivated meat production facilities are minor as the wastewater is treated before being released to water bodies. The magnitude of emissions from the production of culture medium depends on the type of ingredients used. If the culture medium includes substances from agriculture, nitrogen and phosphorus emissions to waterways cannot be avoided as agricultural fields are open systems and 100% nutrient use efficiency is impossible to achieve. As cultivated meat production, however, is likely to require substantially fewer agricultural crops than livestock meat, major reductions in the eutrophication impact can be expected. A study estimated that the eutrophication impact of cultivated meat could be 97% lower compared to beef and 70% lower compared to pork production in the US, whilst poultry production had a similar eutrophication effect with cultivated meat (Mattick et al., 2015).

Other water pollutants from agriculture and livestock farms include pesticide residues, antibiotics and pathogens from manure, all of which affect water quality. These water contaminants would be completely avoided or at least reduced in cultivated meat production systems, assuming that cultivated meat production would not need antibiotics. As cultivated meat production may still require keeping animals for cell donation, not all livestock-related emissions are completely avoided. However, livestock numbers would be just a small fraction of current levels, and the contribution of livestock production to the impacts of the cultivated meat production process would be insignificant. The development of immortalised cell lines for cultured meat production also removes the need to use animals as cell donors.

4.6 Biodiversity

Cultivated meat production would benefit biodiversity due to lower land use requirements, which would reduce the pressure to convert forests to agricultural land and intensify agricultural practices (Machovina et al., 2015). However, the biodiversity benefits would depend on how the land area released from livestock farming is used. If the land is used for intensive production of bioenergy crops, crops for materials or intensive forestry, biodiversity might not improve. Therefore, cultivated meat production would not automatically provide the intended biodiversity benefits and so policies that provide incentives to expand nature conservation areas and to use wildlife-friendly land management practices would be needed.

Complete removal of grazing livestock would not be optimal for biodiversity as many endangered wild plants and insects depend on extensive grazing systems (Alkemade et al., 2013). Furthermore, grazing livestock have a key role in certain areas of landscape maintenance, such as maintaining semi-natural grasslands that provide ecological and cultural benefits (Bengtsson et al., 2019). Therefore, extensive grazing-based livestock production systems have positive impacts on biodiversity, but substantially fewer livestock numbers compared to current levels would be sufficient to provide those benefits.

5 Conclusion

The product-level LCA studies show that cultivated meat could reduce environmental impacts compared to livestock meat, especially beef production, provided that technologies can be scaled up. Culture medium ingredients and bioreactor energy use are key factors affecting the level of environmental impacts. As cultivated meat production technologies are still in the development stage, current estimates of environmental impacts are indicative. Technological development has the potential to substantially reduce the impacts.

To improve our understanding of the wider environmental consequences of cultivated meat, global-level and product-level assessments are needed. Global-level earth system models can be used to estimate the impact of changes in land use dynamics that occur as a result of changes in agriculture (Humpenoder et al., 2022). Some studies have quantified the climate benefits that could be achieved by reforesting land that would be released from livestock production as a consequence of switching to vegetarian diets (Hayek et al., 2021; Sun et al., 2022). These types of global assessments should also consider realistic rates of adoption of cultivated meat products in different parts of the world at different timescales. Furthermore, the assessments need to consider the development of energy technologies as the utilisation of clean energy in cultivated meat production is key to achieving climate benefits.

As the development of cultivated meat technology is still in its early stages, it is unlikely that the technology will be able to help with the urgent environmental targets set for 2030 (e.g. United Nations Sustainable Development Goals and climate targets). While cultivated meat production technologies are being developed, other changes in food systems need to be implemented. These include, for instance, switching towards plant-based diets, improving the sustainability of current agricultural practices, stopping deforestation and reducing food waste. Cell-cultured microbial cells, recombinant proteins and other alternative protein sources, such as algae and insects, may also help improve the sustainability of food systems before 2030.

6 Where to look for further information

Key journal papers:

- Sinke, P., Swartz, E., Sanctorum, H., van der Giesen, C. and Odegard, I. 2023. Ex-ante life cycle assessment of commercial-scale cultivated meat production in 2030. *The International Journal of Life Cycle Assessment* 28, 234–254.
- Tuomisto, H. L., Allan, S. J. and Ellis, M. J. 2022. Prospective life cycle assessment of a bioprocess design for cultured meat production in hollow fiber bioreactors. *Science of the Total Environment* 851(1), 158051.

Key conference:

- The International Conference on the life cycle assessment of food.

Key research groups and organisations:

- Future sustainable food systems -research group (FS)[2], University of Helsinki, Finland: https://www.helsinki.fi/en/researchgroups/future-sustainable-food-systems.

Good Food Institute:

- https://gfi.org/.

7 References

Alexander, P., Brown, C., Arneth, A., Finnigan, J. and Rounsevell, M. D. J. G. E. C. 2016. Human appropriation of land for food: the role of diet. *Global Environmental Change* 41, 88–98.

Alkemade, R., Reid, R. S., Van Den Berg, M., De Leeuw, J. and Jeuken, M. 2013. Assessing the impacts of livestock production on biodiversity in rangeland ecosystems. *Proceedings of the National Academy of Sciences of the United States of America* 110(52), 20900–20905.

Behm, K., Nappa, M., Aro, N., Welman, A., Ledgard, S., Suomalainen, M. and Hill, J. 2022. Comparison of carbon footprint and water scarcity footprint of milk protein produced by cellular agriculture and the dairy industry. *The International Journal of Life Cycle Assessment* 27(8), 1017–1034.

Bengtsson, J., Bullock, J. M., Egoh, B., Everson, C., Everson, T., O'Connor, T., O'Farrell, P. J., Smith, H. G. and Lindborg, R. 2019. Grasslands—more important for ecosystem services than you might think. *Ecosphere* 10(2), e02582.

Boulay, A.-M., Benini, L. and Sala, S. 2019. Marginal and non-marginal approaches in characterization: how context and scale affect the selection of an adequate characterization model. The AWARE model example. *The International Journal of Life Cycle Assessment*, 1–13.

Campbell, B. M., Beare, D. J., Bennett, E. M., Hall-Spencer, J. M., Ingram, J. S. I., Jaramillo, F., Ortiz, R., Ramankutty, N., Sayer, J. A. and Shindell, D. 2017. Agriculture production as a major driver of the Earth system exceeding planetary boundaries. *Ecology and Society* 22(4), 11.

Clark, M., Springmann, M., Rayner, M., Scarborough, P., Hill, J., Tilman, D., Macdiarmid, J. I., Fanzo, J., Bandy, L. and Harrington, R. A. 2022. Estimating the environmental impacts of 57,000 food products. *Proceedings of the National Academy of Sciences of the United States of America* 119(33), e2120584119.

Clark, M. A., Domingo, N. G. G., Colgan, K., Thakrar, S. K., Tilman, D., Lynch, J., Azevedo, I. L. and Hill, J. D. 2020. Global food system emissions could preclude achieving the 1.5° and 2°C climate change targets. *Science* 370(6517), 705–708.

Crippa, M., Solazzo, E., Guizzardi, D., Monforti-Ferrario, F., Tubiello, F. N. and Leip, A. 2021. Food systems are responsible for a third of global anthropogenic GHG emissions. *Nature Food* 2(3), 198–209.

Dudley, N. and Alexander, S. J. B. 2017. Agriculture and biodiversity: a review. *Biodiversity* 18(2–3), 45–49.

FAO 2006. *Livestock's Long Shadow*. Rome: Food and Agriculture Organization.

FAO and UNEP 2020. *The State of the World's Forests 2020. Forests, Biodiversity and People*. Rome.

Filazzola, A., Brown, C., Dettlaff, M. A., Batbaatar, A., Grenke, J., Bao, T., Peetoom Heida, I. and Cahill, J. F. 2020. The effects of livestock grazing on biodiversity are multi-trophic: a meta-analysis. *Ecology Letters* 23(8), 1298–1309.

Giam, X. 2017. Global biodiversity loss from tropical deforestation. *Proceedings of the National Academy of Sciences of the United States of America* 114(23), 5775–5777.

Hayek, M. N., Harwatt, H., Ripple, W. J. and Mueller, N. D. 2021. The carbon opportunity cost of animal-sourced food production on land. *Nature Sustainability*. 4, 21–24.

Humpenoder, F., Bodirsky, B. L., Weindl, I., Lotze-Campen, H., Linder, T. and Popp, A. 2022. Projected environmental benefits of replacing beef with microbial protein. *Nature* 605(7908), 90–96.

IPCC 2018. *Global Warming of 1.5°C - an IPCC Special Report on the Impacts of Global Warming of 1.5°C above Pre-industrial Levels and Related Global Greenhouse Gas Emission Pathways, in the Context of Strengthening the Global Response to the Threat of Climate Change, Sustainable Development, and Efforts to Eradicate Poverty*. Incheon: Intergovernmental Panel on Climate Change.

ISO 2006. *Environmental Management - Life Cycle Assessment - Requirements and Guidelines*. Geneva: International Organization for Standardization.

Järviö, N., Maljanen, N. L., Kobayashi, Y., Ryynänen, T. and Tuomisto, H. L. 2021a. An attributional life cycle assessment of microbial protein production: a case study on using hydrogen-oxidizing bacteria. *Science of the Total Environment* 776, 145764.

Järviö, N., Parviainen, T., Maljanen, N. L., Kobayashi, Y., Kujanpää, L., Ercili-Cura, D., Landowski, C. P., Ryynänen, T., Nordlund, E. and Tuomisto, H. L. 2021b. Ovalbumin production using Trichoderma reesei culture and low-carbon energy could mitigate the environmental impacts of chicken-egg-derived ovalbumin. *Nature Food* 2(12), 1005–1013.

Kyttä, V., Roitto, M., Astaptsev, A., Saarinen, M. and Tuomisto, H. L. 2022. Review and expert survey of allocation methods used in life cycle assessment of milk and beef. *The International Journal of Life Cycle Assessment* 27(2), 191–204.

Lynch, J. and Pierrehumbert, R. 2019. Climate impacts of cultured meat and beef cattle. *Frontiers in Sustainable Food Systems* 3.

Machovina, B., Feeley, K. J. and Ripple, W. J. 2015. Biodiversity conservation: the key is reducing meat consumption. *Science of the Total Environment* 536, 419–431.

Mattick, C. S., Landis, A. E., Allenby, B. R. and Genovese, N. J. 2015. Anticipatory life cycle analysis of in vitro biomass cultivation for cultured meat production in the United States. *Environmental Science and Technology* 49(19), 11941–11949.

McAuliffe, G. A., Takahashi, T. and Lee, M. R. F. 2020. Applications of nutritional functional units in commodity-level life cycle assessment (LCA) of agri-food systems. *The International Journal of Life Cycle Assessment* 25(2), 208–221.

Mekonnen, M. M. and Hoekstra, A. Y. J. E. 2012. A global assessment of the water footprint of farm animal products. *Ecosystems* 15(3), 401–415.

Poore, J. and Nemecek, T. J. S. 2018. Reducing food's environmental impacts through producers and consumers. *Science* 360(6392), 987–992.

Rischer, H., Szilvay, G. R. and Oksman-Caldentey, K. M. 2020. Cellular agriculture – industrial biotechnology for food and materials. *Current Opinion in Biotechnology* 61, 128–134.

Sillman, J., Uusitalo, V., Ruuskanen, V., Ojala, L., Kahiluoto, H., Soukka, R. and Ahola, J. 2020. A life cycle environmental sustainability analysis of microbial protein production via power-to-food approaches. *The International Journal of Life Cycle Assessment* 25(11), 2190–2203.

Sinke, P. and Odegaard, I. 2021. *LCA of Cultivated Meat. Future Projections for Different Scenarios*. Delft: council of Europe.

Smetana, S., Mathys, A., Knoch, A. and Heinz, V. 2015. Meat alternatives: life cycle assessment of most known meat substitutes. *The International Journal of Life Cycle Assessment* 20(9), 1254–1267.

Smetana, S., Schmitt, E. and Mathys, A. 2019. Sustainable use of Hermetia illucens insect biomass for feed and food: attributional and consequential life cycle assessment. *Resources, Conservation and Recycling* 144, 285–296.

Sun, Z., Scherer, L., Tukker, A., Spawn-Lee, S. A., Bruckner, M., Gibbs, H. K. and Behrens, P. 2022. Dietary change in high-income nations alone can lead to substantial double climate dividend. *Nature Food* 3(1), 29–37.

Tilman, D. and Clark, M. 2014. Global diets link environmental sustainability and human health. *Nature* 515(7528), 518–522.

Tuomisto, H. L., Allan, S. J. and Ellis, M. J. 2022. Prospective life cycle assessment of a bioprocess design for cultured meat production in hollow fiber bioreactors. *Science of the Total Environment* 851(1), 158051.

Tuomisto, H. L., Hodge, I. D., Riordan, P. and Macdonald, D. W. 2012. Exploring a safe operating approach to weighting in life cycle impact assessment – a case study of organic, conventional and integrated farming systems. *Journal of Cleaner Production* 37, 147–153.

Tuomisto, H. L. and Teixeira De Mattos, M. J. 2011. Environmental impacts of cultured meat production. *Environmental Science and Technology* 45(14), 6117–6123.

Upcraft, T., Tu, W.-C., Johnson, R., Finnigan, T., Van Hung, N., Hallett, J. and Guo, M. 2021. Protein from renewable resources: mycoprotein production from agricultural residues. *Green Chemistry* 23(14), 5150–5165.

Willett, W., Rockström, J., Loken, B., Springmann, M., Lang, T., Vermeulen, S., Garnett, T., Tilman, D., Declerck, F., Wood, A. J. T. L., Jonell, M., Clark, M., Gordon, L. J., Fanzo, J., Hawkes, C., Zurayk, R., Rivera, J. A., De Vries, W., Majele Sibanda, L., Afshin, A., Chaudhary, A., Herrero, M., Agustina, R., Branca, F., Lartey, A., Fan, S., Crona, B., Fox, E., Bignet, V., Troell, M., Lindahl, T., Singh, S., Cornell, S. E., Srinath Reddy, K., Narain, S., Nishtar, S. and Murray, C. J. L. 2019. Food in the Anthropocene: the EAT–Lancet Commission on healthy diets from sustainable food systems. *Lancet* 393(10170), 447–492.

WRI 2019. *Creating a Sustainable Food Future. A Menu of Solutions to Feed Nearly 10 Billion People by 2050*. World Resources Report. Washington, DC: World Resources Institute.

Xu, X., Sharma, P., Shu, S., Lin, T. S., Ciais, P., Tubiello, F. N., Smith, P., Campbell, N. and Jain, A. K. 2021. Global greenhouse gas emissions from animal-based foods are twice those of plant-based foods. *Nature Food* 2(9), 724–732.

Chapter 12

Consumer attitudes to cultured meat products: improving understanding and acceptance

Christina Hartmann and Michael Siegrist, ETH Zürich, Switzerland

1 Introduction

Meat is one of the most important and valued foods worldwide, and the demand for meat is increasing year by year due to steady growth in both the global population and incomes. However, there are far-reaching consequences of this demand for meat. Meat production and overconsumption in many parts of the world create burdens on the environment, animal welfare and human health (Willett et al., 2019). In a desire to move the current food system towards a sustainable food system, plant-based meat alternatives have flooded the market in the last couple of years to assist in such a nutrition transition. However, consumers' willingness to eat these products on a regular basis seems low, and only a minority consume these products regularly (Hartmann and Siegrist, 2017; Onwezen et al., 2021). This may be due to the fact that meat has a strong symbolic meaning, is nutritious and is a main component of many traditional meals in Western cuisines, making it an indispensable part of most people's diets. It is unreasonable to think that food cultures centred around meat will drastically change and substitute meat with plant-based alternatives. Thus, researchers have questioned the notion that a fundamental reduction in meat consumption and thus production will occur in the near future.

http://dx.doi.org/10.19103/AS.2023.0130.15

Therefore, it has been proposed that if the eating behaviour of consumers cannot be changed, then the way meat is produced needs to change (Cole and McCoskey, 2013; Volden and Wethal, 2021). At first glance, cultured meat - which has fundamentally the same raw materials as conventional meat - seems to be a promising solution for the issues associated with meat overconsumption and production. However, there are also challenges associated with both the technical production of cultured meat (which is not covered here) and consumer acceptance of cultured meat. The present chapter focuses on the consumer's perspective by considering various consumer-related factors for the rejection of cultured meat that have been proposed (Onwezen et al., 2021; Wilks et al., 2019). Given that the acceptance of cultured meat by consumers will determine its commercial success, how consumer perception could be positively influenced is also discussed.

2 Acceptance of novel food technologies and cultured meat

Novel foods and food technologies are important for food safety and the development of a sustainable food system. However, consumers are often conservative in their acceptance of novel food sources and food technologies (Siegrist and Hartmann, 2020a). This is not surprising, given that humans by nature are generally hesitant about accepting unfamiliar foods and food sources. In addition, unlike in other domains, there is considerably less pressure on consumers to accept novel technologies in the food domain, and how consumers perceive novelty determines their expectations of novel and unfamiliar foods (Pliner, 1994; Tuorila and Hartmann, 2020). These observations are also reflected in the low acceptance ratings of cultured meat (Onwezen et al., 2021; Pakseresht et al., 2022; Siegrist and Hartmann, 2020b). Influential factors that impact public acceptance of cultured meat have been reviewed in several publications (Bryant and Barnett, 2018; Onwezen et al., 2021; Pakseresht et al., 2022; Tomiyama et al., 2020), and it has been found that compared to other meat alternatives (e.g. insects and plant-based meat substitutes), cultured meat is the second-least accepted alternative (Onwezen et al., 2021). It is also worth noting that as soon as a comparison with conventional meat is provoked, alternatives (including cultured meat) are evaluated worse (Slade, 2018). These characteristics of being less willing to try and having less interest in consuming cultured meat than conventional meat seem to be stable among consumers to date, as suggested by the findings of a 2-year follow-up study conducted with Belgian consumers (Bryant and Sanctorum, 2021). However, there can be differences in the level of acceptance between countries and cultures (Bryant and Barnett, 2020; Pakseresht et al., 2022; Siegrist and Hartmann, 2020b). For example, consumers in France seem to have a rejection-based attitude, while consumers in Asia appear to be more open to such technological progress in the food sector (the first cultured meat nugget was sold in Singapore) (Bryant et al., 2019). Likewise, consumers in Europe seem to be more open than consumers in the USA (Pakseresht et al., 2022). Nevertheless, while a majority seems to be willing to at least try cultured meat, a much smaller proportion of people would consider buying it regularly (Bryant and Barnett, 2020). Especially, younger, higher educated people and people with higher meat attachment reported higher acceptance ratings (Hocquette et al., 2022; Onwezen et al., 2021; Pakseresht et al., 2022).

However, the concerns about cultured meat expressed by consumers are manifold. Next to less taste and appeal perception, perceived unnaturalness of cultured meat and feelings of disgust towards cultured meat were reported as potential barriers to the uptake of the technology. These product-related factors align with low benefit perceptions and fear of health risks (Hocquette et al., 2015; Pakseresht et al., 2022; Siegrist and Hartmann, 2020b; Wilks and Phillips, 2017). Hence, even though there is potential for cultured meat to prevail as an innovative new technology, personal attitudes and characteristics are the main reasons for rejection.

2.1 Perceived naturalness and disgust

Disgust is a human emotion that prevents us from contacting things that might be pathogenic because it triggers behavioural avoidance of the stimulus (Terrizzi et al., 2013). Disgust is triggered by cues that symbolise hazardous items and the presence of pathogens, such as certain odours (e.g. the smell of decayed food), sights (e.g. mould, a runny nose), tactile cues (e.g. slime) and sounds (e.g. clearing one's throat) (Curtis and Biran, 2001). People can vary in their tendency to react with disgust towards such cues. On the one hand, insensitivity to disgust cues might inhibit the necessary preventive behaviour, and people might expose themselves to higher risks. On the other hand, oversensitivity to cues that are not pathogen related or overgeneralisation based on crudely defined cues can trigger false alarms and might lead to neglecting viable resources. Disgust sensitivity as a personality-like characteristic shapes people's food choices in everyday life and is also a predictor of the acceptance of novel foods and food technologies (Egolf et al., 2019; Siegrist and Hartmann, 2020a; Tuorila and Hartmann, 2020).

A disgust response towards cultured meat is a significant predictor not only for willingness to eat cultured meat but also for risk and benefit perception towards cultured meat. This is important because perceived benefit is a key driver of novel food acceptance among consumers (Bonny et al., 2017). Thus, it was found that the higher the disgust evoked by cultured meat, the higher the risk perception and the lower the benefit perception (Egolf et al., 2019).

In a study by Siegrist and Hartmann (2020b) on the acceptance of cultured meat in ten countries, participants had to indicate their willingness to consume cultured meat, whether they consider cultured meat artificial and whether they consider cultured meat disgusting. The results showed that participants from France, where acceptance was the lowest, reported the highest levels of disgust towards cultured meat. Moreover, it was found that food disgust sensitivity (trait disgust) exerted an effect on disgust evoked by cultured meat (state disgust), which then impacted acceptance. The results also showed that across all included countries (except for China), a lack of perceived naturalness of cultured meat had an impact on how disgust-provoking cultured meat was perceived. Thus, participants who perceived cultured meat as more artificial were more disgusted by cultured meat than participants who perceived it as more natural. In other words, a reason for a disgust response towards cultured meat might be that it is perceived as an artificial and unnatural product.

The naturalness fallacy is based on the consumer's conviction that natural is equal to good and healthy, while unnatural is bad and unhealthy. Thus, naturalness perception is crucial to the acceptance of food technologies (Roman et al., 2017); however, consumers differ in how they perceive cultured meat as natural or artificial (Verbeke et al., 2015). For example, the perceived degree of processing plays a determining role in whether plant-based meat alternatives and cultured meat are described as natural or artificial (Bryant, Anderson, et al., 2019; Hartmann et al., 2022). Unfortunately, it is difficult to manipulate how naturally cultured meat is perceived. In a study with American adults, attempts to manipulate consumers' perceptions by either describing cultured meat as natural, describing conventional meat as unnatural or challenging the appeal of naturalness were not entirely successful (Bryant et al., 2019). The participants could only be persuaded to regard conventionally produced meat as less natural. The findings of another study showed that using technical terms to describe cultured meat leads to lower naturalness perception than using non-technical terms (Siegrist et al., 2018). In the same study, it was found that when participants were informed about cultured meat, a paradoxical effect occurred; they perceived traditionally produced meat as positive as organically produced meat (Siegrist et al., 2018). Thus, it seems that rather than trying to increase the perceived naturalness of cultured meat, messages such as 'conventional meat is unnatural' should be considered (Bryant et al., 2019). Researchers have also suggested highlighting the similarities between cultured meat and conventionally produced meat to lower evoked disgust and increase naturalness perception and consequently positively impact consumers' acceptance (Egolf et al., 2019; Rosenfeld and Tomiyama, 2022). However, more research is needed on the factors associated with cultured meat that influence the level of disgust it can evoke before concrete conclusions can be drawn.

2.2 Trust in the industry and humans tampering with nature

When people lack knowledge about a new technology, such as cultured meat production, they rely on other information sources, such as others' evaluations of the technology. Trust influences the perception and acceptance of novel technologies (Siegrist, 2021), although it is unclear whether trust has a direct or indirect influence on the acceptance of a hazard (or novel technology) (Siegrist, 2021). The simplest model postulates that trust influences risk and benefit perception indirectly and consequently the acceptance of the technology (Siegrist, 2021). In terms of cultured meat, there is limited literature on the role of trust in the acceptance of cultured meat. In one study, researchers found that general distrust in science was not a significant predictor of cultured meat acceptance (Wilks et al., 2019). However, in another study that included samples from ten countries, trust in the food industry was a significant predictor of acceptance (Siegrist and Hartmann, 2020b). Trust in the food industry predicted acceptance not only indirectly via the perceived naturalness of cultured meat but also directly via an effect on willingness to eat cultured meat (Siegrist and Hartmann, 2020b). The latter direct effect was not observed in all countries, and cultural differences were observed in terms of the potential impact of trust on the acceptance of such novel food technologies (Siegrist and Hartmann, 2020b). Nevertheless, trust in the food industry rather than general trust in science seems to be a relevant factor to perceptions of cultured meat, but it is problematic to generalise results from one country to another country (Siegrist and Hartmann, 2020b).

Another obstacle to the introduction of novel technologies is that there can be a lack of acceptance even when the benefits are higher and the risks are lower than those of familiar existing technologies (Siegrist et al., 2016). For instance, a study showed that the risk associated with red meat consumption was more acceptable for familiar, conventional red meat than for cultured meat (Siegrist and Sütterlin, 2017). In a later study, the dimension of tampering with nature was found to be a predictor of perceived risks (Sjöberg, 2000). Tampering with nature is considered a moral heuristic not based on a deliberative processing of information (Sunstein, 2005). It has been found to be the strongest predictor of acceptance of various biotechnologies, including synthetic biology and genetic testing. Technologies that are seen to involve tampering with nature are often perceived as unnatural and morally questionable and linked to concerns about the uncontrollability of (long-term) negative consequences (Hoogendoorn et al., 2021). The latter can be accompanied by a feeling that there is too little knowledge about the negative consequences of a new technology or that people do not have the option to choose between a product produced using conventional methods or one produced using novel technologies (Hoogendoorn et al., 2021). In an early study on the acceptance of cultured meat, the participants stated 'playing god' and 'messing with nature' as underlying reasons for a perceived lack of naturalness and thus rejection of cultured meat (Verbeke et al., 2015). Tampering with nature and perceived naturalness are linked but conceptually distinct constructs (Hoogendoorn et al., 2021). The perception that cultured meat technology involves tampering with nature and how this influences acceptance needs to be further researched above and beyond naturalness perception. Having greater insight into these aspects will support the development of targeted communication and persuasion strategies for increasing acceptance and tackling consumer concerns and risk perceptions. Hence, the role of trust and humans interfering with something that is traditionally considered natural should be researched.

3 Consumers' motivation to substitute conventional meat

For consumers to have any interest in cultured meat, it is critical that they are motivated to find a substitute for conventional meat. Several potential factors exist that might function as incentives and convey additional benefits of cultured meat consumption. First and foremost is price. If cultured meat was available at a lower price than conventional meat, then meat consumers might be motivated to consume it. However, production costs are still very high, and it is unclear when a market promotion stage will be reached.

Another motivating factor for substituting meat with cultured meat is an environmental concern. However, environmental friendliness is not a vital driver of daily food choices (Bauer et al., 2022). Many consumers also lack basic knowledge of environmentally friendly food selection and tend to underestimate the environmental impact of meat production and overestimate the environmental impact of meat-substitute products (Hartmann et al., 2022; Hartmann et al., 2021). Thus, it is questionable whether consumers see the benefits of cultured meat over traditional products when it comes to environmental friendliness. It is worth noting that cultured meat production is very energy intensive, and life cycle assessment data is not currently available. When the production technology facilitates market access and all food processing stages are determined, the environmental friendliness of cultured meat production can be assessed.

Health considerations might also drive consumers to test meat substitutes. For example, overconsumption of red meat poses a health risk (Rohrmann et al., 2013). However, the nutritional profile of cultured meat is still unknown, and it cannot be expected that high consumption of cultured red meat will be healthier than high consumption of conventionally produced red meat. Consumers seem to question the health benefits too (Mancini and Antonioli,

2019), and the safety concerns of consumers must be addressed first. To date, no adequate product risk assessment is available (Guan et al., 2021).

Lastly, concerns about animal welfare in highly industrialised animal husbandry systems are a driving factor for the rejection of conventionally produced meat and the willingness to substitute (Hartmann and Siegrist, 2020).

While willingness to reduce meat intake or substitute meat with other foods seems low, the proportion of people who have reduced or stopped their meat intake has slightly increased in the last couple of years in high-income countries (Hartmann and Siegrist, 2017). In fact, people living in societies with meat-centred food cultures are socialised into the habit of eating meat during childhood, and there is little reflection on that behaviour in everyday life (Bastian and Loughnan, 2017). When confronted with the current situations in industrialised animal husbandry systems, consumers do not consider them to be morally justified (Hartmann and Siegrist, 2020), but this does not automatically imply that meat consumption is reduced. Simultaneously caring for animals and eating meat that requires the slaughter of animals creates an ambivalence that is referred to as the 'meat paradox' (Loughnan et al., 2014). Accordingly, one's behaviour is not in line with one's ethical beliefs and can cause negative emotions and moral concerns. While some people reduce their meat consumption to counteract their psychological discomfort (i.e. cognitive dissonance), others maintain their behaviour but (unconsciously) disengage from their moral principles. This disengagement from moral principles is accompanied by a diverse set of strategies and beliefs, such as denial of animal suffering in meat production systems, dissociation of the animal–food connection and avoidance of negative thoughts about animal husbandry and slaughtering (Rothgerber, 2013). In their 2015 study, Rothgerber (2015) identified a relatively new consumer segment that consists of people who have turned towards 'ethical' meat eating – 'conscientious omnivores'. They consume meat only if its production meets certain moral standards. Compared to vegetarians and vegans, conscientious omnivores were found to score higher on a meat-eating justification scale even though they care for animal welfare issues (Rothgerber, 2015). It is imaginable that this consumer segment is more willing to adopt cultured meat since two motivational aspects, enjoyment of meat and animal welfare, are fulfilled. Consequently, meat-eating justification could be reduced since cognitive dissonance should no longer be an issue when cultured meat is consumed instead of conventional meat. It also fits with the observation that people with higher meat attachment are more willing to consume cultured meat.

Interestingly, even though most consumers are calling on farmers to pay more attention to animal welfare, the ongoing consumption of conventionally produced meat products supports the continuation of intensive livestock farming. This contrasts with consumer survey findings in which animal welfare is given much more weight than when making decisions at the point of purchase (also referred to as the attitude–behaviour gap). A major reason for this is that consumers prioritise product price over animal welfare. If cultured meat cannot be produced at lower costs than conventional meat, it is questionable whether cultured meat can enter the mass market.

4 Overcoming the rejection of cultured meat

Current research is focusing on determining how reluctance towards trying cultured meat can be overcome. The impact of informational claims on acceptance (Onwezen et al., 2021) and how cultured meat should be named to prevent negative associations and rejection were among the first subjects to be investigated (Bryant and Barnett, 2019). It was found that the terms 'clean meat' and 'animal-free meat' provoked more positive attitudes than the more technological 'lab-grown meat' (Bryant and Barnett, 2019), which likely provokes concerns about naturalness and healthiness (Siegrist et al., 2018). Baum et al. (2022) investigated the effect of counter-messaging on the acceptance of cultured meat and found that when the benefits of cultured meat relate to problems of conventional meat, acceptance can change in a positive direction, independent of whether animal welfare or environmental impact was stressed. Even though the effect sizes were small, these results gave the first indication of how focusing on societal benefits (rather than individual benefits) could impact acceptance ratings.

Piochi et al. (2022) tested the effect of information about animal welfare, environmental impact and human safety on willingness to substitute meat with cultured meat, willingness to try and willingness to purchase with a sample of Italian consumers. Compared to the control group, willingness to substitute was higher among people who received additional information about cultured meat. However, no effect was found on willingness to try and willingness to purchase. In addition, Van Loo and colleagues (Van Loo et al., 2020) found that information about animal welfare and environmental benefits had the largest effect on consumers with preferences for meat alternatives, but overall the impact was small. In the same study, most consumers preferred conventional beef (i.e. labelled farm-raised beef) to alternatives (i.e. plant-based with pea protein, plant-based with animal-like protein and cultured meat) even when prices were constant. Cultured meat was the least-preferred product in that study. Notably, they also found that over 70% of their American survey participants preferred that plant- and lab-grown meat not be labelled with the term 'beef' and that only products from traditional meat sources should be allowed to be labelled with 'beef'. Nevertheless, there is a strong argument for referring to cultured meat and cultured fish as meat and fish when providing consumers, such as those with meat allergies, with information about the nature of the product.

A positive effect of information on acceptance was also found by Rolland et al. (2020). Information about personal benefits increased acceptance even more than information about the quality and taste of cultured meat. No effect was found for societal benefits (animal welfare, environment), however. The initial acceptance of cultured meat was already rather high in that sample and therefore also prior awareness (Rolland et al., 2020), which influenced the results of the study. Furthermore, provided information about personal benefits conveyed the idea that cultured meat is superior to traditional meat in terms of healthiness, while information about taste highlighted similarities with traditional meat. Societal benefits, however, did not include information about product attributes. Thus, the findings of the presented studies suggest that information provision can impact willingness to try or acceptance, but only if certain (personal) benefits are emphasised.

5 A synthesis of possibilities

Considering the factors discussed in the current chapter that are associated with the acceptance of cultured meat, it is questionable whether cultured meat will disrupt the meat industry in the near future, as consulting companies have suggested (ATKearney, 2020). For sure, there are many consumer issues to address before cultured meat achieves market adoption. However, the benefits of cultured meat that are derived from its similarity to conventional meat indicate that this new food technology has great potential. Existing survey data suggest that interest in cultured meat varies considerably between countries and is dependent on the sample's characteristics. The lowest acceptance rates have been observed in France; however, approximately 20% of participants in France still answered that cultured meat is 'fun and/or promising' (Hocquette et al., 2022). The average European consumer does not appear to be a good target at the moment because of the negative attitudes towards novel food technologies. However, populations with relatively positive attitudes towards technological innovations in food systems, such as the Singaporean population, may be good starting points to gain early adopters of cultured meat. In fact, Singapore was the first country to officially approve cultured meat. Nevertheless, there are many unresolved issues concerning cultured meat production, and it will take some time until products are officially approved by authorities in Europe.

The topics outlined in this chapter lead to several open questions about cultured meat acceptance. First and foremost, how can producers and scientists effectively communicate the benefits and risks of cultured meat? This has important implications for public acceptance. To accelerate commercialisation, it is crucial to determine the optimal ways to name and advertise cultured meat. Positive impressions and expectations need to be built.

Providing informational claims might be one strategy for shaping perceptions (Onwezen et al., 2021). The similarity of cultured meat to conventional meat needs to be emphasised, as well as the risks associated with the consumption of conventionally produced meat (even though these might be different from the risks associated with cultured meat). The media also plays a crucial role in this process because it can shape consumers' risk perception. People whose concerns about eating meat are driven primarily by health and environmental reasons are also more likely to value the naturalness of products and see risks in consuming products developed using novel food technologies. Thus, technologically processed products, such as many alternative protein products (including cultured meat), are more likely to be rejected due to being unnatural. Additionally, it can be assumed that people with a high meat attachment, but who have moral concerns about (conventional) animal husbandry, are more likely to be interested in organic meat consumption. It is imaginable that they are also more open to cultured meat than people with low meat attachment. In this regard, the superiority of cultured meat over traditional meat is probably a selling point and needs to be emphasised.

It will not be possible to develop a single product that satisfies all consumer segments and needs. Thus, several approaches and products will coexist in the future and need to be promoted differently. From a nutritional perspective, cultured meat has great potential to outperform other plant-based alternatives that are additional carriers of salt, sugar and fat in a westernised diet (Macdiarmid, 2021). Moreover, consumers' concerns about the high degree of processing of meat-mimicking products and the many additives used in their production could have a positive impact on the attractiveness of cultured meat.

Notwithstanding all of the abovementioned discussion points, the acceptance of cultured meat may depend on whether the technology can be developed to produce textured meat that resembles whole cuts. This is often considered the holy grail of the alternative protein industry because of the challenges associated with generating products that replicate the texture of fillets, steaks and other pieces of intact meat tissue. It remains to be seen to what extent cultured meat products will resemble whole cuts. To date, most processed meat products, such as sausages and burgers, can be replicated with alternatives. These products are traditionally made from slaughter by-products, co-products and leftovers after premium pieces and whole cuts have been removed. Hence, they play a crucial role in the utilisation of the whole animal, from nose to tail. If alternative proteins substitute these processed meat products and consumers still seek premium cuts, then the co-products and inferior meat cuts will no longer be used. This might create the problem that animals are raised and slaughtered for whole cuts and the environmental impact of meat will further increase if not all products arising from the slaughter are used. Thus, it is of central importance that cultured meat products that are suitable substitutes for whole cuts are developed instead of cultured meat that can only substitute for processed products. The latter would reduce the share of an animal's body that is used for food production.

Today's marketing strategies and product recognition processes are very different to those of the nineteenth century. Beyond Meat is a prime example of how the viral digital spread of information about a product can lay the foundation for a successful market launch. Business leaders also play a key role as ambassadors and supporters. The promotion of certain lifestyles on social media platforms, such as Instagram, appeals primarily to the younger generations, and they will presumably be the main target group for cultured meat products. However, it is currently unclear whether people are interested in regularly consuming cultured meat products or are only driven by the excitement of just trying it. Thus, it remains to be seen how preferences will develop when products are available on the market.

6 Conclusion

The world population is growing and so too is the demand for meat (Godfray et al., 2018). However, there are planetary boundaries that limit food production via traditional methods; thus, novel agricultural solutions are needed that have a low environmental impact. If cultured meat is to be part of the solution, it must be accessible, affordable and sensory attractive. It needs to be seen whether this will be achieved in the near future.

Many factors related to the perception of cultured meat and the personality-like characteristics of consumers that shape acceptance have been identified (Bryant and Barnett, 2020; Onwezen et al., 2021; Pakseresht et al., 2022; Siegrist

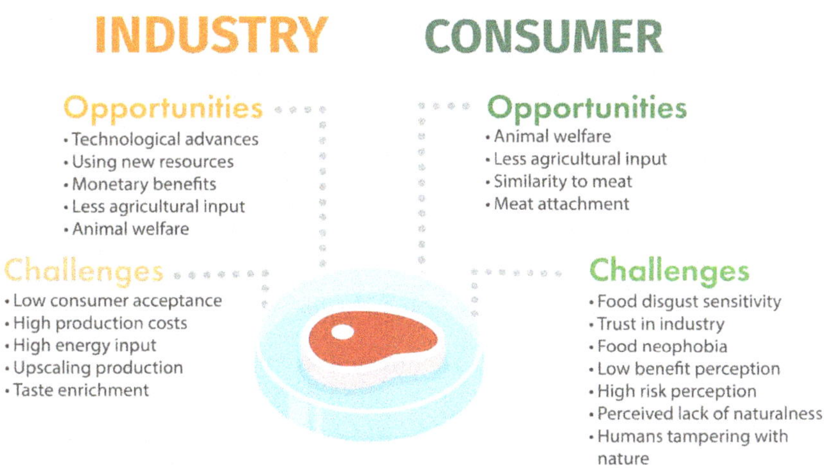

Figure 1 Opportunities and challenges associated with the acceptance of cultured meat for the industry and the consumer.

and Hartmann, 2020b). There is also research in progress that is focused on how to shift perceptions of cultured meat in a positive direction. The resultant insights can be used in efforts to promote cultured meat when it is ready for the mass market.

There are many opportunities and challenges for the cultured meat industry and the consumer (Fig. 1). Notably, the impact of the production and consumption of cultured meat on multiple Sustainable Development Goals has yet to be determined (Nobre, 2022).

7 Where to look for further information

The following review articles provide good overviews of the relevant predictors of acceptance of cultured meat:

- Bryant, C. and Barnett, J. (2018). Consumer acceptance of cultured meat: A systematic review. *Meat Science* 143, 8–17.
- Bryant, C. and Barnett, J. (2020). Consumer acceptance of cultured meat: An updated review (2018–2020). *Applied Sciences* 10(15), 5201.
- Onwezen, M. C., Bouwman, E. P., Reinders, M. J. and Dagevos, H. (2021). A systematic review on consumer acceptance of alternative proteins: Pulses, algae, insects, plant-based meat alternatives, and cultured meat. *Appetite* 159, 105058. https://doi.org/10.1016/j.appet.2020.105058.
- Pakseresht, A., Kaliji, S. A. and Canavari, M. (2022). Review of factors affecting consumer acceptance of cultured meat. *Appetite* 170, 105829.

Publications that highlight predictors of novel food technology acceptance:

- Siegrist, M. and Hartmann, C. (2020a). Consumer acceptance of novel food technologies. *Nature Food* 1(6), 343–350. https://doi.org/10.1038/s43016-020-0094-x.
- Tuorila, H. and Hartmann, C. (2020). Consumer responses to novel and unfamiliar foods. *Current Opinion in Food Science* 33, 1–8. https://doi.org/10.1016/j.cofs.2019.09.004.

A publication that focuses on meat culture:

- Volden, J. and Wethal, U. (2021). What Happens When Cultured Meat Meets Meat Culture? In Hansen, A. and Lykke Syse, K. (Eds.), *Changing Meat Cultures: Food Practices, Global Capitalism, and the Consumption of Animals* (1, 185). Rowman & Littlefield Publishers.(Volden and Wethal, 2021).

8 References

ATKearney (2020). When consumers go vegan, how much meat will be left on the table for agribusiness? Available at: https://www.kearney.com/consumer-retail/article/?/a/when-consumers-go-vegan-how-much-meat-will-be-left-on-the-table-for-agribusiness-.

Bastian, B. and Loughnan, S. (2017). Resolving the meat-paradox: A motivational account of morally troublesome behavior and its maintenance. *Personality and Social Psychology Review* 21(3), 278–299.

Bauer, J. M., Nielsen, K. S., Hofmann, W. and Reisch, L. A. (2022). Healthy eating in the wild: an experience-sampling study of how food environments and situational factors shape out-of-home dietary success. *Social Science and Medicine* 299, 114869.

Baum, C. M., Verbeke, W. and De Steur, H. (2022). Turning your weakness into my strength: how counter-messaging on conventional meat influences acceptance of cultured meat. *Food Quality and Preference* 97. https://doi.org/10.1016/j.foodqual.2021.104485.

Bonny, S. P., Gardner, G. E., Pethick, D. W. and Hocquette, J.-F. (2017). Artificial meat and the future of the meat industry. *Animal Production Science* 57(11), 2216–2223.

Bryant, C. J., Anderson, J. E., Asher, K. E., Green, C. and Gasteratos, K. (2019). Strategies for overcoming aversion to unnaturalness: the case of clean meat. *Meat Science* 154, 37–45.

Bryant, C. and Barnett, J. (2018). Consumer acceptance of cultured meat: A systematic review. *Meat Science* 143, 8–17.

Bryant, C. J. and Barnett, J. C. (2019). What's in a name? Consumer perceptions of in vitro meat under different names. *Appetite* 137, 104–113.

Bryant, C. and Barnett, J. (2020). Consumer acceptance of cultured meat: an updated review (2018–2020). *Applied Sciences* 10(15), 5201.

Bryant, C. and Sanctorum, H. (2021). Alternative proteins, evolving attitudes: comparing consumer attitudes to plant-based and cultured meat in Belgium in two consecutive years. *Appetite* 161, 105161. https://doi.org/10.1016/j.appet.2021.105161.

Bryant, C., Szejda, K., Parekh, N., Deshpande, V. and Tse, B. (2019). A survey of consumer perceptions of plant-based and clean meat in the USA, India, and China. *Frontiers in Sustainable Food Systems* 3. https://doi.org/10.3389/fsufs.2019.00011.

Cole, J. R. and McCoskey, S. (2013). Does global meat consumption follow an environmental Kuznets curve? *Sustainability: Science, Practice and Policy* 9(2), 26–36. https://doi.org/10.1080/15487733.2013.11908112.

Curtis, V. and Biran, A. (2001). Dirt, disgust, and disease: is hygiene in our genes? *Perspectives in Biology and Medicine* 44(1), 17–31.

Egolf, A., Hartmann, C. and Siegrist, M. (2019). When evolution works against the future: disgust's contributions to the acceptance of new food technologies. *Risk Analysis* 39(7), 1546–1559.

Godfray, H. C. J., Aveyard, P., Garnett, T., Hall, J. W., Key, T. J., Lorimer, J., Pierrehumbert, R. T., Scarborough, P., Springmann, M. and Jebb, S. A. (2018). Meat consumption, health, and the environment. *Science* 361(6399). https://doi.org/10.1126/science.aam5324.

Guan, X., Lei, Q., Yan, Q., Li, X., Zhou, J., Du, G. and Chen, J. (2021). Trends and ideas in technology, regulation and public acceptance of cultured meat. *Future Foods* 3, 100032. https://doi.org/10.1016/j.fufo.2021.100032.

Hartmann, C., Furtwaengler, P. and Siegrist, M. (2022). Consumers' evaluation of the environmental friendliness, healthiness and naturalness of meat, meat substitutes, and other protein-rich foods. *Food Quality and Preference* 97. https://doi.org/10.1016/j.foodqual.2021.104486.

Hartmann, C., Lazzarini, G., Funk, A. and Siegrist, M. (2021). Measuring consumers' knowledge of the environmental impact of foods. *Appetite* 167, 105622.

Hartmann, C. and Siegrist, M. (2017). Consumer perception and behaviour regarding sustainable protein consumption: A systematic review. *Trends in Food Science and Technology* 61, 11–25. https://doi.org/10.1016/j.tifs.2016.12.006.

Hartmann, C. and Siegrist, M. (2020). Our daily meat: justification, moral evaluation and willingness to substitute. *Food Quality and Preference* 80, 103799. https://doi.org/10.1016/j.foodqual.2019.103799.

Hocquette, A., Lambert, C., Sinquin, C., Peterolff, L., Wagner, Z., Bonny, S. P. F., Lebert, A. and Hocquette, J.-F. (2015). Educated consumers don't believe artificial meat is the solution to the problems with the meat industry. *Journal of Integrative Agriculture* 14(2), 273–284.

Hocquette, É., Liu, J., Ellies-Oury, M. P., Chriki, S. and Hocquette, J. F. (2022). Does the future of meat in France depend on cultured muscle cells? Answers from different consumer segments. *Meat Science* 188, 108776. https://doi.org/10.1016/j.meatsci.2022.108776.

Hoogendoorn, G., Sütterlin, B. and Siegrist, M. (2021). Tampering with nature: A systematic review. *Risk Analysis* 41(1), 141–156.

Loughnan, S., Bastian, B. and Haslam, N. (2014). The psychology of eating animals. *Current Directions in Psychological Science* 23(2), 104–108.

Macdiarmid, J. (2021). The food system and climate change: are plant-based diets becoming unhealthy and less environmentally sustainable? *Proceedings of the Nutrition Society*, 1–13.

Mancini, M. C. and Antonioli, F. (2019). Exploring consumers' attitude towards cultured meat in Italy. *Meat Science* 150, 101–110. https://doi.org/10.1016/j.meatsci.2018.12.014.

Nobre, F. S. (2022). Cultured meat and the sustainable development goals. *Trends in Food Science and Technology* 124, 140–153.

Onwezen, M. C., Bouwman, E. P., Reinders, M. J. and Dagevos, H. (2021). A systematic review on consumer acceptance of alternative proteins: pulses, algae, insects, plant-based meat alternatives, and cultured meat. *Appetite* 159, 105058. https://doi.org/10.1016/j.appet.2020.105058.

Pakseresht, A., Kaliji, S. A. and Canavari, M. (2022). Review of factors affecting consumer acceptance of cultured meat. *Appetite* 170, 105829.

Piochi, M., Micheloni, M. and Torri, L. (2022). Effect of informative claims on the attitude of Italian consumers towards cultured meat and relationship among variables used in an explicit approach. *Food Research International* 151, 110881. https://doi.org/10.1016/j.foodres.2021.110881.

Pliner, P. (1994). Development of measures of food neophobia in children. *Appetite* 23(2), 147–163. https://doi.org/10.1006/appe.1994.1043.

Rohrmann, S., Overvad, K., Bueno-de-Mesquita, H. B., Jakobsen, M. U., Egeberg, R., Tjønneland, A., Nailler, L., Boutron-Ruault, M.-C., Clavel-Chapelon, F. and Krogh, V. (2013). Meat consumption and mortality-results from the European Prospective Investigation into Cancer and Nutrition. *BMC Medicine* 11(1), 1–12.

Rolland, N. C. M., Markus, C. R. and Post, M. J. (2020). The effect of information content on acceptance of cultured meat in a tasting context. *PLoS ONE* 15(4), e0231176.

Roman, S., Sánchez-Siles, L. M. and Siegrist, M. (2017). The importance of food naturalness for consumers: results of a systematic review. *Trends in Food Science and Technology* 67, 44–57.

Rosenfeld, D. L. and Tomiyama, A. J. (2022). Would you eat a burger made in a petri dish? Why people feel disgusted by cultured meat. *Journal of Environmental Psychology* 80. https://doi.org/10.1016/j.jenvp.2022.101758.

Rothgerber, H. (2013). Real men don't eat (vegetable) quiche: masculinity and the justification of meat consumption. *Psychology of Men and Masculinity* 14(4), 363.

Rothgerber, H. (2015). Underlying differences between conscientious omnivores and vegetarians in the evaluation of meat and animals. *Appetite* 87, 251–258.

Siegrist, M. (2021). Trust and risk perception: A critical review of the literature. *Risk Analysis* 41(3), 480–490.

Siegrist, M. and Hartmann, C. (2020a). Consumer acceptance of novel food technologies. *Nature Food* 1(6), 343–350. https://doi.org/10.1038/s43016-020-0094-x.

Siegrist, M. and Hartmann, C. (2020b). Perceived naturalness, disgust, trust and food neophobia as predictors of cultured meat acceptance in ten countries. *Appetite* 155, 104814.

Siegrist, M., Hartmann, C. and Sutterlin, B. (2016). Biased perception about gene technology: how perceived naturalness and affect distort benefit perception. *Appetite* 96, 509–516. https://doi.org/10.1016/j.appet.2015.10.021.

Siegrist, M. and Sütterlin, B. (2017). Importance of perceived naturalness for acceptance of food additives and cultured meat. *Appetite* 113, 320–326. https://doi.org/10.1016/j.appet.2017.03.019.

Siegrist, M., Sütterlin, B. and Hartmann, C. (2018). Perceived naturalness and evoked disgust influence acceptance of cultured meat. *Meat Science* 139, 213–219.

Sjöberg, L. (2000). Perceived risk and tampering with nature. *Journal of Risk Research* 3(4), 353–367.

Slade, P. (2018). If you build it, will they eat it? Consumer preferences for plant-based and cultured meat burgers. *Appetite* 125, 428–437.

Terrizzi, J. A., Shook, N. J. and McDaniel, M. A. (2013). The behavioral immune system and social conservatism: A meta-analysis. *Evolution and Human Behavior* 34(2), 99–108.

Tomiyama, A. J., Kawecki, N. S., Rosenfeld, D. L., Jay, J. A., Rajagopal, D. and Rowat, A. C. (2020). Bridging the gap between the science of cultured meat and public perceptions. *Trends in Food Science and Technology* 104, 144–152.

Tuorila, H. and Hartmann, C. (2020). Consumer responses to novel and unfamiliar foods. *Current Opinion in Food Science* 33, 1–8. https://doi.org/10.1016/j.cofs.2019.09.004.

Van Loo, E. J., Caputo, V. and Lusk, J. L. (2020). Consumer preferences for farm-raised meat, lab-grown meat, and plant-based meat alternatives: does information or brand matter? *Food Policy* 95, 101931.

Verbeke, W., Marcu, A., Rutsaert, P., Gaspar, R., Seibt, B., Fletcher, D. and Barnett, J. (2015). 'Would you eat cultured meat?': consumers' reactions and attitude formation in Belgium, Portugal and the United Kingdom. *Meat Science* 102, 49–58. https://doi.org/10.1016/j.meatsci.2014.11.013.

Volden, J. and Wethal, U. (2021). What happens when cultured meat meets meat culture? In: Hansen, A. and Lykke Syse, K. (Eds). *Changing Meat Cultures: Food Practices,*

Global Capitalism, and the Consumption of Animals (vol. 1). Rowman & Littlefield Publishing Group Publishers.

Wilks, M. and Phillips, C. J. (2017). Attitudes to in vitro meat: A survey of potential consumers in the United States. *PLoS ONE* 12(2), e0171904.

Wilks, M., Phillips, C. J. C., Fielding, K. and Hornsey, M. J. (2019). Testing potential psychological predictors of attitudes towards cultured meat. *Appetite* 136, 137–145. https://doi.org/10.1016/j.appet.2019.01.027.

Willett, W., Rockström, J., Loken, B., Springmann, M., Lang, T., Vermeulen, S., Garnett, T., Tilman, D., DeClerck, F., Wood, A., Jonell, M., Clark, M., Gordon, L. J., Fanzo, J., Hawkes, C., Zurayk, R., Rivera, J. A., De Vries, W., Majele Sibanda, L., Afshin, A., Chaudhary, A., Herrero, M., Agustina, R., Branca, F., Lartey, A., Fan, S., Crona, B., Fox, E., Bignet, V., Troell, M., Lindahl, T., Singh, S., Cornell, S. E., Srinath Reddy, K., Narain, S., Nishtar, S. and Murray, C. J. L. (2019). Food in the Anthropocene: the EAT–Lancet Commission on healthy diets from sustainable food systems. *The Lancet* 393(10170), 447–492.

Chapter 13

Techno-economic analysis for cultivated meat production

John P. Ellersick, Next Rung Technology, LLC, USA; Elliot Swartz, The Good Food Institute, USA; and Reina Ashizawa, Next Rung Technology, LLC, USA

1 Introduction

The cultural, regulatory, and technical aspects of cultivated meat production have been presented in previous chapters as critical areas of focus required to achieve commercialization and to meaningfully address global sustainability objectives. This chapter addresses financial evaluations and modeling, which combine technical and economic factors to create the thorough business understanding required to successful scale the cultivated meat value chain.

http://dx.doi.org/10.19103/AS.2023.0130.16

Understanding of production economics is essential throughout the business lifecycle - from launch, through funding, technology development, and commercial scaling. Given the critical role cultivated meat plays in achieving sustainability and climate goals, this chapter also demonstrates how the same material inputs and technical drivers can be used to quantify key impacts such as water and land use, waste production, and carbon impacts.

The elements of cost, technology, external inputs and outputs, and sustainability are integrated into a techno-economic model (TEM) based on the proposed processing equipment and technology platform. The use of this model, in concert with financial parameters and evaluation activities such as sensitivity studies or business pro forma development, constitute the set of activities commonly referred to as techno-economic analysis (TEA).

Throughout this chapter, the required activities to develop a comprehensive TEM are presented in a stepwise manner, in sufficient detail to support the reader in developing their own working model. The goal of any model is to simplify underlying complex processes into a format that can be clearly understood and used to communicate and develop core areas of technology, business, markets, and execution plans.

This chapter first discusses how to develop and build models (Sections 2 through 9) with a focus on the fundamental objectives of creating a TEM, as well as the major inputs and outputs of the model. Effective TEMs are not static and continue to develop throughout the life cycle of a business to track technology changes and improvements and support effective responses to changes in the external economic environment and to business opportunities. Impactful uses of the TEM model for critical business objectives are further presented in Section 10 discussing technical and business uses of the model, including the capability to model sustainability impacts such as water consumption, land use, and carbon emissions.

While this chapter does not aim to develop a cultured meat TEA, Sections 11 through 14 do provide:

- a review of five significant cultivated meat industry models that have been published to date;
- a simple case study on the development of a model for a representative, simplified cultivated meat process and approach; and
- a final summary of the topic as well as guidance on where to look for further information.

The TEA process and techno-economic modeling highlight areas of scale, cost reduction, necessary technological advances, and business and sustainability objectives that require addressing. These models are core to the development and growth of cultivated meat processes and the industry, as we seek more sustainable and cost-effective ways to supply the world's protein and meat in a matter the public will accept and savor in the decades to come.

2 Objectives of techno-economic analyses

TEA of agricultural technologies seeks to achieve multiple fundamental objectives:

- To develop an understanding of the complete financial elements of any given endeavor in a clear, simple, and flexible manner;
- To capture the key costs of inputs to the process, the value of outputs including primary products and co-products, as well as the expenses associated with wastes, emissions, and other impacts;
- To integrate the model with a meaningful technical representation of the underlying processes;
- To integrate the model with the key economic parameters of importance, including financial parameters and supply chain cost parameters; and
- To present the process and these results clearly and simply for multiple audiences including key C-level members of any organization (CEO, CFO, and CTO), technical audiences, leaders seeking to use the results to determine how process changes may impact economics, financial professionals, and importantly (potential) investors to see the impacts of financial parameters.

A high-quality TEM presents complex information in a simple and interpretable manner. This is achieved by developing a tool that allows for easy adjustments to key parameters and a clear presentation of economic impacts. Understanding and focusing on major cost drivers is key for the cultivated meat industry to achieve a delivery price that is affordable, which in combination with higher quality products can lead to mass adoption.

3 Basic structures and philosophy of techno-economic analysis models

This chapter will focus on the elements to build a business-focused TEM, oriented to capture required capital investment, operating costs, and potential production improvements to create a complete financial picture. Financial results and costs of goods are presented with simple and clear economic results in the language of the key users (e.g. return on investment) to drive smart, effective technical, business, and investment decisions.

The model structure, as presented, is applicable to any basic spreadsheet tool. Microsoft Excel was used for this example, with the following critical areas developed and discussed in separate spreadsheet tabs:

- Faceplate;
- Technical, facility, and operating parameters;
- Material balance;
- Plant design and equipment list; and
- Cost references

Further sections are commonly developed as spreadsheet tabs to capture the results of tool use, allowing for case studies and sensitivity analyses to be carried out, as presented in later sections of this chapter. Process design or simulation software may also be used to generate economics for the process areas being modeled, with the caveat that cost information in any third-party provided model should be independently verified for accuracy and applicability.

Developing clear and usable models can be assisted through good 'book-keeping' practices. In this chapter, the following conventions are used: any user-changeable inputs are shaded in light blue, calculated values are unshaded, and units of measure are always explicitly shown.

TEM practitioners are encouraged to emphasize clarity over complexity and a philosophy of 'showing your work.' The best models make the model development approach clear for use by others. Buried formulas and complex calculations limit the usability of a model by other parties and should be avoided.

4 The faceplate: a control panel for the techno-economic analysis model

The primary objective of the modeling activity is to understand the complete financial elements of the target business in a clear, simple, and flexible manner. The modeling process begins with the translation of critical inputs and outputs into an easily interpretable dashboard called the faceplate (Fig. 1). The reader should know that the figures and tables included throughout this chapter do not show a real TEA study of cultivated meat, and numbers included are simply representative values from the industry.

The key input elements typically included in the faceplate are listed subsequently. These elements serve to allow streamlined communication between technical, commercial, project, and financial teams. The information presented in the faceplate is thus interpretable by broad audiences and can be used to demonstrate impacts such as economies of scale or economic impacts based on productivity improvements:

- **Scenario modeling:** Includes information about the facility size or production volume output of the facility. This information can be adjusted

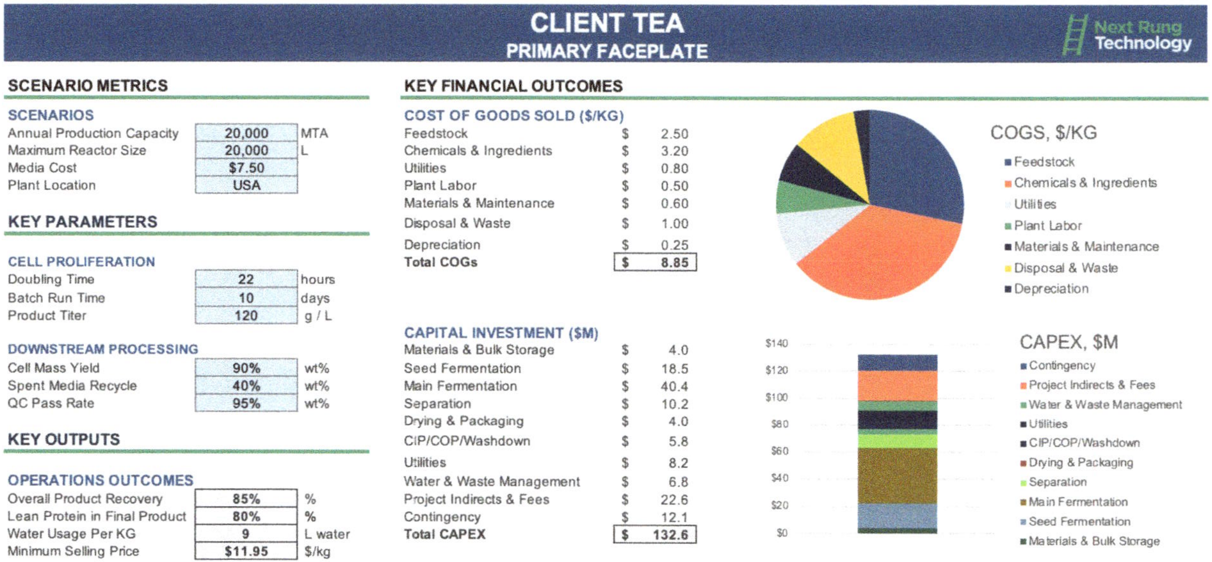

Figure 1 A representative faceplate of a business-focused TEM for the cultivated meat industry.

to analyze costs at different scales or opportunities associated with economies of scale.
- **Major ingredients and other inputs:** Pricing of major ingredients or inputs that drive process economics. Optionality for different ingredients may also be included for case study analyses.
- **Utilities.** Pricing for key utilities, which varies significantly by locality.
- **Process parameters.** Key process parameters that drive process efficiency and economics. For cultivated meat, this may include cell density, doubling time, and other productivity or quality-related metrics.
- **Financial parameters.** Key economic inputs that can be used for financial evaluation. Examples may include depreciation parameters, pricing options, scenarios for interest and inflation, and insurance and maintenance.

The key output elements typically include:

- **Cost of goods sold (COGS).** An easily understood parameter that accounts for the direct costs incurred for producing a given quantity or volume of product.
- **Profitability metrics.** This parameter can be expressed in various terms such as net present value (NPV) over a specified timeframe or absolute margin (sales over total costs).
- **Economic returns on investment.** Calculations for determining the overall return on investment (ROI) or when return on initial investment may be reached.

5 Major inputs to the techno-economic model

The most common major inputs to the model will typically be captured in a separate tab centered on technical, facility, and operating parameters. This tab will be referred to as the 'calculator' section. These parameters include:

- production facility parameters and scale;
- ingredients and other inputs supporting production;
- processing parameters, conditions, chemicals;
- product information, recipes, co-products;
- targeted markets, and market information; and
- production platforms and requirements - farm, plant, processing to final product.

5.1 Facility parameters and scale

Plant size and operating parameters are key drivers in the development and commercialization of a technology. To properly understand and plan for these factors, one must consider:

- understanding the economies of scale, often for the initial pilot plant, first commercial plant, and the ultimate commercial plant;
- understanding how a plant may be developed in phases;
- recognizing how the overhead expenses to bring a facility into operations can be spread to reduce costs as plant operations expand; and
- thinking about overlaying a process onto an existing operating partner's facility, such that only a portion of capital and or operating expenses may be borne by the next added capacity.

A typical facility parameters section as shown in Table 1 is simple; however, it has significant impacts on results and is an especially useful tool to model both scale and how a production plant may operate.

5.2 Ingredients and other inputs supporting production

Rudimentary TEMs often start simply with just the costs of inputs and ingredients to check the basic feasibility of a process and product. It is recommended that the ingredients are managed carefully to achieve both accuracy and clarity by distributing the location of inputs in three locations:

- Major inputs of importance for study evaluations should be in the faceplate, with the values then carried back to the calculator section and material balance sections. In the cultivated meat industry, this may be used to study various growth media options, proportions, or approaches to the use of fat additives, scaffolding, microcarriers, etc.
- General inputs that contribute to operating costs are often more fully articulated in the operating costs section of the calculator tab. This section will pull together pricing as well as usage rates for the process and

Table 1 Example high-level facility parameters typically included for commercial bioprocessing facilities

High-Level Facility Parameter	Typical Unit	Details
Annual production target	MTA	Annual production target for production plant
Annual operating basis	hours	Annual scheduled runtime for production plant
Plant operating availability	%	Ratio of actual runtime to scheduled runtime, accounting for downtime
Plant performance	%	Accounts for batch failures
Product quality	%	Ratio of approved product to total product made
Incoming material hold time	days	Storage of incoming materials
Outgoing product hold time	days	Storage of outgoing product

conditions being modeled and convert them to annual costs. As desired, major categories of costs can be pulled out and used for case studies.

- Complex inputs are generally simplified by creating separate tabs specifically for their formulas, as many media may have 15 or more ingredients. These are often the focus of cost reductions. Simplifying strategies, such as creating media formulas for years 1, 2, and 5, along with their associated costs, can be used to clarify what is typically a complex area of cultivated meat development.

5.3 Processing parameters, conditions, chemicals, utilities, waste management and recovery

Processing parameters are the core of any TEM; they are the transcription of how a given production process is translated into the model to analyze and evaluate approaches. A step-by-step description of the process - from the moment materials arrive to the production facility, to how they are stored, mixed, processed, and used to create raw products, and eventually methods by which they are purified and formulated into final products - can be extensive but is critical. Guidance for this section is to organize in blocks of processing, such as the following:

- Material receipt and storage;
- Media and chemicals preparation;
- Processing of inputs to proliferation and differentiation;
- Proliferation;
- Differentiation;
- Intermediate product processing;
- Final product(s) processing;

Table 2 Example process parameters typically included for cultivated meat production

Process Parameter	Typical Unit	Details
Seeding density	%	Volume of seed inoculated at start of production run
Final cell concentration, wet cell basis	g WCW/L	Final concentration of cell mass at end of production run
Maximum size, main reactor	L	Maximum allowable reactor size
Working volume, main reactor	%	Working volume of main reactor
Production time	hours	Production time in main reactor
Turnaround time	hours	Turnaround time in main reactor, accounting for transfer and cleaning
Aeration requirements	vvm	Aeration required for main reactor

- Product storage and transport;
- Waste collection and media reuse;
- Cleaning systems; and
- Waste treatment

Within each of these blocks, the full operating processes and parameters of importance should be simply and clearly mapped out for both technical and nontechnical team members to understand.

5.4 Product information, recipes and co-products

A facility or proposed process will likely have multiple potential end products or markets or require some additives or other ingredients brought in to make finished products. For a TEA, this can be modeled by a series of relevant recipe parameters, as in the following case, where a pure meat product is adjusted with ingredients to make a product that is more like the ground beef products found on grocery shelves today. Other examples that could be listed here include the valorization of co-products derived from ammonia or lactic acid collection from spent cell culture media.

5.5 Market information

Just as different product formulations are required to produce one main product or multiple products, the products may also be sold into different markets at different price points. Collectively, the recipes, formulas, products and co-products, and market opportunities can be combined with expected pricing for these various products to project the revenue for a business model.

Modeling the revenues is not always necessary if an economic model is focused on the costs of production. However, it is generally good practice and can provide an important mechanism for including the various grades of a product or incorporating co-product credit into the overall costs of production and goods.

Table 3 Example product formulation parameters typically included for cultivated meat production

Product Formulation Parameter	Typical Unit
Cell mass	wt. %
Scaffolding	wt. %
Fat source 1	wt. %
Fat source 2	wt. %
Coloring	wt. %
Flavoring	wt. %

Table 4 Example operating revenue and gross profit margin block

Product Line	Target Pricing	Product Allocation	Production	Revenues
Product 1	$12 / KG	40%	8 000 000	$96 000 000
Product 2	$10 / KG	60%	12 000 000	$120 000 000
Total Plant Revenues				$216 000 000
Annual Cost of Production				$150 000 000
Gross Profit				$66 000 000
Gross Margin				30%

Table 5 Example operating expenses associated with a commercial cultivated meat production facility

Expense Category	Operating Expense
Media Ingredients	Basal media, dry ingredients Glucose Protein hydrosylates Growth factors Supplementary amino acids
Product Ingredients	Fat source 1 Fat source 2 Colorings Flavorings
Chemicals	Caustic Ammonia Antifoam
Utilities	Electricity Natural gas Water
Plant Labor	Leadership Technical Operators
Plant Material Costs	Replacement filters Spares
Waste Disposal	Wastewater BOD surcharge Solid Waste

5.6 Production platforms and requirements – farm, plant and processing to final product

This section serves as a catch-all to note and highlight that all other parameters within the scope of modeling inputs should be identified and set up to completely map the entirety of the endeavor. As a starting point, all costs at sources, transport, and storage requirements, as well as final processing, water recovery reuse, and waste treatment parameters, should be included. Additionally, key plant items such as staffing and labor requirements, materials usage, and equipment rentals should be modeled here. The unit pricing and usage rates are converted to annual cost and ultimately costs per pound or kilogram of product.

6 Major outputs of a techno-economic model

The plant design itself is built up from input technical parameters, sizing and sparing decisions, and the material balance as developed in Section 7. The material balance then feeds both the operating costs, as discussed later, as well as being used for further plant equipment design. The sized equipment list is developed in its own tab, and with the sized equipment come other key elements such as total project capital cost, power requirements, and other elements of operating costs that are further incorporated in all sections of the operating costs modeling.

Many economic models will differentiate costs as either fixed (capital investment or other costs that do not vary with operating throughput) or variable costs which vary with throughput in a facility (such as media ingredients). While this is a common way to organize cost information, in dynamic techno-economic models, fixed costs will typically change with processing capacity, and thus, that terminology can be ambiguous and is not presented here.

6.1 Operating costs

In a simple view of TEMs, the inputs and outputs can be expressed as materials costs in, products cost out, operating costs for other supplies and utilities, and capital costs for the facility itself. Within operating costs are the following categories of information that are of varying levels of difficulty to estimate and model well:

- **Chemicals**. These are practical aspects of operations, most often cleaning-in-place chemicals (CIP) that are typically caustic or food-grade formulas as well as any chemical additions for pH adjustment of streams or food formulations.
- **Utilities**. There is consistently the same set of plant utilities required for most operating facilities, which include the following:
    - o Electricity - for plant power users;
    - o Steam - for sterilization of bioreactors and other equipment, this may be purchased as steam from a site or generated using a fuel (commonly natural gas);

- o Cooling water - for cooling of bioreactors and other cooling uses that can be supported by temperatures achieved by atmospheric cooling towers;

Chilled water - for chilling of intermediate or product materials to colder temperatures often 40 degrees Fahrenheit or below;

- o Raw water treatment and water purification [reverse osmosis (RO), deionization] - preparing incoming water for use in contact with food;
- o Plant and instrument air - used in bioreactors as well as for plant uses such as control valve actuation; and
- o Wastewater collection and treatment.

Although many models will seek to charge utilities out in a simplified format (e.g. cost per thousand gallons of treated water, cooling water), in practice, these utilities can be further reduced to the type of utilities an operating facility buys every day: electricity, raw water, and natural gas.

To be consistent, the capital cost to produce these utilities (e.g. a cooling tower) is carried out as part of the capital plant design. Utilities are often found to be impactful, and these insights can allow for the evaluation of different siting locations for a facility, where water availability and energy costs may be vastly different.

- **Plant setup and staffing**. Staffing of facilities is typically a key element of cost, especially in the initial stages and/or smaller scales of plant operations. Regardless of size, plants will require plant leadership, lead technical staff, mechanical staff and capabilities, and operations teams. Geographic location is a prime determinant of labor costs, and often a faceplate variable is used to compare project economics with labor rates at various locations compared to a basis project located (i.e. % labor rate relative to a reference location).
- **Wastes, wastewater, and other materials disposal**. This is another major category and can also vary sufficiently by location. Understanding these elements of plant production is critical to understand the true economics of any undertaking, and several simple rules can be kept in mind:
    - o A complete accounting of everything that comes in and goes out, all water flows, and all waste flows is critical for true economic evaluations.
    - o Determination of whether a product becomes a co-product to be monetized or becomes a waste with associated costs for disposal in landfills. Key waste categories include simple materials and disposables (gloves, filters, etc.), solid wastes separated from

wastewater and washings, and treated wastewater, which generally involves a charge or surcharge for flow and biological oxygen demand (BOD). The costs associated with wastewater treatment are site specific and depend upon what treatment is required to be provided by the production facility (commonly wastewater source collection and equalization in a holding tank, solid removal, and pH adjustment) and further biological treatment (typically aerobic and/or anaerobic waste treatment) which may either be provided by the production facility or external parties (a municipality or a site host).

- **Supplies and consumables**. Many plant operations will involve consumable materials such as filters, gloves, sample jars, packaging, etc.
- **Other**. Other operating costs may include equipment maintenance and spares (typically estimated as a percentage of plant capital cost), insurance, taxes, royalties, and fees.

Together, these operating costs are summarized in Table 6, converting unit pricing and usage rates to an annualized cost.

6.2 Capital costs and investment (depreciation)

While the plant design itself is built using input technical parameters, sizing decisions, and the material balance (as developed in Section 7), the sized equipment list is developed in its own tab. Other key elements are related to the sized equipment, including total project capital cost, power requirements, and other elements of operating costs and overhead. Examples of how the capital cost is built up are shown subsequently, following cost estimating methodology based on pricing equipment and estimating the installation costs as a multiplier of the equipment cost to arrive at the project's direct cost. Further costs are

Table 6 Equipment items and typical sizing considerations

Equipment Line Items	Equipment Sizing Considerations
Tanks	Mass balance flow, residence time
Agitators	Tank size, mixing requirement
Pumps	Mass balance flow, transfer time
Bioreactors	Designated maximum bioreactor size, mass balance flow, production time, headspace
DSP Equipment	Mass balance flow
CIP Skid	Equipment CIP requirements
Boiler	Steam usage
Air Compressors	Process aeration requirements, equipment air requirements
Chillers	Process cooling requirements

incurred during project execution, commonly referred to as indirect costs, which are related to engineering design, project management, and the many fees and services paid by project owners in the execution and delivery of a project.

A sized and priced equipment list is developed as discussed in Section 7, with line-item equipment costs and requirements developed for each area, as shown in Table 8. The area-by-area equipment costs are then totaled for

Table 7 Typical expenses included in a plant capital cost summary

Project Capital Costs	Method of Estimation
Building fit out cost	Space estimate from equipment
Site/shell cost	Space estimate for enclosed areas
Process equipment cost	Individually sized and priced equipment items
Equipment installation	Installation factors applied for each equipment item
Unpriced equipment	Factored estimate on total installed equipment cost
Project indirects	Factored estimate; includes engineering, procurement, management
Owner's costs	Factored estimate; includes commissioning, startup
Escalation	Factored estimate on total project cost
Contingency	Factored estimate on total project cost

Table 8 Example summary of product demand and revenues

Product volumes (MM lbs)	Year 1	Year 2	Year 3	Year 4	Year 5
Product 1	23.3	100.0	233.3	300.0	666.7
Product 2		6.3	50.0	125.0	250.0
Product 3			4.3	57.1	171.4
Total	**25.3**	**133.3**	**418.3**	**872.1**	**2133.8**
Product Revenue	**Year 1**	**Year 2**	**Year 3**	**Year 4**	**Year 5**
Product 1	$1.40	$6.00	$14.00	$18.00	$40.00
Product 2		$0.50	$4.00	$10.00	$20.00
Product 3			$0.30	$4.00	$12.00
Total	**$1.60**	**$9.20**	**$30.60**	**$66.50**	**$162.00**

Table 9 Example of equipment sizing parameters to convert a material balance to sized equipment list. Knowledge of similar applications is often required to estimate motor size for mixing

Equipment type	Typical sizing parameter	Typical design allowance/ margin
Pump	Forward flow rate	110% of flow
Tank	½ to 1 h for process tanks12 h to 3 days for storage tanks	90% full
Mixer	Sized to match tank being mixed, with motor size related to mixing requirements	None

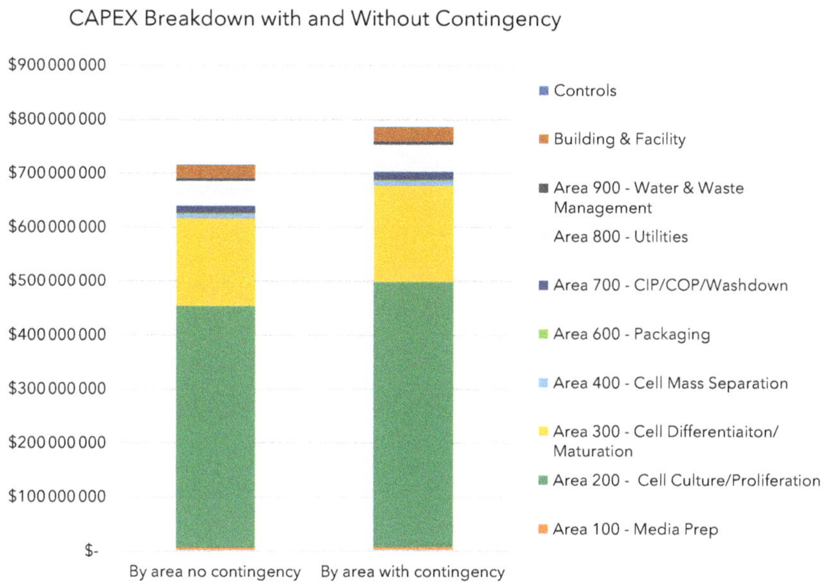

Figure 2 Simplified summary of project capital cost by area with and without contingency.

the entire project, as shown in Table 9 to develop a full capital cost estimate including a contingency allowance.

Capital costs are often further simplified and visualized for broader audiences. In Fig. 2, the calculated values are presented in the left case study, and these same values with a 10% contingency incorporated (a general good practice for capital management) are provided in the right case study.

For use in follow-on COGS evaluations, capital cost information is further simplified using relevant financial parameters to convert capital cost investments into an annual depreciated value that may be applied over the production capacity to convert it to represent the capitalized costs for any facility's operations.

While there are many more sophisticated and in-depth ways to view the cost of capital and investments over time, most analyses reduce the cost of capital to a straight-line depreciated value over a specified duration, most commonly 10, 15, or 20 years. These simplifications support clear and usable model creation.

6.3 Cost of goods sold

The economic picture of the costs of production can now be brought into simpler and clearer forms of definition through the COGS metric. Typical units for the cost of goods are $ or €/lb. or kilogram, which represent all the fixed and variable costs divided by the production rate of the sellable product.

For audiences with an interest in the detailed breakdown of costs, COGS are presented in visualizations on the main faceplate. For financial audiences and for the use in side-by-side case studies, a stacked bar approach is commonly used. A stacked bar visualization continues the objective of simplifying complex information and can support financial audiences who may be considering

- what needs to be financed versus being an operating expense;
- what are the labor requirements, and how might that change with location and scale?; and
- how do different sites with differing utility costs impact the overall costs, etc.?

6.4 Revenues

The revenue output comes as a combination of all goods produced, pricing, and markets. As with other parameters, the sources of revenues can be provided with varying levels of detail to provide a snapshot in time with some detail or simplified for broader audiences and for comparisons in case studies of approaches to market and changes over time.

6.5 Overall financial and key business parameters (rate of return, cash flow, margin and sustainability metrics)

Once the modeling results and outputs are fully defined, key business parameters such as margin, rate of return on investments, and cash flow projections over time can be calculated. Cash flow modeling can further support major financial investment decisions including calculating NPV or net present cost, which summarize the enterprise value and profitability of the project. Another common advanced application is to develop manufacturing models in support of 'make versus buy' decisions for partial or complete process areas, and for evaluating the use of a contract manufacturing organization to outsource production versus developing and making one's own production facilities. Finally, as many businesses incorporate and clearly track sustainability metrics, outputs such as water and land use, or carbon footprint have become common outputs.

7 The underlying model: mapping the process and facility – onsites

For a TEM to be effective, the model must be able to convert technical approaches into a process flow through a facility at various scales and then represent the equipment and capital investments required, operating costs, especially for utilities and process-related items, and the true costs of facilities, plants, or farms. These are achieved through the development of a material balance that is kept simple, yet adjustable to any key inputs or variable setup in the faceplate or technical input section.

7.1 Keys to setting up a material balance

Material balances are best kept simple and clear, with ingredients limited only to what must be modeled and total flows tracked to make it easy to cross-check numbers in useful units. Cross references may be provided to clearly indicate where the information came from, to correlate streams to a PFD or other pictorial reference, and to identify key factors that may be used to calculate a stream's composition or flow. Examples include the solids concentration (wt%) that may be expected out of a filter press. Being explicit and clear with formulas supports use by third parties.

In the subsequent figures, typical material balances are shown for a limited number of components, and often below the mass flows, other key process parameters may be carried such as the mass fraction composition, calculated densities, or process conditions. Understanding flows in this section supports all the key economic parameters related to product production and equipment sizing. Tracking the flows as mass flows of individual components and streams, combined with the knowledge of the density of both streams and components, allows for the subsequent use of flow information in either mass or volumetric flow as is required for the follow-on calculations of equipment sizing and utility usage.

7.2 Creating a sized and scaling equipment list for onsite equipment

With the foundation of a material balance, the process of equipment sizing and then costing can be conducted. Equipment sizing can primarily be developed based on an understanding of the flows through the equipment, the time-related sizing parameter that is used to size equipment, and any adjustments to capacity for working capacity/design margin and sparing approach.

Sizing calculations do not need to be rounded to a more typical nominal volume, as sizing will adjust as the inputs to the model are changed to understand the economic impacts of size and process conditions.

7.3 Capturing utility impacts from equipment

The primary philosophy for estimating utility impacts is to focus first on major utility users and then on minor users. Finding relevant utility guidelines can be difficult without access to true plant design information. As an example, the model is developed in this section, assuming this information has been made available. However, utility usage can also be built up from basic process fundamentals (e.g. pump power requirements and mixer requirements). In the following section, the utility usage will be quantified on a plantwide basis as part of utilities and 'offsites'.

The key utilities to be tracked in a TEM are generally the fundamental input values one would bring to a site to have it operate - power, natural gas (typically) for steam production, and water (if not created onsite by a well). The most complete models will then have cost and processing information incorporated for major utility systems that will create the utilities used at processes such as softened or RO water, steam of different pressure levels and cleanliness categories, CIP system, as well as wastewater treatment operating costs.

Utility estimates come from several sources, building off the material balance and sized equipment list:

- The sized equipment list includes motors.
- Utility summaries to address intermediate plant utilities such as steam usage or CIP solutions on a user basis.
- Rough estimates that can be used to scale from other similar facilities (CIP, steam out, etc.) or from government or nonprofit-funded studies and resources for overall plant use are discussed.

8 The underlying model: mapping the process and facility - offsites and utilities

For a model to be truly effective in understanding the production costs of cultivated meat, the plant and facility's entire requirements must be understood and built into a cost model that expands beyond the core processing areas. Generally called offsites, these areas of a facility typically include the following: shipping and receiving, incoming materials storage, outgoing product materials storage, major utilities, site infrastructure, water, and wastewater treatment.

For every facility, a general philosophy is typically developed for the quantity of incoming materials that will be stored onsite (commonly one to two weeks) and the quantity of outgoing product. These basic plant decisions are commonly passed through into the model for related tanks, equipment, and utilities sizing. The basis then sizes equipment such as storage tanks, their mixers, and pumps, as well as facility and building requirements for dry, refrigerated, or frozen storage of ingredients and product.

Table 10 Example of development of utilities usage from typical plant users and material balance

Utility	Individual users	Facility users
Power	Sized equipment list	Summary of process users, with added allowance for buildings or facility usage
Energy - typically Natural gas	Steam usage, any power usage by turbine or other fired	Summary of plant users, with general allowances for intermittent steam usage, which may be a function of vessel sizes that are CIP'ed
Water	Water used in mixing tanks, recipe preparation, to wash product intermediate or streams	Cooling towers, boilers, water softeners, RO water generators all use raw water or treated water sources in the production of a plant-used utility
Wastewater	The material balance will be used to identify waste streams created in the process; this may include the filtration step permeate or filtrate (the unused portion), product washing or rinsing steps	Plant utility systems will generate wastewater streams such as utilities blowdown streams, water treatment reject streams

Table 11 Studies included in the analysis.

Authors	Year	Title
Vergeer, R., Sinke, P. and Odegard, I.	2021	TEA of cultivated meat: future projections for different scenarios
Humbird, D.[a]	2021	Scale-up economics for cultured meat
Risner, D., Li, F., Fell, J. S., Pace, S. A., Siegel, J. B., Tagkopoulos, I. and Spang, E. S.	2021	Preliminary Techno-Economic Assessment of Animal Cell-Based Meat
Negulescu, P. G., Risner, D., Spang, E. S., Sumner, D., Block, D., Nandi, S. and McDonald, K. A.	2022	Techno-Economic Modeling and Assessment of Cultivated Meat: Impact of Production Bioreactor Scale
Garrison, G. L., Biermacher, J. T. and Brorsen, B. W.	2022	How much will large-scale production of cell-cultured meat cost?

[a]The Humbird study included in this chapter is the version that has been shortened for peer review. An extended analysis has also been published as a preprint (Humbird, 2020). There are a few minor differences between the two studies

8.1 Capturing equipment impacts from tankage, shipping and receiving

A typical top-level input template is shown subsequently, along with the corresponding impact on several storage tanks and storage areas for incoming materials.

All incoming materials should then be assessed based on the type of storage that may be required. Dry goods with no special storage requirements will be allocated to a basic storage area, refrigerated dry goods will require a conditioned space with general dry storage equipment support, and refrigerated liquid goods may be stored either indoors if the material arrives in totes or in larger cooled or chilled tanks if the material is used in bulk and may be stored in an agitated indoor or outdoor located tank. Other common storage approaches include bulk tankage for nonrefrigerated liquid goods, mixed tankage to make up daily batches of materials if constituted from dry ingredients, and heated tankage for liquids such as sugars, which are commonly heated to support keeping them both as a liquid and sterile.

In addition to storage vessels, TEAs should track impacts related to warehouse space for the ingredients noted previously and for chemicals that may be used in utility or wastewater areas of the plant. Outdoor space for bulk tankage, truck loading and unloading racks, and the roads and weigh scales that may be used to support transport and receipt can also contribute significantly to a project's storage cost and should be accounted for.

8.2 Major utilities

The final major category of equipment to be modeled is the utilities themselves. Nearly all facilities will be expected to have the following major utilities:

- Significant water treatment is a function of the water quality that is available to the facility. This may begin with untreated raw water from wells that require management of sediments, or potable water may be available to the plant. Further treatment commonly includes softening, the creation of RO water, and even deionization (IX water). The economics will show that the highest grades are best used in limited quantities, and each plant will develop a philosophy of when to use lower or higher grades of water. In all cases, for food products, the water choice should be compliant with food safety standards.
- Air compressors to supply plant instrument air, and any processes that are aerated including the cultivation processes.
- Other industrial gasses including oxygen, nitrogen, and CO_2 may be generated onsite, or purchased by a major supplier who can provide capital and operating costs for establishment.
- Steam is used throughout most cultivated meat facilities for heating and for managing sterility and cleanliness, whether in support of a CIP system, steam in place (SIP) system, or clean out of place system. The steam itself

is derived from fuel, and that cost is carried typically as natural gas or whatever other energy sources may be used (e.g. electric and oil).

- Wastewater is perhaps the most important and often overlooked component of the plant infrastructure. Countries have national standards that set the guidelines for what can be discharged to either a local collection system or a receiving body of water. Wastewater sources are best understood as lower-flow, high-strength sources that resemble general food or brewery waste streams and higher-flow, lowly contaminated (generally clean) sources that may be good candidates for minimal treatment and possible reuse, cleaning wastes, and other sources such as rinsing wastes. The preferred approach for source management is to start by understanding each source separately in a source segregation table, with minimum/normal/maximum flows and composition (pH, BOD, and COD) defined and then the system and approach managed as a whole.

Of all these utilities the management of water is one of the least sustainable aspects of the early generation of cultivated meat production. It is becoming critical for all facilities to closely study their water usage and reuse requirements, including investment in membrane-based recycling and reuse systems to capture clean water and often unused media components to improve economics and reduce waste stream strength. Technologies employed may include various forms of pretreatment, ultrafiltration (UF), RO, and other types of treatment including vibrating membranes (e.g. VSEP) and highly cleanable emerging element technologies such as Super Filtration.

9 Overview of a step-by-step approach to techno-economic analysis development

The TEA development process starts with having an underlying approach and technology to be evaluated that is well understood. A block flow diagram that tracks all the major processing units is then generated to give an overview of the facility. This diagram identifies all the major ins and outs to the battery limits of the process being studied – ingredients and products, major utilities, waste, and wastewater.

With a major facility concept developed, the processing steps should be further mapped out to the level of a process flow diagram (PFD) showing each piece of equipment that will be used in the plant, its function, and the order it is used in processing. This diagram will then be used to serve as the basis for an equipment list, identification of the streams to include in the material balance, and to construct utility usage estimates.

Major streams in the PFD are then numbered, and these numbered streams are tracked in the material balance section, with the flow rates used to set the equipment size in the equipment list and to track the ins and outs used for overall economic evaluations.

Finally, with the underpinnings of a model in place, the technology itself should be expressed in the TEA calculator tab. This is the place for an understandable translation of the technical process from the key technical parameters of the plant design itself. The TEA calculator section has both input values (shaded) and calculated values, all seeking to be as explicit as possible in the calculations being performed so third-party users can follow the model or use it themselves.

The faceplate should be developed in close connection with the end users who are more financially and less technically focused. Faceplate calculations and graphical results are those that are most impactful to the plant economics and will be connected directly back into the TEA calculator page. The faceplate's purpose is to achieve clarity and simplicity and to answer the primary users' questions, which are commonly centered around asking what are the economics of the approach if performed today, and what improvements can be made to make the process more economical or sustainable in the future. These end uses of the model, and where its power truly lies, are discussed in Section 10.

10 Beyond the basics

With the completion of a first comprehensive model, the model's use can expand beyond the basic cost of goods type evaluations to provide insights into areas that are key to business and technical success. In this section, we will highlight three of the primary uses of TEA models for research and development programs, business development and funding activities and for sustainability considerations.

10.1 R&D, process development and optimization

If the fundamental structure and buildup of the TEA model is thorough and correct, the most impactful process input parameters should be prominently located in the faceplate, or in a dedicated block in the TEA calculator page. The model can be used to conduct a series of case studies on a single parameter (such as titer in the case shown later), or on multiple parameters, perhaps representing a programs' progress over time as multiple aspects of a process are improved.

Because the model is bottom-up and complete, changing these input parameters will impact all key aspects of the plant model – CAPEX will adjust with sizing, flow stream, and utility usage will change and often hidden or secondary impacts of process changes on economics can be understood. In the example of product titer, for a given production target, an increase there would effectively reduce most of the supporting plant size and large number of other parameters that changed. As a word of note and caution, when changing parameters, the inputs to the model should be complete and consistent - to get higher yields, if higher materials inputs are also needed, those input values should all be changed in one consistent case study that can realistically be achieved.

In Fig. 7, an entire program's life cycle and goals are represented in case studies representing years of progress in technology development. Improvements do not happen in a vacuum; in this case, titer, yield, processing time, and media cost are all shown to improve. As plans develop and are implemented, the model should be as well. Process learnings and advances are updated so that the next set of goals has a proper starting point and a new projection, and accurate focus is incorporated into the R&D planning cycle.

10.2 Sustainability - water, land use and carbon emissions

A new and important addition to the key output results of techno-economic modeling is the tracking of sustainability impacts of different processes and production approaches. Because the model is a fundamentally complete facility design, it is already tracking water and energy uses in wastewater and waste production outputs. Additionally, facility footprint, carbon usage, workforce, and other inputs to detailed life cycle analysis models are also being modeled. The primary inputs for developing simple or complex Life Cycle Analyses are already present.

What is not tracked in a typical TEA are the more detailed secondary impacts of feedstock sourcing and supply, alternative means of production, and the many deeper evaluations that go into a true lifecycle analysis that are beyond the scope of this section and model.

As a primary recommendation, two approaches are recommended:

- First, the top inputs for the LCA process should be brought to the faceplate as four to six key output results related to land use, water use, waste, and wastewater production, carbon impacts, and energy use.
- If the desire is to fully integrate the model further, this is a tremendous approach, and a separate tab or set of tabs/workbook section is recommended to integrate the models fully without compromising the basic function and, importantly, clarity of the TEA model and its goal.

Figure 5 shows the addition of key sustainability and environmental parameters to the faceplate output.

11 A review and gap analysis of published techno-economic analyses of commercial cultivated meat production

11.1 Overview of studies included in analysis

As of October 2022, there have been five publicly-available studies that have examined the techno-economics of a scaled, commercially cultivated meat production process (Table 1). Because cultivated meat is still a nascent industry, each of the studies examines a hypothetical production process based on certain assumptions, often taking into account knowledge or best practices employed in the fermentation and biopharmaceutical industries and considering how this may be applied to cultivated meat production. Each study also takes a different approach when considering different sensitivity or scenario analyses to identify cost hotspots. Later, we summarize each study, compare assumptions, examine their key findings and areas of overlap, and make recommendations for future techno-economic analyses of cultivated meat.

11.2 Humbird (2021)

The first study by Humbird was released in late 2020 and later published in 2021. Humbird takes a bootstrapped approach to first establish an optimized cellular metabolism and derive biomass equations based on Chinese hamster ovary (CHO) cells and other available literature. These equations were then leveraged for analysis of nutrient composition and imposed limitations for achievable cell density by catabolites such as ammonia and CO_2. The software was used to estimate equipment costs, which were combined with other assumptions to estimate other capital costs of the facility. TEA models were created in Excel for both fed-batch and perfusion-based production, and sensitivity was assessed by varying global production volume, the number of bioreactors and production rate per facility, the volume of the largest production bioreactor, and retained cell density for perfusion culture.

The TEA model suggested that a metabolically optimized fed-batch process would cost \$37/kg but could be lowered to \$22/kg if amino acids were sourced from hydrolysates at \$2/kg. For a perfusion process, the cost was found to be even higher at \$51/kg, owing to increased capital costs of bioreactors and perfusion equipment, including consumables such as filters.

11.3 Risner et al. (2021)

The second study by Risner et al. was published in 2021. Informed by standard process and chemical engineering methods, the authors modeled a semi-continuous batch production process for cultivated meat. An economic model

Company X TEA
Technical Case Studies
Cellular Agriculture TEA Example

Technical Case Study - Impact of Titer Isolated

Key Parameters	Base Case	Titer 80	Titer 120	Titer 160	Titer 200	Titer 240
Target Annual Production, MTA	20 000	20 000	20 000	20 000	20 000	20 000
Maximum fermenter size, l	30 000	30 000	30 000	30 000	30 000	30 000
Product titer, g/l	160	80	120	160	200	240
Batch time, days	14	14	14	14	14	14
Basal Media Cost, $/kg	$4.000	$4.000	$4.000	$4.000	$4.000	$4.000

COGs Breakdown– CASES	Base Case	Titer 80	Titer 120	Titer 160	Titer 200	Titer 240
Main Ingredients	$1.13	$2.48	$1.58	$1.13	$0.86	$0.68
Chemicals	$0.13	$0.13	$0.13	$0.13	$0.13	$0.13
Utilities	$0.02	$0.02	$0.02	$0.02	$0.02	$0.02
Staff & labor	$0.03	$0.03	$0.03	$0.03	$0.03	$0.03
Materials. Maint. & spares	$0.01	$0.01	$0.01	$0.01	$0.01	$0.01
Disposal and waste treatment	$0.08	$0.17	$0.11	$0.08	$0.06	$0.05
Depreciation	$0.02	$0.02	$0.02	$0.02	$0.02	$0.02
Total COGS	$0.79	$1.55	$1.04	$0.79	$0.64	$0.54
COGs vs Basecase	100%	197%	132%	100%	81%	68%

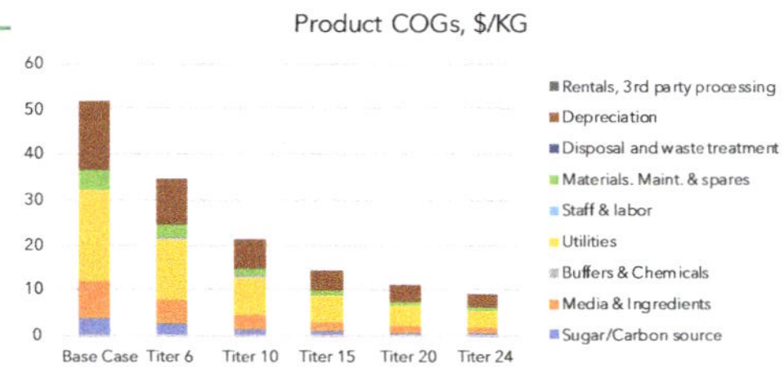

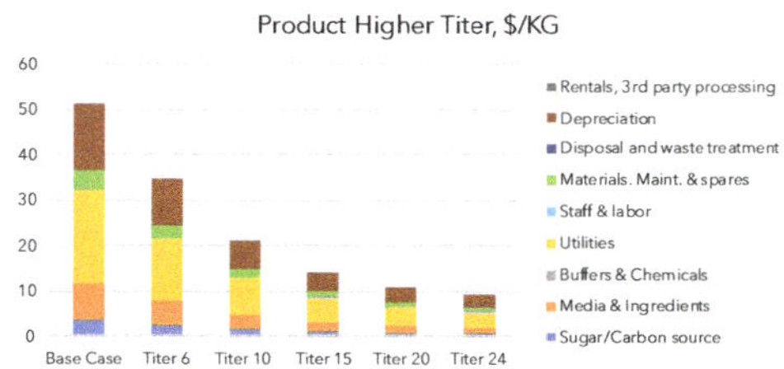

Figure 3 Case study tab to compare impacts of technical parameters.

Company X TEA
Facility Ramp Up Case Studies
Cellular Agriculture TEA Example

Modeling Scaleup Scenarios Years 1 to 5

Key Parameters	Year 1	Year 2	Year 3	Year 4	Year 5
Target Annual Production, MTA	5000	7500	10000	15000	20000
Maximum fermenter size, l	10000	20000	30000	30000	30000
Product titer, g/l	120	160	200	220	240
Batch time, days	20	20	14	24	14
Basal Media Cost, $/kg	$10.00	$8.00	$6.00	$4.00	$4.00

COGs Breakdown– CASES	Year 1	Year 2	Year 3	Year 4	Year 5
Main Ingredients	$7.90	$5.56	$4.16	$3.62	$3.62
Chemicals	$2.66	$1.78	$1.33	$0.89	$0.67
Utilities	$0.66	$0.44	$0.33	$0.22	$0.17
Staff & labor	$1.21	$0.80	$0.60	$0.40	$0.30
Materials. Maint. & spares	$0.44	$0.29	$0.22	$0.15	$0.11
Disposal and waste treatment	$1.07	$0.77	$0.59	$0.52	$0.52
Depreciation	$0.92	$0.61	$0.46	$0.31	$0.23
Total COGS	$14.85	$10.25	$7.69	$6.11	$5.62

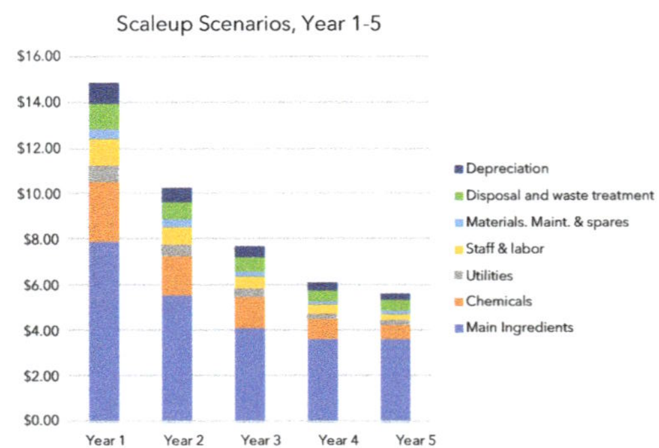

Figure 4 Case study tab to analyze COGs during the first 5 years of startup.

written in Python and based on 67 different parameters related to operations, cellular attributes, finance, media, utilities, and labor was then created, with values for parameters determined by literature, equations, and other assumptions. Six different sensitivity analysis algorithms were used to identify the parameters that had the highest influence on CAPEX and OPEX, including cell density, growth factor and glucose concentrations, cell doubling and maturation times, and glucose consumption rates. Based on these parameters, four different hypothetical production scenarios were assumed, with each scenario adjusting the previously identified parameters. The maximally optimistic and likely implausible scenario, where cost drivers are dramatically reduced or even eliminated, suggested a COGS of $2/kg for cultivated meat, with a range of $2/kg to $437 000/kg.

11.4 Vergeer et al. (2021)

The third study by Vergeer et al. was published as a white paper in 2021. Based on interviews and data collected from over 15 different companies involved in cultivated meat development, the authors constructed a baseline semi-continuous process with a given set of process assumptions that involved both fed-batch and perfusion operations. Media composition was informed by collected data, and pricing for equipment and other inputs was sourced from literature, current suppliers, or calculations in consultation with study partners. The TEA model was built in Excel and first assessed baseline costs under three scenarios for medium-use efficiency. Then, additive scenarios for cost reduction were analyzed, including reduced costs for growth factors, extended investment payback periods, variable production run times, and larger cell volumes. Sensitivity analysis was also conducted for the investment payback time, production run time, and cell volume. Assuming that all optimistic scenarios could be achieved, the COGS for cultivated meat was $6.43/kg. The same bioprocess model and cell culture media formulation was also used in an anticipatory life cycle assessment to examine the environmental impacts of cultivated meat production (Sinke et al., 2021).

11.5 Negulescu et al. (2022)

The fourth study by Negulescu et al. was published as a preprint in 2022. The authors created three process models in SuperPro, with each model containing a different maximum-size production bioreactor (42 m^3 and 210 m^3 stirred tanks and a 260 m^3 air-lift reactor) operating in batch mode. Many of the biological, engineering, and media cost assumptions were taken from Humbird and supplemented with values from current suppliers or literature. The total number and cost of facilities were calculated based on a total market size of 100 kt of annual production, which resulted in a COGS range of $13 to $30.40/kg, depending on the facility type. Sensitivity analyses were performed for media cost and cell doubling time, which demonstrated further cost reduction opportunities, predominantly in media costs.

11.6 Garrison et al. (2022)

The fifth study by Garrison, Biermacher, and Brorsen was published in 2022. The study assumes a facility size significantly smaller than others analyzed, with an annual output of 560 tons. The cell culture media costs and bioprocess metrics and design were based on assumptions by Specht (Specht 2018). The authors include additional costs not examined in previous studies, such as transportation, packaging, insurance, communications, and cold storage, but include fewer details overall. The facility was planned to be located in San Francisco, California, and thus the labor, utilities, and transport costs correspond to numbers acquired for this region. A baseline scenario arrives at a production cost of $63/kg. Scenario analyses examined the impact of 0% to 10% facility downtime and 0% to 30% changes in specific input costs such as growth medium, salaries, and maintenance. The study found that the media, bioreactors, and labor are expected to be the largest cost drivers.

11.7 Key assumptions for bioprocess design, costs and other metrics

The results of a TEA are dependent on the assumptions that are used in the model. In Table 12, a nonexhaustive list of assumptions is provided for the four published TEA studies on cultivated meat. Note that some assumptions are similar across studies, some have a higher degree of variance, and others were included or excluded from certain models.

11.8 Cost drivers and areas for innovation

Although each study assembles a different TEA model, they cumulatively identify several cost drivers that the cultivated meat industry can use as a roadmap for innovation and cost reduction opportunities.

Cell culture media is a major cost driver for cultivated meat production. The cost drivers within the cell culture media are primarily growth factors and recombinant proteins and secondarily amino acids. The cultivated meat industry will need to derive ways to use fewer of these ingredients and source them at lower price points. The studies and other analyses suggest that tractable paths exist for growth factors, but there is less certainty for amino acids (Swartz, 2021; O'Neill et al., 2020). The Humbird, Negulescu, and Vergeer studies highlight that sourcing will be especially important for amino acids, where large cost discrepancies are observed based on scenarios that model amino acids sourced predominantly from fermentation compared to those sourced from hydrolysates of raw materials such as soy. It remains unclear which scenario will accurately map best onto the cultivated meat industry or which raw materials will be used. In either case, new scaled and efficient supply chains will need to emerge to accommodate their use in cultivated meat production. A driving factor influencing overall cell culture media usage and associated costs is cell metabolism and, to a lesser extent, the overall production time. Cell line optimization, including reduced doubling times and metabolic engineering, will be key areas of focus for reducing costs. Even with an extremely metabolically efficient cell line, the Humbird study shows that approaching cost parity with conventional meats will be challenging.

Cell density is also an important metric for the cultivated meat industry to optimize. The studies demonstrate that growing cells at higher densities will result in lower CAPEX, as this will reduce the number of bioreactors needed in a facility, and lower OPEX, as this will reduce the labor and overhead costs associated with bioreactor equipment. The Humbird TEA highlights that maximum achievable cell densities will be limited by the buildup of catabolites such as NH_3 and CO_2. Discovering ways to reduce or overcome their accumulation will therefore be crucial to scaling economically.

Lastly, capturing just a fraction of a percent of the global meat market will require constructing many new facilities that house many large bioreactors. While the studies are aligned in identifying the CAPEX of bioreactors and buildings as a cost driver, there is disagreement on how their costs are analyzed. As one example, a large cost discrepancy in bioreactors is noted between the Humbird and Negulescu studies, as the former assumes pharma-grade 316L stainless steel will be used, as opposed to food-grade 304 stainless steel in the latter. Other discrepancies are noted in the installation (Lang) factor, where divergence is again represented by assumptions between food versus pharmaceutical production. As cultivated meat manufacturers move beyond the pilot scale in the coming years, narrowing the uncertainty in projected costs for bioreactors and buildings will become increasingly important by sourcing primary data and information directly from industry manufacturers.

12 Future directions

Cultivated meat is a nascent industry, and the findings and assumptions from these studies can be expected to change over time as the industry matures. The five studies outlined in this chapter should be viewed as a starting point for understanding cost drivers within the cultivated meat industry. Many more studies will be required to fully dissect the costs for the full range of products,

Company X TEA
Media Recycle Case Study
Cellular Agriculture TEA Example

PARAMETERS OF CASE STUDY

	Input	Units
Spent media to be processed	1,000,000	L / day
% recovered water from spent media	50%	%
% recovered media to WWTP	25%	%
Flux	10	LMH
Concentration factor	2	

PROCESS OUTPUT VALUES

	Input	Units
Water recovered	500,000	L / day
Concentrated media to WWTP	125,000	L / day
Concentrated media recovered	375,000	L / day
Total recycled media (effective)	750,000	L / day

ECONOMIC VALUES

	Input	Units
Cost to Process Spent Media at WTP	0.22	$ / L
Cost of Fresh Media	2.50	$ / L
Value of Recovered water	0.015	$ / L
Power	0.12	$ / kWh

REFERENCE

	Input	Units
Typical BOD fee	0.40	$ / lb BOD
Typical media bod	50000	ppm
Media fee calculated	0.42	$ / gal
Media fee calculated	0.11	$ / L

OPERATING COSTS, SCENARIO COMPARISON

Operating Cost / Savings		0% recycle	50% recycle	Value of recycle
Media to WTP, Processing Cost	$/day	-$110,185	-$27,546	
Recovered water value	$/day		$7,500	
Recovered media value	$/day		$1,875,000	
Power use	$/day		$756.33	
		0% recycle	0% recycle	Value of recycle
Total	$/day	-$110,185.19	$1,855,710.04	$1,965,895.22
Investment Summary				
Years to recover investment (payback)				13.4
Investment return				7.5%

EQUIPMENT SIZING & CAPEX

Equipment	Size	Units	Price	Reference System Information Size	Power (kWh/day)	Count	Install Factor	Total Cost
Spent media tank	1,000,000	L	$225,000	500,000	min	2	4	$2,728,289.82
Pump	694	LPM	$72,000	658	50	2	4	$594,934.91
Membrane system	4,167	m2	$1,300,000	1,000	1,500	5	1.5	$22,955,030.32
System Total					6,303			**$26,278,255.04**

Figure 5 Media recycle example case study.

Table 12 A nonexhaustive list of assumptions compared across the five published cultivated meat TEA models. OTR = oxygen transfer rate; OUR = oxygen uptake rate; FTE = full-time employee. * Some numbers are taken from Humbird, 2020

	METRICS				
Key assumption	**Vergeer**	**Humbird***	**Risner**	**Negulescu**	**Garrison**
Facility output per year	10 kt (combined semi-continuous fed-batch and perfusion process); No specific market size set	6.8 kt (fed-batch), 6.9 kt (perfusion); Market size set at 100 kt annually	Single facility output not stated (semi-continuous fed-batch process); Market size set at 121 kt annually	4 kt for the 42 m^3 stirred-tank facility (batch process) 20 kt for the 210 m^3 stirred-tank facility (batch process) 25 kt for the 260 m^3 air-lift facility (batch process) Market size set at 100 kt annually	548 tons Semi-continuous fed-batch process
Cell doubling times analyzed (range)	28 h (baseline); 21 to 35 h for sensitivity analysis	24 h	8 to 24 h	23 h (baseline); 12 to 72 h for sensitivity analysis	28 h
Maturation or differentiation times analyzed (range)	10 days	N/A	1–10 days	10 days	N/A
Cell densities analyzed (range)	50×10^6	Max cell densities are dependent on other model limitations; 36.6×10^6 cells per mL (110 g/L) in stirred-tank bioreactors < 20 m^3; limitation imposed by NH_3 65×10^6 cells per mL (195 g/L) in perfusion bioreactors, limitation imposed by the cell retention device and perfusion rates	10×10^6 to 20×10^7 cells per mL	33×10^6 cells per mL (100 g/L)	40×10^6 cells per mL

(Continued)

Table 12 (*Continued*)

Cell volumes analyzed (range)	3500 μm³/cell	2903.5 μm³/cell	5000 μm³/cell	2903.5 μm³/cell	5000 μm³/cell
Cell mass (stated or calculated based on density of meat assumptions)	3084 pg	3000 pg	5300 pg	3000 pg	4405 pg
Cell culture type and culture conditions	Unstated non-GM mammalian cells grown in suspension (no microcarriers); no antibiotics; 37°C	CHO cell or similar grown in suspension (no microcarriers); no antibiotics; 37°C	Bovine mesenchymal stem cell or myoblast grown in suspension (no microcarriers); no antibiotics; 37°C	Unstated mammalian cells grown in suspension (no microcarriers) and differentiated into muscle, fat, and connective tissue; no antibiotics; 37°C	Unstated non-GM mammalian cells grown in suspension (no microcarriers)
Product being modeled	1 kg of wet animal cell slurry (70 to 77% water) that can be used in unstructured products	Wet animal cell slurry (70% water)	Undifferentiated cell slurry	Wet animal cell slurry (70% water)	Not stated

CAPEX

Key assumption	**Vergeer**	**Humbird***	**Risner**	**Negulescu**	**Garrison**
Lang factor	~3	7.2 (from Humbird, 2020)	2 but the authors acknowledge that 3-5 may be appropriate	Not stated	Not stated
Size of largest proliferation vessel	10 m³	20 m³	20 m³	42 m³ stirred-tank; 210 m³ stirred-tank; 260 m³ air-lift reactor	20 m³

Bioreactor cost assumptions	\$325 000 per 10 m^3 stirred-tankbioreactor; \$600 000 per 2 m^3 perfusion bioreactor	For a 20 m^3 stirred-tank bioreactor, \$330 000 per bioreactor for bare costs, \$1.5M/bioreactor with piping, instrumentation, and other internals and externals ~\$860 000 per 2 m^3 perfusion bioreactor Bioreactors are made of pharma-grade 316L stainless steel	\$50 000/m^3 (food-grade bioreactor, scaling factor of 0.6); \$778 000 per 20 m^3 bioreactor	\$902 000 per 42 m^3 stirred-tank\$2380000per 210 m^3 stirred-tank \$312 000 per 260 m^3 air-lift reactor Bioreactors are made of food-grade 304 stainless steel	Specific bioreactor costs were not provided. All equipment including four bioreactor lines, preparation, harvesting, and waste inactivation tanks was estimated to be \$60M
Total number of large bioreactors per facility	130 × 10 m^3 stirred-tank bioreactors;430 × 2 m^3 perfusion bioreactors in baseline scenario	24 × 20 m^3 stirred-tank bioreactors in the fed-batch facility96 × 2 m^3 stirred-tank bioreactors in perfusion mode in the perfusion facility	Between 50 × 20 m^3 and 5205 × 20 m^3 stirred-tank bioreactors depending on scenario assumptions	50 × 42 m^3 stirred-tank bioreactors 50 × 210 m^3 stirred-tank bioreactors 50 × 260 m^3 air-lift reactors	4 × 20 m^3 stirred-tank bioreactors
Total estimated facility cost	\$450M under baseline scenarios	\$328M for the fed-batch facility\$663M for the perfusion facility	N/A	\$431M for 42 m^3 stirred-tank facility \$1158M for 210 m^3 stirred-tank facility \$352M for the 260 m^3 air-lift facility	A warehouse building was leased for \$36 000/year

(Continued)

Table 12 (*Continued*)

Annual capital charge	$110M/year to $15M/year depending on payback time. A payback time range of 4 years to 30 years was included for sensitivity analysis.	$48M/year for the fed-batch facility $96.5M/year for the perfusion facility	N/A	Not stated	N/A
FIXED OPEX					
Key assumption	**Vergeer**	**Humbird***	**Risner**	**Negulescu**	**Garrison**
Labor and FTEs	100 000 US$/yearFTEsalary; 200 FTEs in facility	50 000 US$/year FTE salary, supervisor salary at 140% of FTE salary.A labor burden of 100% added to the total labor cost. 95 FTEs in fed-batch facility, 132 FTEs in perfusion facility	$13.68/h, based on meat packer wagesAssumes one laborer per full-scale bioreactor	Upstream operators earn $20/h; total labor cost = $46/h Downstream operators earn $25/h; total labor cost = $57/h	Baseline scenario assumes facility operates 24 h/day, 365 days per year 1 laborer per bioreactor. Average salary is $75868basepayand $103 00 including benefits (range is $32 000 to $245 000 base pay for all positions)

Overhead	Maintenance as 5% of bare equipment costs charged yearly; no insurance	4% of total capital investment (TCI) for maintenance and 5% of TCI for insurance annually	Fixed manufacturing costs were estimated as a percentage of the fixed equipment costs except loan and equity interest; these costs include equipment maintenance, insurance, taxes, and royalties' costs	Fixed facility-dependent costs included maintenance, depreciation, insurance, local taxes, and factory expense. The exact maintenance cost is dependent on equipment and was not stated	5% for repairs and maintenance 6% interest on loans utilized for operating inputs

OTHER NOTABLE ASSUMPTIONS

Key assumption	Vergeer	Humbird*	Risner	Negulescu	Garrison
Media: amino acid costs and sourcing	70% of amino acids will be sourced from plant protein hydrolysates (soy is analyzed at $2.00/kg to $3.50/kg). The other 30% will come from fermentation sources ($2.50/kg to $50/kg); all nutrients are considered to be food-grade	Individual amino acids will be sourced primarily from fermentation processes or as bulk amino acids from food-grade soy hydrolysate at $2/kg. Cost of amino acids are derived from a regression model based on economies of scale. At 100 kTA scale of cultivated meat production, prices range from $40/kg to $146/kg	Not listed	The costs of amino acids are based on the anticipated yearly demand and calculated with the regression model used by Humbird.	The total media cost is assumed to be $3.74 per liter in the baseline scenario (taken from Scenario 5 from Specht)

(Continued)

Table 12 (*Continued*)

Media: growth factors and growth factor costs	Based on Essential 8; includes insulin, transferrin, albumin, FGF2, and TGFßScenarios assume that growth factor costs will decrease by 100-1000X of current prices and that other innovations will permit the reduction of albumin, insulin, and transferrin by a net factor of 5	Based on Essential 8; includes insulin, transferrin, FGF2, and TGFßThe cost of growth factors is derived from a regression model based on economies of scale for recombinant protein production via fermentation	Based on Essential 8; includes insulin, transferrin, FGF2, and TGFßGrowth factor costs are set at current market rates. In the most optimistic scenario, growth factor costs are eliminated entirely	Based on Beefy-9 formulation (Stout et al., 2022); includes insulin, transferrin, albumin, FGF2, NRG1, and TGFß The costs of growth factors are based on the anticipated yearly demand and calculated with the regression model used by Humbird	
Media sterilization	90% of the medium is heat-sterilized and the remaining medium that contains concentrated heat-sensitive substances is sterilized via filtration	Media tanks are attached to high-temperature short-time sterilizers (HTST) to provide preinoculation fill; a smaller dedicated tank (or tanks) that contains premixed media to be added during batches is attached to sterile retention filters	Details not included	The entire volume of media is sterilized using HTST	Details not included
Infrastructure: clean rooms	Class 8 clean room environment in the production area	Class 8 clean room for cell culture areas, Class 6 for laboratory areas	Details not included		Details not included

Imposed constraints	Bioreactor volume fraction is < 0.25 Constrains for oxygen and metabolite inhibition were not considered.	• Bioreactor volume fraction is < $0.25k_La$ such that OTR = OUR at u_s = 0.006 m/s and $\lambda_k \geq 20$ μm; mixing time is < $1/k_La$ Inhibition by cell metabolites leads to limitations in achievable densities and production scales:5 mM NH_3 • 50 mM lactate • < 100 mbar CO_2	Not examined	Bioreactor volume fraction is < 0.25 Calculations based on literature values were performed to ensure OTR was sufficient to meet the OUR of cells at their maximum concentration. Constraints for metabolite inhibition were not considered.	Not examined

manufacturing approaches, and new technologies that may be pursued in the developing industry. It will be important for future TEMs to incorporate data and practices from cultivated meat manufacturers or other research institutions that accurately reflect the state of the art (Swartz, 2021). Later, a set of ideas that future TEA practitioners may consider incorporating into future studies, scenarios, or sensitivity analyses is provided.

1. **Hybrid products.** Because the costs of cultivated meat production are high, many manufacturers believe that hybrid products - those that contain a mixed percentage of cultivated animal cells with other plant-based ingredients - may offer the fastest pathway to market. TEA can be useful to determine the costs of hybrid products with varying percentages of cultivated animal cells and different types of plant-based or fermentation-derived ingredients.
2. **A full spectrum of products.** How might costs differ between the production of cultivated muscle vs. fat? What are the costs of differentiating cells? How do the production processes and costs compare when modeling a chicken nugget vs. a marbled steak? These may be important questions to dissect in future TEAs.
3. **Bioreactor selection.** The Negulescu study is a starting point for understanding how bioreactor selection influences costs. Future TEAs may build on this idea, extending analyses to other bioreactor designs, adjusting bioprocess parameters, and incorporating computational fluid dynamics models.
4. **Clean rooms**. The Humbird study found that beyond a certain number of large proliferation reactors, the clean room costs outpace the productivity from added reactors. Larger reactors can also increase clean room costs because the facility needs more levels to accommodate taller vessels. Thus, the environmental controls and assumptions related to contamination events within the facility will be important areas to explore in future TEAs.
5. **Scope.** Only the Garrison study presented in this chapter contains some of the elements of a full production process, especially downstream processing steps such as dewatering, drying, flavoring, chopping, texturizing, packaging, storage, labeling, and shipping. Other considerations for inclusion in future TEAs are full accounting for solid and liquid wastes, potential co-product creation, geographic variability, projected revenues, and other financial parameters.

13 A cultivated meat techno-economic analysis case study

In this section, a simple case study is presented, which brings together the basic concepts of building a techno-economic model for a limited portion of a processing facility. The case study considers the collection and recycle of spent media, an important portion of cultivated meat production processes, and an electronic file has been provided for users to experiment with and build off for their own modeling efforts.

The model includes the following sections:

- Parameters of case study: Key defining user inputs for conducting case studies. The effectiveness and economics of the proposed system can be evaluated by adjusting typical key parameters - what level of recovery the system can achieve, what portion of spent media may require disposal (vs. reuse), the expected flux achieved by the system, and the concentration factor achieved (calculated from water recovery).
- Process output values: Process results used to support model equipment and economic calculations.
- Economic values: The user entered economic values that will be different for any locale and process.
- Equipment sizing and CAPEX: The model works with the flows to size physical equipment - a spent media collection tank and pump and a membrane system that can process media. This area also models representative power use, the estimated cost of reference equipment, and an estimate of the cost of the bare equipment installed in a plant (Installation Cost). As the model is used to change size, the equipment sizing, cost, and power use also change correspondingly. The predicted cost of the recovery equipment is determined by comparing the model size to the reference size, and scaling cost proportionally to the size ratio raised to the 0.6 factor (a standard cost estimating practice).
- Operating costs, scenario comparison: Water treatment costs are calculated based on the input economic values, the model's flow predictions, and power usage.
- Investment summary: The bottom-line evaluation of economics for adding media recycle is determined by comparing the total value of implementing recycle with the required investment to calculate a simple payback period in years, as well as an ROI based simply on savings achieved divided by the required investment.

The purpose of this tool is to provide a starting point for readers to build their own tools, get comfortable with the concepts in this chapter, and consider this important aspect of cultivated meat production. All variables and results are based on user inputs, and the values used as a starting point are for example purposes only.

14 Conclusion

This chapter presents the central and critical role of techno-economic modeling for process and business understanding, development, improvements, and execution. The TEA model is something investors will often ask for as a first item before considering funding of an enterprise, and it creates and tracks the information that is needed for the development of a business plan as well as any technical or process improvements program or plan. No business can be launched or function well without this understanding.

We have shared a step-by-step approach to building the plan, moving from basic models to advanced models and their use for key technical and business activities. Effective models translate technical and economic impacts into true plant designs. Capital projects and plant operations are understood in all their resource and cost impacts. Finally, we have shared the concept that these plans should also consider the sustainability impacts of the endeavor being modeled, to achieve true triple bottom-line objectives.

Behind a strong techno-economic model, there must of course be a fundamentally sound process and commercialization program. A primary model used is to provide business answers for potential investors, especially to demonstrate commercial potential and sufficiently low costs of goods. While models effectively convey the value of technology improvements, for example, the reduction of media costs to a target price, these target improvements are only achieved when a corresponding technology program is planned and effectively executed as a result of the model guidance.

Similarly, as discussed in the case studies section, the inputs must form one consistent whole and realistic set of simultaneously achievable results. The inputs should not be unrealistically optimistic or have simultaneously unachievable results that constitute what is often referred to as a 'green light schedule' (the time it would take to cross a major city if every traffic signal turned green when encountered).

Models do, however, give you the power to understand the impacts of achieving a goal, and from there, teams and an execution program can be set to work on the key activities – they may be supply chain or co-location opportunities, site selection, process improvements, partnering, or third-party manufacturing agreements.

With this guidance, we strongly encourage the reader to move forward with their own model development, starting with a simple model and layering on additional information, inputs, and outputs in a progressive elaboration manner. In support of this effort, we offer the final chapter section, providing the reader with guidance on where to look for further information in support of successful model development and implementation.

15 References

Humbird, D. (2020). *Scale-Up Economics for Cultured Meat: Techno-Economic Analysis and Due Diligence*.

Humbird, D. (2021). Scale-up economics for cultured meat. *Biotechnol. Bioeng.* 118(8), 3239–3250. .

Negulescu, P. G., Risner, D., Spang, E. S., Sumner, D., Block, D., Nandi, S. and McDonald, K. A. (2022). Techno-economic modelling and assessment of cultivated meat: Impact of production bioreactor scale. bioRxiv. https://doi.org/10.31224/2414.

O'Neill, E. N., Cosenza, Z. A., Baar, K. and Block, D. E. (2020). Considerations for the development of cost-effective cell culture media for cultivated meat production. *Compr. Rev. Food Sci. Food Saf.* 3, 1.

Risner, D., Li, F., Fell, J. S., Pace, S. A., Siegel, J. B., Tagkopoulos, I. and Spang, E. S. (2021). Preliminary techno-economic assessment of animal cell-based meat. *Foods*. 10(1), 3. https://doi.org/10.3390/foods10010003.

Sinke, P., Vergeer, R. and Odegard, I. (2021). *LCA of Cultivated Meat: Future Projections for Different Scenarios (CE Delft).*

Stout, A. J., Mirliani, A. B., Yuen, J., White, E., Kaplan, D. L. (2022). Simple and effective serum-free medium for sustained expansion of bovine satellite cells for cell cultured meat. *Commun. Biol.* 5(1), 466.

Swartz, E. (2021). *Anticipatory Life Cycle Assessment and Techno-Economic Assessment of Commercial Cultivated Meat Production: A Summary of Key Findings and Future Directions (The Good Food Institute).*

Vergeer, R., Sinke, P. and Odegard, I. (2021). *TEA of Cultivated Meat: Future Projections for Different Scenarios (CE Delft).*

Index

www.ingramcontent.com/pod-product-compliance
Ingram Content Group UK Ltd.
Pitfield, Milton Keynes, MK11 3LW, UK
UKHW021825100926
8556IPUK00003B/162

9 781801 463768